The 2024 Declaration of Helsinki

Chieko Kurihara • Dirceu Greco • Ames Dhai
Editors

The 2024 Declaration of Helsinki

Global Efforts Towards the Highest Ethical Standards

Springer

Editors
Chieko Kurihara
Kanagawa Dental University
Yokosuka, Kanagawa, Japan

Ames Dhai
School of Clinical Medicine
University of the Witwatersrand
Johannesburg, Gauteng, South Africa

Dirceu Greco
School of Medicine
Federal University of Minas Gerais
Belo Horizonte, Brazil

ISBN 978-981-96-9293-4 ISBN 978-981-96-9294-1 (eBook)
https://doi.org/10.1007/978-981-96-9294-1

This Springer imprint is published by the registered company Springer Nature Singapore Pte Ltd.
The registered company address is: 152 Beach Road, #21-01/04 Gateway East, Singapore 189721, Singapore

Foreword by Lujain Alqodmani

The Declaration of Helsinki (DoH) stands as a cornerstone of our decisiveness to act together, which crystallizes our shared commitment and vision to uphold the highest ethical standards in medical research discussed and agreed among all the interested parties from a number of regions, including conflict-affected countries, through regional consultations and global dialogue. The World Medical Association (WMA), as the custodian of the Declaration, is honored to have stewarded this milestone through eight regional meetings in in Tel Aviv; Sao Paulo; Copenhagen; Tokyo; the Vatican; Johannesburg; Munich; and finally, Washington, DC, in collaboration with National Medical Associations and external stakeholders.

The revised Declaration reflects the diversity and depth of this process—integrating insights from frontline clinicians, ethicists, community advocates, and research participants themselves. The adoption of this new version marks not an end, but a beginning. Now comes the real work: dissemination, education, implementation, and preparing for the next iteration through continued global engagement.

The DoH is not a mere document for physicians alone. It is a guiding light for all healthcare professionals, researchers, and, needless to say, for research participants, patients, and communities.

We also need to address the humanitarian crises, armed conflicts, and the erosion of health systems. Throughout my presidency, there has been immense danger in health care all over the world. We are now seeing devastating situations of tremendous attacks against the healthcare systems, such as in Ukraine, Sudan, Gaza, Lebanon, and many other regions. Health professionals must be safeguarded, and the rights of health for all preserved—even amid war and disaster.

In addition, we bear another moral duty. We are now still away from fulfilling "sustainable development goals" (SDGs). Goal 3 of the SDGs (good health and well-being) aims to ensure healthy lives and promote well-being for all, at all ages, which might be accomplished by establishing universal health coverage all over the world. The health of the people and the rights of the professionals that provide healthcare must be at the top of our political agenda. The ultimate prerequisite is to ensure lasting peace for the sake of health, human rights, and dignity of all people. Also, it is important to remember Goal 4 of the SDGs (quality education), which aims to ensure inclusive and equitable quality education and promote lifelong learning opportunities for all, to provide equal capacity-building opportunities.

This book offers a window into the 2024 revision of the DoH, a tool to guide collaborative action, identify social determinants of health, and amplify the voices too often unheard. Our joint, continuous discussions of the DoH could ensure that medical research shall contribute to realize the equitable global health. The revised Declaration places new emphasis on social value, community engagement, and the co-creation of research with participants and their communities.

This Declaration must be more than a framework; it must be a shared promise to shape a future in which ethics, inclusion, and justice define the heart of medical and global health research.

The Immediate Past President (2023–2024) Lujain Alqodmani
of the World Medical Association
Ferney-Voltaire, France

Foreword by Ramin W. Parsa-Parsi et al.

A strong foundation in globally applicable ethical principles is key to ensuring the protection and respect of human research participants across vast cultural contexts. Cognizant of the vital importance of codifying such a framework for the global medical profession, and in response to the dehumanizing and unethical practices of Nazi physicians during World War II, as well as other failures to respect the well-being, rights, and safety of research participants, the World Medical Association (WMA) adopted the Declaration of Helsinki (DoH) in 1964.

The DoH outlines the ethical principles for medical research involving human participants. While the Declaration has stood the test of time—providing guidance and consistency to physicians and other researchers worldwide and fostering trust in research and the medical profession—it has maintained its relevance through a series of regular comprehensive revisions.

In the six decades since its adoption, the DoH has been revised multiple times—most recently by a broad international workgroup chaired by the American Medical Association—to meet the challenges and seize the potential of a changing research environment. While not altering the document's fundamental framework, these revisions have acknowledged the DoH as a living document. As language, science and technology, global collaboration, and social expectations develop and evolve, the DoH must adapt to answer new and, in some cases, previously unexplored questions.

The broad spectrum of international perspectives found in the chapters that follow is a reflection of the foremost topics addressed during the most recent DoH revision process, which concluded with the unanimous adoption of the policy by the WMA's General Assembly in October 2024 in Helsinki.

The most notable of these topics were accorded more extensive attention in a series of expert conferences held throughout the WMA's official regions, followed by global comment periods open to all stakeholders and the public. This is standard procedure for the revisions of WMA's core ethical documents, and it serves to ensure an inclusive and comprehensive global evaluation of the document in question.

The first such conference—focused on big data and artificial intelligence—was held in December 2022 in Tel Aviv, Israel, at the invitation of the Israeli Medical Association. Stressing new challenges related to consent for the collection, storage, and secondary use of patient data, the preliminary discussions held during this meeting led the workgroup to expand on provisions in the 2013 DoH focused on identifiable human material and data by adding a direct reference to the WMA's Declaration of Taipei on the ethical considerations regarding health databases and biobanks.

This meeting was followed two months later by a conference in São Paulo, Brazil, on the use of placebo in clinical trials. Organized by the Brazilian Medical Association (Associação Médica Brasileira), the program addressed how this topic was approached in previous revisions of the DoH and explored how the use of placebo is regulated in various parts of the world and by other guidelines (e.g., those of the Council for International Organizations of Medical Sciences (CIOMS)). In contrast to previous revisions, during which this issue was a source of great contention, the workgroup ultimately decided to largely retain the existing language on placebo, which makes allowances for its use in strictly defined cases.

During the subsequent expert meeting on emerging trial design hosted by the Danish Medical Association in Copenhagen, Denmark, a crucial question emerged about whether to expand the DoH to directly address non-physician researchers. The final version of the DoH reflects the WMA's view of these principles as an essential guide for prioritizing the interests, rights, and safety of all research participants in that it directly addresses not only physicians, but also "all individuals, teams, and organizations involved in medical research." Furthermore, the discussions in Copenhagen shed light on the complexities of multinational studies and the importance of bolstering the role of research ethics committees. The outcome of this meeting and other ensuing debates held throughout the revision process led to the inclusion of further protections against exploitation of research participants in the DoH. One example of this is the new requirement for those carrying out multinational studies to consult with an ethics committee in every country where a given study is being carried out, including the country of the sponsor. In addition, the revised DoH no longer refers to "research subjects" but rather "research participants," thus linguistically reflecting the indispensability of patient autonomy in medical research and acknowledging the growing understanding of the role of patients as partners and co-creators in the research enterprise.

Attendees of the Japan Medical Association's regional meeting in Tokyo, Japan, delved into the especially timely ethical questions surrounding medical research on humans in times of public health emergencies and disasters. The unprecedented rise in fast-track trials during recent public health crises led the workgroup to carefully consider the challenges of maintaining the usual rigorous ethical standards in situations where there is a need for rapid results—most notably during the COVID-19 pandemic. The DoH was revised to reaffirm that the validity of these core principles must not be compromised or undermined by such situations. The outcome of the meeting also led the workgroup to further emphasize the ethical importance of environmental considerations in research in the revised DoH.

A growing awareness of the impact of research gaps and disparities in clinical trial participation led to a more in-depth assessment of existing language on and approaches to vulnerability—a topic so pertinent it was the subject of three expert meetings held in the Vatican City (hosted by the Pontifical Academy for Life); in Johannesburg, South Africa (hosted by the South African Medical Association); and in Munich, Germany (hosted by the German Medical Association and the Leopoldina National Academy of Sciences).

Earlier versions of the DoH had stipulated that research should first be conducted on nonvulnerable groups. However, during the most recent revision process, there was much deliberation about the needs and priorities of individuals and groups in situations of more vulnerability. The discussions held during these meetings contributed to a shift from an understanding of intrinsic vulnerability to one of contextual vulnerability—that is, heightened susceptibility caused by specific, though mutable social, environmental, or cultural factors. They also shed light on the detrimental effects of exclusion from clinical research. The outcomes of these meetings led the workgroup to develop new wording for the DoH paragraphs focused on vulnerability. While there is no question that the protection of individuals and groups in a situation of more vulnerability remains a priority, the revised DoH now adds that the harms of exclusion, e.g., delayed access to new therapies, must be considered and weighed against any harms of inclusion. For medical research with individuals, groups, or communities in situations of particular vulnerability, the DoH declares that medical research "is only justified if it is responsive to their health needs and priorities and the individual, group, or community stands to benefit from the resulting knowledge, practices, or interventions." Furthermore, "researchers should only include those in situations of particular vulnerability when the research cannot be carried out in a less vulnerable group or community, or when excluding them would perpetuate or exacerbate their disparities."

On the subject of research in limited-resource settings, the workgroup introduced language underscoring the need to strive for global justice, fair benefits for all participants, and meaningful community engagement before, during, and after conducting research. The revised DoH also strengthened existing language on the provision of post-trial treatment access.

These and other changes integrated into the 2024 revision of the DoH all reflect the WMA's unwavering commitment to the policy's guiding principle of safeguarding the interests, rights, and safety of the research participant. This was one main impetus behind the workgroup's decision to strengthen the requirement of advanced assurance of post-trial provision of demonstrably beneficial interventions to trial participants.

The enduring impact of the DoH as one of the most authoritative documents in the field of research ethics was further cemented during this most recent revision process through the efforts of a representative international workgroup, a series of expert conferences in five world regions, an extensive online public consultation and its unanimous acceptance by the WMA's constituent members, which represent 10 million physicians worldwide.

This interdisciplinary book will serve as an instrumental advocacy tool and contribute to the ongoing in-depth global analysis of this living document.

The World Medical Association
Ferney-Voltaire, France

German Medical Association
Berlin, Germany

Ramin W. Parsa-Parsi

Workgroup Chair of the 2024 revision
of the Declaration of Helsinki
American Medical Association
San Francisco, CA, USA

Jack Resneck Jr

Secretary General, The World Medical Association
Ferney-Voltaire, France

Otmar Kloiber

Contents

DoH, publishing a substantial number of peer-reviewed papers, interview articles, and proceedings after forming various domestic and international meetings involving experts, patients, and the public. She took the role of leading editor of a book titled *Ethical Innovation for Global Health: Pandemic, Democracy and Ethics in Research* (Springer; 2023), collaborating with two other editors and some chapter authors of this book to discuss the 2024 DoH.

Dirceu Greco, MD, PhD Professor Emeritus of Infectious Diseases and Bioethics at the School of Medicine, Federal University of Minas Gerais (UFMG) in Belo Horizonte, Brazil. He received his MD degree and his PhD from UFMG, with further specialization in Clinical Immunology at State University of New York, Buffalo, and the University of London, England. Member (2018–2024), Brazilian National Health Council (CNS), representing the Brazilian Society of Bioethics (SBB). Member (2018–2025) and one of the vice-Chairs (2020–2023) of UNESCO's (Paris) International Bioethics Commission (IBC). Chair (2018–2024), Brazilian Society of Bioethics (SBB). Member (2023 on), WHO Expert Group on ethics and governance of infectious disease outbreaks and other emergencies of public health importance. Member (2020–2023) WHO Access to COVID-19 Tools Ethics Working Group. Director (2010–2013) Department of STI, AIDS and Viral Hepatitis (Ministry of Health, Brazil). Member (2007–2010), Brazilian Research Ethics Commission (CONEP). Founding member and chair, UFMG Research Ethics Committee (1994–1998). Dean for Graduate Affairs (1994–1998), UFMG. Temporary consultant for several national/international institutions such as the Brazilian AIDS Programme, WHO, UNITAID, UNAIDS, UNICEF, UNESCO, CIOMS, World Medical Association (WMA), and the United States Presidential Commission for the Study of Bioethical Issues. Main topics of interest include infectious and parasitic illnesses, bioethics, public health, human rights, and clinical immunology.

Ames Dhai, MASSAf, PhD, MBChB, FCOG, LLM Founder and Past Director of the Steve Biko Centre for Bioethics at the Wits Faculty of Health Sciences (2007–2019), Professor of Bioethics and Health Law at the Wits School of Clinical Medicine, Specialist Ethicist at the Office of the President and CEO of the South African Medical Research Council (SAMRC), member of the Academy of Science South Africa (ASSAf), and Chairperson of the UNESCO International Bioethics Committee. Some of her current activities include: Chairperson of the SAMRC Bioethics Advisory Panel, the SA Human Sciences Research Council Research Ethics Committee and the SA National Blood Services Research Ethics Committee, and the ASSAf Biosafety and Biosecurity Committee. She is a board member of the SA Medical Association; Vice-Chair on the Ministerial Advisory Committee for COVID-19 Vaccines; member of the Ministerial Advisory Committee on Unrelated Organ Transplants; and a member of the Advisory Committee to the SA Law Commission Project on Medical Litigation. She is editor-in-chief of the *South African Journal of Bioethics and Law* and Associate Editor of the *South African Medical Journal*. She started her career as a medical doctor, specialized as an obstetrician and gynecologist, and then moved into the disciplines of bioethics and health law.

About the Contributors

Foreword 1 From the World Medical Association

Lujain Alqodmani, MD is a global health and medical professional, currently holds the prestigious position of Immediate-Past President at the World Medical Association (WMA). Renowned for her leadership, she spearheads the WMA's initiatives, advocating for global healthcare excellence and ethical medical practices worldwide. Alongside her role as WMA President, she serves as the Healthcare Professionals Network Lead of Healthcare Without Harm. Previously, Dr. Alqodmani contributed significantly to various health sectors. Her experiences range from serving as an emergency physician in Kuwait to holding vital roles such as the co-chair of the WMA Environment Caucus and the International Relations Director at Kuwait Medical Association. Moreover, she held influential positions at EAT, Women in Global Health and the International Federation of Medical Students' Associations (IFMSA), where she represented the Eastern Mediterranean Region and led crucial internal affairs portfolios. With a medical degree from Kuwait University and a Master's in international healthcare management, economics, and policy from SDA Bocconi, Dr. Alqodmani remains dedicated to elevating healthcare standards worldwide.

Foreword 2 From the World Medical Association

Ramin W. Parsa-Parsi, MD, MPH Head of the Department for International Affairs at the German Medical Association (GMA) in Berlin, Germany. Prior to joining the GMA, he worked with Harvard Medical International in Boston, USA, as the Director of Health Policy. He has been a member of the Board of the European Doctors (CPME), the Council of the World Medical Association (WMA), and the WMA Medical Ethics Committee. He has chaired international workgroups for the 2013 revision of the WMA Declaration of Helsinki, the 2017 revision of the WMA Declaration of Geneva, and the 2022 revision of the WMA International Code of Medical Ethics. He holds a master's degree in public health from Harvard University, obtained a doctoral degree at the Institute of Clinical Pharmacology at the University of Heidelberg, Germany, and received his MD from the University of Cologne in Germany, where he also did his postgraduate medical education in Hematology/Oncology. He received additional training at medical schools and hospitals in the USA, France, and Switzerland.

Jack Resneck, MD Chair and Bruce U. Wintroub Endowed Professor of Dermatology at the UCSF School of Medicine in San Francisco, CA, and holds a joint appointment as affiliated faculty at the Philip R. Lee Institute for Health Policy Studies. He serves as the Chair of the World Medical Association, where he recently oversaw its inclusive, global 30-month process to revise and modernize the Declaration of Helsinki, the seminal document defining the ethical principles and protections governing medical research with human participants. Dr. Resneck is a passionate advocate for physicians and patients, a prominent spokesperson for thoughtful, well-tested healthcare innovation, and a champion for a more equitable health care system. Whether testifying about removing healthcare dysfunction and burdens, advocating for improved patient access, or leading his department and caring for patients at UCSF, he channels his collaborative, inclusive leadership to improve the lives of patients and the working environment for physicians to help build stronger, healthier communities. Active in health services research, his studies on patient access to care, health care delivery and economics, digital medicine, and public health have been published in prominent journals. Dr. Resneck received his BA in public policy from Brown University and his MD from UCSF in San Francisco, California —where he also completed his internship in internal medicine, residency in dermatology and fellowship in health policy.

Otmar Kloiber, MD, PhD has served as Secretary-General of the World Medical Association since 2005, after he left the German Medical Association as Deputy Secretary General. He holds an MD (1984) and PhD (1986) from the University of Cologne, was a postdoctoral fellow in the Department of Biochemistry at the University of Minnesota, Duluth, and a scientific research fellow at the Max Planck Institute for Neurological Research. He holds an honorary doctorate from the Victor Babes University, Timisoara, Romania, and was appointed Clinical Professor in Health Administration at the Brooks College of Health, University of North Florida, from 2009 to 2013. He is interested in the development of deontology under the influence of health system organization and its relation to the provision of medical care. He provided advice to numerous governments on medical ethics and socio-medical issues. His advocacy focus is on equitable access to quality healthcare for all people.

Part I Research Ethics Perspectives from the Global South and Asia

Chapter 1

Bioethics in Brazil to Overcome Pandemics: Striving for the Highest Ethical Standards to Prevent Exploitation and Achieve Health for All —A Case Study of the AIDS and COVID-19 Pandemics—

Dirceu Greco (See "About the Editors")

Chapter 2

Ubuntu Principles: Solidarity to Include Vulnerable Communities for Equitable Access

Ames Dhai, Chieko Kurihara (See "About the Editors")

Chapter 3

Strengthening Ethical Safeguards Protecting Vulnerable Populations in Egyptian: Role of Research Ethics Committee

Hany Sleem, MD Egyptian physician and bioethics leader currently serving as Head of the Department of Dermatology and Andrology at the National Hepatology and Tropical Medicine Research Institute (NHTMRI) in Cairo and General Manager of the Clinical Research Center at the Air Force Specialized Hospital. Dr Sleem graduated from the Faculty of Medicine, Cairo University, and specialized in Dermatology and Andrology, he further honed his expertise through certifications in research ethics from the University of Maryland and Harvard Medical School's Clinical Scholar Research Training program. Dr. Sleem's career is marked by pioneering contributions to ethical research governance in low- and middle-income countries. He played a pivotal role in drafting Egypt's first national laws for Animal Care and Use in Research and Education (2015–2018) and Clinical Trials (2016–2018), while coordinating the Egyptian Network of Research Ethics Committees (ENREC) to strengthen ethical standards nationwide. Moreover, he was the Co-Principal Investigator of the NIH-funded Middle East Research Ethics Training Initiative (MERETI) and advanced regional capacity for ethical review systems. His foundational works included Research Ethics in Africa (2014) and Ethical

Issues in Research in the Arab Region (2018), alongside peer-reviewed publications in journals such as Developmental World Bioethics and the European Journal of Public Health. Dr Sleem also participates as a chairman or a member in over a dozen institutional review boards in Egypt. He was recognized with multiple Egyptian Awards for scientific and community achievements.

Chapter 4

The DoH, the DoT, and the Duty to Participate in Data Research: A Case Reflection from Taiwan

Daniel Fu-Chang Tsai, MD, PhD is a family physician and bioethicist. He graduated from the National Taiwan University College of Medicine (NTUCM) and earned his PhD in bioethics from the University of Manchester, UK. He is the founding professor of the Graduate Institute of Medical Education and Bioethics, a joint professor in the Department of Family Medicine and the Graduate Institute of Clinical Medicine, and the past Director of the Department of Social Medicine at the NTUCM. He is an attending physician in the Department of Medical Research, the Director of the Ethics Center, the Chairman of the Research Ethics Committee, and the Executive Secretary of the Clinical Ethics Committee at National Taiwan University Hospital. He is also the Director of the Center for Biomedical Ethics at National Taiwan University, the past Vice President of the International Association of Bioethics, and the past president of the Taiwan Association of Institutional Review Boards. He is a member of the Merk Ethics Advisory Panel and was awarded Honorary Membership by the UNESCO Chair of Bioethics in 2015, the Goldman-Berland Lectureship in Palliative Medicine in 2019, and elected as Hastings Center Fellow in 2021. His research interests include research ethics, clinical ethics, cross-cultural bioethics, and medical ethics education.

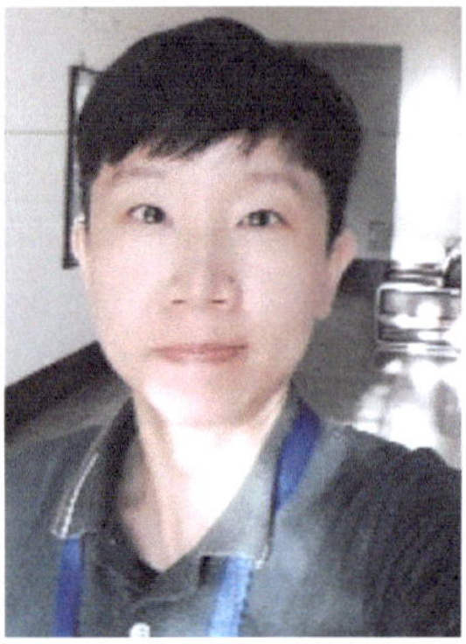

Yu-Chen Juang, MA is currently a PhD student at the Institute of Health Policy and Management, National Taiwan University. Juang holds a bachelor's degree in medical biotechnology and a master's degree in technology law. Juang's expertise lies in interdisciplinary research on topics related to medicine, ethics, and law, with a particular interest in developing popular science knowledge to promote health. Since 2016, Juang has been a member of Professor Daniel Fu-Chang Tsai's research team in the Graduate Institute of Medical Education and Bioethics at the National Taiwan University College of Medicine, conducting research in medical ethics and related legal issues.

Chapter 5

Ethical, Legal, and Social Issues of Biobanks, Including Brain Banks: Experiences in South Korea, Taiwan, and Japan, and Future Perspectives for Global Principles

Young-Joon Ryu, MD, PhD Professor in the Department of Medical Humanities and Medical Ethics, as well as the Department of Pathology at Kangwon National University College of Medicine. He earned his MD degree from Kosin University and his PhD from Seoul National University. He witnessed research mis- conduct by his advisor, Hwang Woo-Suk, while study- ing at Seoul National University as a PhD graduate student, and reported it to MBC TV. Afterwards, he was forced out of the hospital where he worked and spent 2 years struggling on the streets. Even in oppression and hardship, he did not lose faith in science and continued his research. He continued his journey as a clinical specialist in pathology at Korea University and Asan Medical Center. He also obtained a degree in medical humanities, broadening his research interests to include ethics. He has served as an expert member of the fifth Presidential National Bioethics Committee, as well as the president of Korea Human Resources Bank and Korean Human Brain Bank. He is currently the research director of the Neuroethics Project of the National Research Foundation of Korea.

Daniel Fu-Chang Tsai (See Chapter 4)

Takeo Saio, MD Department of Internal Medicine and Psychiatry, Fuji Toranomon Orthopedic Hospital; Chief, K&S Consulting Office for Occupational Mental Health. SMBC Nikko Securities Inc. Corporate Occupational Health Manager; Certified Specialist and Supervisor of the Japanese Society of Psychiatry and Neuropsychiatry (JSPN). Member of the JSPN Committees of Pharmaceutical Affairs; Suicide Prevention; Editorial Committee member of the Japanese journal *Clinical Evaluation.* After graduating from Gunma University (Japan) School of Medicine in 1989, he has practiced in psychiatry, gastroenterology, occupational health, and general internal medicine. He engaged in introducing the basic philosophy behind evidence-based medicine (EBM), public understanding of science (PUS), as well as ethics and regulatory science to Japanese people by way of publishing Japanese translation of books on background information of these areas, including *Evidence-Based Medicine Toolkit* (Badenoch D & Heneghan C, 2002.), *The Institutional Review Board Member Handbook* (Amdur RJ, et al., 2002), *The Resourceful Patient* (Gray JAM, 2002), *The Truth About the Drug Companies: How They Deceive Us and What to do About It* (Angell M, 2004), *Inside the FDA: The Business and Politics Behind the Drugs We Take and the Food We Eat* (Hawthorne F, 2005).

Chieko Kurihara (See "About the Editors")

Chapter 6

From COVID-19 to Global Standards: Redefining Korean Bioethics and Research Ethics

Ock-Joo Kim, MD, PhD is Professor and Chair of the Department of History of Medicine and Medical Humanities at Seoul National University College of Medicine, Korea. She also serves as Director of the Center for Human Research Protection at Seoul National University Hospital. She received her M.D. (1989) and M.S. (1992) from Seoul National University, and her Ph.D. in the history of medicine from the University of Minnesota in 1998. She held a postdoctoral fellowship at Harvard University until 2001. Professor Kim has been a leading figure in the fields of biomedical ethics, history of medicine, and research governance. She has served on the Korean National Bioethics Committee and its Expert Committees since 2005, and was a member of the UNESCO International Bioethics Committee (2016–2019). She served as President of the Korean Bioethics Association (2020–2021). Her recent scholarship has focused on international research ethics, data governance, and the evolving role of ethics in pandemics. Her publications include analyses of the 2024 revision of the Declaration of Helsinki, ethical challenges during the COVID-19 pandemic, and comparative governance of research ethics systems. She is also co-author and editor of several books on Korean and global bioethics, and has actively contributed to aligning Korea's Bioethics and Safety Act with international ethical standards. Through her interdisciplinary work, she continues to shape ethical innovation and policy discourse in global health research.

Chapter 7

Scientific Misconduct and the Waste of Research Funding: Can Existing Ethical Principles and Legal Frameworks Be Sufficient to Prevent Intentional Violations?

Young-Joon Ryu (See Chapter 5)

Chapter 8

From Professional Freedom to Professional Autonomy: Ethical Principles for the Use of Unproven Interventions

Takeo Saio (See Chapter 5)

Chieko Kurihara (See "About the Editors")

Part II Forward-Looking Perspectives for International Ethical Principles

Chapter 9

Bridging Between the CIOMS and the WMA: Common Ground, Similarities, Differences, and Next Steps

Dominique Sprumont Professor of Health Law, Institute of Health Law, University of Neuchâtel (http://www.unine.ch/ids), WMA academic partner. President of the Research Ethics Committee of Vaud (www.cer-vd-ch), Vice-President of CIOMS and past vice-director of the Swiss School of Public Health. Collaborated in the drafting of numerous laws and regulation in health law at the Swiss, European and international levels, including the WMA Declaration Helsinki (2008, 2013, 2024) or the Declaration of Taipei (2016), as well as the 2023 CIOMS International Guidelines on Good Governance Practice for Health-Related Research Institutions. Founder in 2002 – and member of the Board since 2021 – of the European Network of Research Ethics Committees (www.eurec-net.org). Since 2006, coordinator of the initiative Training and Resources in Research Ethics Evaluation (http://elearning.trree.org). He has written and edited more than 150 scientific publications, articles and book chapters in the field of public health law, research ethics and regulation, and patients' rights.

Chapter 10

Justice and Equity of Placebo-Controlled Trials: What Has Changed?

Chieko Kurihara, Dirceu Greco, Ames Dhai (See "About the Editors")

Takeo Saio (See Chapter 5)

Chapter 11

Post-trial Access as a Fundamental Right to Health: What Has Changed in the Implementation Principles?

Chieko Kurihara, Dirceu Greco, Ames Dhai (See "About the Editors")

Takeo Saio (See Chapter 5)

Chapter 12

Universities as Guardians of Ethics and Equity in Medical Research: Protecting the Rights of Vulnerable Populations, Ensuring Post-trial Access, and Advancing Transparency in Clinical Trials

Sarai Mirjam Keestra is a MD/PhD student within the Departments of Epidemiology and Data Science and Paediatric Endocrinology at the Amsterdam University Medical Centre, the Netherlands. A medical anthropologist by training, she previously conducted research at the University of Oxford and the London School of Hygiene and Tropical Medicine. She has been active in global health advocacy organisations such as Universities Allied for Essential Medicines and the People's Health Movement World Health Organization Watch since 2016. Her work focuses on applying principles of global health equity throughout the research process, with a specific focus on equitable research & development, technology transfer, clinical trial transparency, and post-trial access.

Yara de Koning is an independent global health researcher and access to medicines advocate. She works with various civil society organizations, including Universities Allied for Essential Medicines and the Pharmaceutical Accountability Foundation, focusing on equitable partnerships, access to essential medicines, and socially responsible licensing. She holds an MSc in International Public Health and through her work investigates equity, transparency, and accountability in pharmaceutical policy and industry practices, with a commitment to ensuring health innovations serve the public good.

Maurice Luca Remy is a PhD student at the Athena Institute of the Vrije Universiteit Amsterdam, the Netherlands, investigating the epistemic value of patient involvement in biomedical research. Prior to that, he obtained an MSc in Global Health. Since 2019, he has been active on various issues of global health advocacy, mainly with Universities Allied for Essential Medicines (UAEM). At UAEM, he held multiple leadership positions, among others serving on the European Coordinating Committee and as UAEM's representative to the World Health Organization's Youth Council during its inaugural term.

Molly Pugh-Jones currently works as Advocacy Manager (Digital Health and Human Rights) at STOPAIDS, a UK-based membership network working on HIV and other health rights and equity issues. Prior to this, she worked in research and advocacy roles focused on equitable access to medicines. She has an MSc in Global Health and Development, and previously held leadership roles in health justice groups including Youth Stop AIDS and Universities Allied for Essential Medicines.

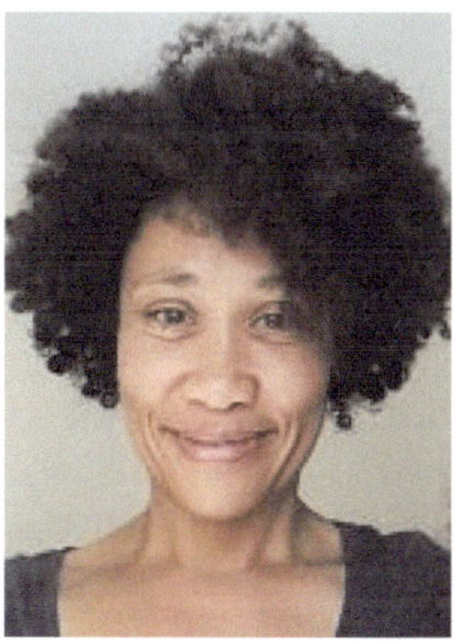

Lauren Paremoer is an Associate Professor at the Department of Political Studies, University of Cape Town. Informed by a feminist political economy approach, her research focuses on health activism, conceptions and enactments of solidarity in global governance for health, and political mobilisation aimed at realising social citizenship in the Global South. She is a member of the People's Health Movement and is active in leading its Democratising Global Health Governance programme.

Till Bruckner is the founder of TranspariMED, a campaign that works to end evidence distortion in medicine. Over the past eight years, TranspariMED has successfully driven improvements in clinical trial reporting in the UK, European Union and United States through research and analysis, coalition building, political advocacy, and public pressure. Till currently works as a postdoctoral research fellow at Karolinska Institutet in Sweden. He previously worked with the QUEST Center for Responsible Research and various nonprofit groups. He holds a PhD in politics from the University of Bristol, UK.

Toby Pepperrell is a medical doctor with a background in regenerative economics. He is currently undertaking a PhD in Environmental Sciences and Human Health at the University of Stirling, investigating the health impacts of community urban agroecology via the microbiome as a part of ecological reparative justice. He is an active member of the People's Health Movement (PHM) and has worked closely with Universities Allied for Essential Medicine (UAEM) UK. Having worked clinically and in trade and health policy, his research has predominantly focused on the Social and Commercial Determinants of Health.

Chapter 13

Patient Public Declaration of Research Ethics: Opinions and Proposals from Patients and the Public on Research Ethics

Chieko Kurihara (See "About the Editors")

Yoshiko Saito, Master of Social Design is a breast cancer survivor and 2018 Master's graduate in Social Design at Rikkyo University. She completed her research on expanding peer support networks for cancer survivors in hospitals throughout Japan, and now dedicates her time to helping others in her community. After suffering from breast cancer in 2015, Saito participated in the Breast Cancer Survivor Coordinator Program sponsored by NPO Cancer Net Japan to learn more about the medical care system and the mental implications of breast cancer to serve others in similar situations. In 2016, she completed a training course at

NPO Cancer Patients Support Organization. With her two qualifications, she now works as a peer supporter at both private and government-funded medical institutions in the greater Tokyo area. Over the last 10 years, Saito has moved around the globe, teaching Japanese in Asia before returning to Japan. Saito graduated from Rikkyo University in 1986 with a BA in Christian Ethics and Middle and High School Education in Social Studies.

Hiroto Kai is a patient of rare blood disease and is presently an undergraduate student at Aichi University of Education. The affliction has necessitated medical intervention for a duration exceeding 5 years. He is pursuing his academic interests in the field of bioethics, drug development processes, and science pedagogy. He has received fundamentals of the process of pharmaceutical medicine at the Japanese Institute for Public Engagement (Ji4pe). Furthermore, he has augmented his knowledge base via coursework offered in the EUPATI Open Classroom, where he focused on medical research and development. He holds memberships with the Chemical Society of Japan and the Japanese Society on Thrombosis and Hemostasis. In the spring of 2023, he commenced his studies at Aichi University of Education, where he is conducting research on bioethics and pharmaceuticals from an educational and young perspective.

Yoshikazu Funabashi, BEng graduated from Keio University in 1976 with a Bachelor of Engineering. After graduation, I worked for the Japan branch of a major American IT company as an SE/IT Specialist and Marketing Specialist. However, during my employment, I had bladder cancer and left the company 2 years later. I also recently had prostate cancer. As a peer supporter, I had joined the NPO Cancer Patients Support Organization in 2014 and received training as a peer supporter. Furthermore, in order to improve the quality of peer support, I studied to become a Japan Society of Clinical Oncology Cancer Network Navigator in 2020 and received certification as a result. In 2023, the Certified Network Navigator Committee appointed me to serve as an external member of the Navigator Working Group. Currently, as a Cancer

Survivor, I pro- vide peer support at hospitals in Tokyo and Yokohama, as well as at a hospital in Kanagawa Prefecture as a part-time employee.

Keiko Inoue, Associate degree in Law is a breast cancer patient and a victim of medical malpractice. She serves as a board member of the Association for Medical Malpractice Victims, providing counseling and support for victims and raising issues for society. She has served as a member of Relief Operations Committee of PMDA (Pharmaceuticals and Medical Devices Agency), and several third-party committees on research misconduct. She currently serves on Ethics Review Committees and a Certified Review Board at a national university and a national research institute. From her own experience, she realized that the general public without a medical background needs basic knowledge to participate in various fields in medicine, and is currently an instructor at the Japanese Institute for Public Engagement (Ji4pe) study course where she studied.

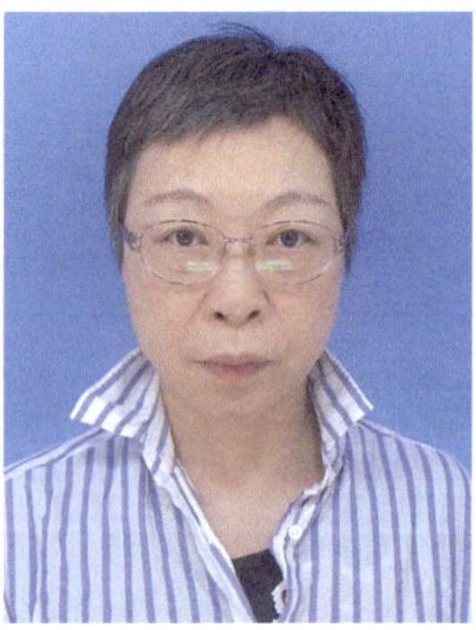

Noriko Kishi, BA Executive Director of CMT Association JAPAN (J-CMT). She was diagnosed with Charcot-Marie-Tooth disease in 1993 and joined the founding J-CMT in 2008. After graduating from the Department of Literature at Rikkyo University in 1983, she became the editor-in-chief at the Japan Finance News until 2014, then served as a senior researcher at the Japan Credit Information Reference Center Corp. from 2014 to 2020. Since 2020, she has been promoting project planning to improve the QOL of all intractable disease and disabled parties. Her academic affiliation includes the Japan Society for Disability Studies, the Japan Society of Social Design Studies, and the Japan Society for Business Ethics. The main organizations she belongs to as a member are the Japanese Institute for Public Engagement (Ji4pe), the Japan Fundraising Association (JFRA), and the Social Impact Management Initiative (SIMI).

Akemi Kuge, MA in Social Welfare Freelance social worker. After completing graduate school at Osaka Women's University, she worked as a law clerk and a writer for a local government women's policy public relations magazine and a consumer cooperative's bulletin. From her own experience, she became interested in the welfare of children who have not attended school. While working as a support worker for the mentally disabled and as a committee member of local government and social welfare councils, she obtained qualifications as a certified social worker and a mental health social worker. She has worked as a school social worker for 6.5 years. Participated in the establishment of a family peer support group for children and adolescents with mental disorders and served as its president for 5 years. She focuses on adolescent mental health welfare and suicide prevention education. Currently, she aims for the social restoration of all people with difficulties in life and their families, and the realization of empowerment, liberation, and well-being, including those who support them. She is also trying to promote networking and advocacy in education, welfare, healthcare, medical care, labor, justice, and community life as the environment for this, through activities that are not bound by position or affiliation.

Toshie Murakami, Bachelor of Psychology I work part-time as a peer supporter at Sagamihara Kyodo Hospital. As a survivor of cervical and breast cancer, I have been dedicated to peer support since 2007, grateful to be alive. In 2017, I obtained certification as a Senior Navigator from the Cancer Medical Network of the Japan Society of Clinical Oncology (JSCO), feeling the importance of ensuring the quality of peer support. Since then, I have served as an external committee member of Cancer Care Network Navigator Certification Committee of JSCO. I have also had the opportunity to present the need for peer support at JSCO annual meeting for three consecutive years. Related Committees: External Committee Member, Cancer Care Network Navigator Certification Committee of the Japan Society of Clinical Oncology (JSCO); Committee Member, Kanagawa Prefecture Cancer Control Review Board. Patient Groups: Representative of facilitators, Sagamihara Kyodo

Hospital Cancer Patients' Association Fukiiso; Vice Representative, Cancer Support Sagamihara; Representative, Cancer Philosophy in Sagamihara F.

Katsura Suzuki, Pharmacist Katsura Suzuki, Medical and life sciences researcher and patient with advanced-stage Parkinson's disease, involved in drug idea generation and discovery research for 27.5 years in ASKA Pharma until 2021. She is still employed as of 2025 and her current duties include ensuring legal guidelines for the provision of information on the company's medicines, and conducting and supervising medical reviews.

Member of the Japan Institute for Medical Development (Ji4pe).

Engaged in facilitation of shared decision-making. Member of an international patient network, PD Avengers, research committee. Founder of the Life Design Institute for PwP Japan, the aim that patients themselves develop treatments and take charge of their lives.

Since 2022, she has been a PhD student at the Juntendo University Graduate School of Medicine under. She is belong to the Department of Neurologic Sciences. Japanese Pharmacist License holder, Master of Pharmacology.

Hiroko Takahashi, Bachelor in Management After a family member was diagnosed with hereditary hemorrhagic telangiectasia (HHT), she became deeply aware of the importance of shared decision-making through the treatment process. To support patients and their families, she now serves as a board member of Hereditary Hemorrhagic Telangiectasia (HHT) Patients Association JAPAN, engaging in peer support and collaboration with healthcare professionals. As HHT is a genetic condition and was passed on to her own child, she began to consider what could be done for future generations. This led her to study basic medical knowledge and Patient and Public Involvement (PPI) through study sessions at the Japanese Institute for Public Engagement (Ji4pe). Based on this experience, she currently serves as a member of an ethics review committee at a public hospital.

Eiko Uchida President, Breast Cancer Patient Support Group BOUGAINVILLEA (in charge of breast cancer), Facilitator, Patient Voice Council (in charge of Basic Act on Medical Care), Vice President, Nangan Net (in charge of all cancer and intractable disease patients), Town Net Cafe (engaged in Local Civic Activities). In January 1994, while in Singapore, I experienced total breast cancer resection, chemotherapy, and breast reconstruction. Having encountered such medical care in Singapore that emphasizes patient self-determination and information disclosure, promoting informed consent, I established the Breast Cancer Patients' Association in January 1998 after returning back to Japan. It was out of gratitude to and requital for medical providers who saved my life. In our activities, we set the goal of the International Federation of Patient Organizations as the five principles for measuring patient centricity in healthcare, supporting the final outcome of patient and family to be their satisfaction of feeling of "living my own life even if having cancer." I am also working for the enactment of the "Basic Act on Medical Care," which should be the parent law of the healthcare system. Award: Received a letter of appreciation from the 23rd Annual Meeting of the Japan Breast Cancer Society. Received the Excellence Award at the Tokyo Metropolitan Women's Advancement Awards.

Kyoko Imamura, MD, DrMedSci, PhD is the past president of IFAPP, after completed her presidency in 2018–2020. She is an orthopedic surgeon and holds MD, Dr of Medical Science (clinical pharmacology), and PhD (outcomes research). Since joining the pharmaceutical industry in 1995, she has built up her career in R&D and Medical Affairs and held management positions in international pharmaceutical companies. At the same time, she joined Japanese Association of Pharmaceutical Medicine (JAPhMed), the Japanese member association of IFAPP, and served as the president from 2009 to 2016. Since 2017, she has been engaged in academic career, first as the project professor of Pharmaco-Business Innovation, and later as the professor of Social Cooperation Program of IT Healthcare, at Graduate School of Pharmaceutical Sciences, The University of Tokyo. She has led research projects to promote innovative clinical trials including virtual (site-less) clinical trials, promoting pharmacists' leadership in community, patient and public involve- ment, and real-world data studies. In 2020, she devel- oped Japanese Institute for Public Engagement (Ji4pe) as the non-profit organization and has promoted social engagement in drug development and health policy projects as the founding President.

Part III Alternative Frameworks for Ethical Innovation Towards Global Health

Chapter 14

Medicines Development for Global Health: The Evolution of IFAPP's Ethical Framework in the Ecosystem of the Life Sciences in a New World Order for Health

Varvara Baroutsou, MD, PhD, GFMD, EMAUD Consultant in Internal Medicine; Consultant in Pharmaceutical Medicine; Research & Experimental Development in Medical Sciences Expert; CIOMS Executive Committee Member; IFAPP President (2022-2025). With almost 40 years of experience in the practice of Internal Medicine and over 20 years experience as a Pharmaceutical Medicine Expert, she has led scientific research and development (R&D) departments, focusing on medicines development, through translational research and clinical trials to document innovative treatments. She started her career as an academic research fellow, hospital physician and clinical researcher at N.N.Th.A (Sotiria) and Am. Fleming she has held senior management positions in Clinical Development and Medical Groups of major pharmaceutical and biotech companies (BIANEX, MSD, Abbott, Sanofi, Novartis) in Greece and as Head of Western Europe. As President of the International Federation of Associations of Pharmaceutical Physicians and Pharmaceutical Medicine (IFAPP), she collaborates with non-profit scientific organisations (CIOMS, WMA, PharmaTrain, Faculty of Pharmaceutical Medicine -FPM), universities (ECPM-Switzerland, King's College, UK), patient associations and international organisations to implement initiatives of excellence in clinical research with patient involvement (Stavros Niarchos Bioethics Academy Alumni). Her contributions include reviewer and editor scientific journals, publications, scientific book chapters, teaching in postgraduate programmes and advisory contributions to educational, technical and regulatory standards committees. She is a graduate of the Medical School of the University of Athens, a Certified Consultant, Global Fellow in Medicines Development (GFMD), with postgraduate studies in Health Economics from the Stockholm School of Economics and a European Market Access Diploma from the University of Lyon (EMAUD).

Chapter 15

Ethical Principles for the Sustainable Development of Drugs for Neglected Tropical Diseases in Neglected Populations: Facilitating Meaningful Participation of Vulnerable People and Communities

Chieko Kurihara (See "About the Editors")

Takeo Saio (See Chapter 5)

Dirceu Greco, Ames Dhai (See "About the Editors")

Chapter 16

Reformation of Global Research Ethics Principles from the Perspective of Patient Public Involvement

Kotone Matsuyama, BPharm, RPh President-Elect, The International Federation of Associations of Pharmaceutical Physicians and Pharmaceutical Medicine (IFAPP), since Apr 2025, Former Chair of the Ethics Working Group June 2021 to Apr 2025; Contents Committee Lead since 2025, Community Lead of Bioethics since 2022, Drug Information Association (DIA) Japan. She is the Director of the Center for Clinical Research and Development, National Center for Child Health and Development (NCCHD) since June 2025. She graduated from Kyoto University Faculty of Pharmaceutical Science and obtained a license as a pharmacist. She started her Research and Development (R&D) career as a data manager in 2003 and has worked as a project manager since 2005. She spent 12 years at the Translational Research Informatics Center, Foundation for

Biomedical Research and Innovation, Kobe, 2.5 years as a Lecturer at the Center for Quality Assurance in R&D, Kyoto Prefectural University of Medicine, and Professor at the Department of Health Policy and Management at Nippon Medical School since 2017. She has managed or supported many R&D projects initiated by academia, including product seeds management, clinical trial protocol development, not only for drugs but also for medical devices and regenerative medicines. She specializes in clinical pharmacology and project management. She works with the research governance and the activities on patient public involvement.

Chapter 17

Global Efforts Towards the Highest Ethical Standards: Horizon for the Future

Chieko Kurihara, Dirceu Greco, Ames Dhai (See "About the Editors")

Part I
Research Ethics Perspectives from the Global South and Asia

Chapter 1
Bioethics in Brazil to Overcome Pandemics: Striving for the Highest Ethical Standards to Prevent Exploitation and Achieve Health for All

A Case Study of the AIDS and COVID-19 Pandemics

Dirceu Greco

Abstract The experiences in Brazil in responding to the AIDS and COVID-19 pandemics underscore the critical importance of bioethics as guiding principles in confronting public health crises. While Brazil's response to the AIDS pandemic, marked by a strong ethical framework promoting and protecting human rights, serves as a notable example of successful intervention, the COVID-19 pandemic revealed significant shortcomings in the application of these principles, particularly concerning social equity, international solidarity and the role of scientific evidence.

This chapter emphasizes the importance of a universal public health system and a national research ethics framework for the protection of human research participants. It also examines how the highest research ethics principles, promulgated by the Declaration of Helsinki (DoH) by the World Medical Association in 2000, have been incorporated into and have contributed to the Brazilian research ethics framework, which has been maintained until now. The analysis compares not only Brazil's responses to the AIDS and COVID-19 pandemics, but also the DoH 2000 and the 2024 versions, to be shared with the global community of research ethics for deliberations.

Keywords AIDS · COVID-19 · Human rights · Health for all · Bioethics · Universal public health system

---

D. Greco (✉)
School of Medicine, Federal University of Minas Gerais, Belo Horizonte, Brazil

C. Kurihara et al. (eds.), *The 2024 Declaration of Helsinki*,
https://doi.org/10.1007/978-981-96-9294-1_1

1 Introduction

1.1 Role of Bioethics and Research Ethics Principles: Comparative Analysis Between the AIDS and COVID-19 Pandemics

Bioethics plays a vital role in navigating the complex moral and ethical dilemmas that invariably arise during public health emergencies. These crises often force societies to confront difficult questions at the intersection of individual rights and collective well-being, where established norms and values may be challenged by the urgency and scale of the situation.

Brazil, a nation marked by unacceptable social and economic disparities, was confronted with two major pandemics in the last four decades: the AIDS pandemic, which began to spread in the early 1980s and disproportionately affected marginalized communities, and the more recent COVID-19 pandemic, which placed an unprecedented strain on the national health-care system and resulted in over 700,000 deaths and also affected disproportionately vulnerable individuals. A robust ethical framework is fundamental for confronting such crises, ensuring that efforts to control disease outbreaks are not only effective but also just and equitable.

This chapter begins with a review of the 2024 revision of the Declaration of Helsinki (DoH) by the World Medical Association (WMA) [1], and followingly, provides reasoning for this evaluation through a comparative analysis of the role of bioethics in confronting the two health crises: AIDS and COVID-19. Bioethics has contributed to Brazil's efforts to prevent the exploitation of vulnerable populations and to promote equitable access to health for all during both pandemics. Also, by examining the ethical dimensions of these experiences, this analysis seeks to identify lessons that can help in preparing and establishing response strategies for facing future pandemics and similar challenges in Brazil and other nations. Comparison between the versions of the 2000 [2] and the 2024 DoH [1] would present significant implications in the context of research, as part of the whole picture of the role of bioethics in these health crises.

1.2 Role of Bioethics for Human Rights, Equality, and Sustainable Ecosystem

The emergence of bioethics as a discipline was partly motivated by the recognition of the need to associate scientific advancements with humanistic values, as this confrontation demands both scientific understanding and adequate ethical considerations regarding human life and dignity. According to Potter [3], the risks to the ecosystem's survival are due to the widening gap between scientific and humanistic knowledge. This remains highly relevant when addressing pandemics, where scientific data on disease transmission and treatment must be carefully balanced with ethical considerations of individual rights, social justice, and the allocation of scarce resources.

Brazil's experiences with both the AIDS and COVID-19 pandemics, in a context of marked social and economic inequality, underscore the critical need for bioethical frameworks that help address systemic vulnerabilities. The huge social disparities in Brazil explain the nonuniform impact of pandemics, with marginalized communities bearing a disproportionate burden of illness and mortality. The application of bioethical principles that explicitly aim to mitigate social inequalities is essential for ensuring a just and effective pandemic response, although by themselves they are not sufficient. This will be shown by the analysis of the differences in the responses to the AIDS and the COVID pandemics, led by the deliberation of the implication of the 2024 revision of the DoH.

2 Implication of the 2024 Revision of the DoH from the Perspective of Justice in Bioethics

2.1 Pros and Cons of the 2024 Revision of the DoH

The new version of the DoH was approved in Helsinki in October 2024, receiving various changes, many of which are real progress to its scope. The following points are especially notable ("para." means "paragraph"):

- Recognition of "*various structural inequalities*" in research and ways to deal with them (para. 6).
- That meaningful engagement with research participants and their communities is fundamental and should be required at all stages of the research (para. 6).
- The proposition to strengthen research ethics committees, which "*must have independence and authority to resist undue influence from the researcher, the sponsor, or others*," keeping qualification and diversity by education and training, and having "*sufficient familiarity with local circumstances and context, and include at least one member of the general public*" (para. 23).
- That the fixed, contextual, and dynamic nature of vulnerability must be recognized and that medical research in situations of particular vulnerability is only justified if it is responsive to their health needs and priorities. They could be included only when the research cannot be carried out in a less vulnerable group or when excluding them would perpetuate or exacerbate their disparities (para. 19, 20).
- Reinforces that the principles of the Declaration must also be upheld during public health crises (para. 8).

However, the fact that the following were rejected during the revision process has raised special attentions [4, 5] and needs further deliberations in the global community:

1. The "social value" of research has been agreed upon internationally as an established prerequisite condition of research [6] and has already been incorporated into policy in several jurisdictions and guidelines [7]. Although it was included in the proposed draft for the WMA's first public consultation, it was deleted in the final version adopted in 2024. Social value is in accordance with the core, fundamental principle of the DoH that the purpose of research "*never takes precedence over the rights and interests of individual participants.*" Thus, it must be reinforced to implement relevant research and to ensure benefit sharing through co-creation with the community, respecting its value.
2. The use of a placebo in clinical trials can be acceptable only when there is no proven effective and safe comparator. This text was in the 2000 DoH but has been downgraded since the 2002 Note of Clarification to the DoH, and the compromised principle has been maintained in 2024, with a very high level of risk to the research participants ("*participants who receive any intervention other than the best proven one(s), placebo, or no intervention will not be subject to additional risks of serious or irreversible harm as a result of not receiving the best proven intervention*" (para. 33).
3. Post-trial access: The 2000 version stated that at the end of a trial, participants who still need interventions that have been shown to be safe and effective must have access to them ensured by the sponsor(s). This has also been downgraded since the 2004 Note of Clarification, and the 2024 DoH is much weaker because instead of guaranteeing access, it just says it "*must be arranged*" (para. 34).
4. Additionally, the opportunity was missed to extend post-trial access beyond drugs or devices to include tools or systems utilizing artificial intelligence (AI), particularly those developed using data from studies and other relevant data from low- or middle-income countries (LMICs) [8]. This would ensure more equitable access to these tools, fostering fairness in global collaborative studies in the field of digital health. This is also a missed opportunity to capitalize on the inclusion of the WMA Declaration of Taipei (DoT) on health databases and biobanks [9], incorporating in the DoH (para. 32), to expand the ethics of data-driven research for equitable global health.

2.2 *Implication of the 2024 DoH from a Perspective of Global Justice*

The author cohosted online meetings twice (October 15 and 19–20, 2024) during his stay in Helsinki to participate in the WMA General Assembly (October 16–19, 2024) that adopted the new DoH, with international contributors present in person and others connected globally. The "Helsinki Statement" [4, 5] was issued in the above context, one day before the adoption of the new DoH. Particularly with regard to placebo and post-trial access, the Uruguayan Medical Union submitted two motions [10] to reinstate the principles of the 2000 DoH, and their representatives attended one of our meetings. Although the motions were voted down, they represent the positions of Latin America, the Caribbean [11] and some countries in Southwestern Europe [12]. Furthermore, our Helsinki Statement was endorsed by 125 individuals and 5 organizations, mainly from the Global South and Asian countries. It is worth noting that at the General Assembly, 50 national medical associations were present, out of a total of 115 members; and only national medical associations have the right to vote on the DoH.

The DoH has been considered one of the most recognized ethical principles for research involving humans. The 2000 DoH defined strict limits to placebo use and added the guarantee of the right to post-trial access, as the result of international controversies to avoid exploitation of vulnerable research participants [13–16]. However, soon after its adoption, the strong pressure exerted mainly by the US researchers and US agencies watered down these principles of protection in the 2000 version [17], and this threatened its authority. Some communities (represented by the US) abandoned such ethically high standards, which are difficult to follow [18, 19], while others (represented by Brazil and other Latin American and Caribbean countries) rejected the weakened protection included in the main texts of the 2008 DoH [20, 21].

Such reactions may be seen as a testament to the importance of the DoH as a benchmark and should be read as people's aspirations and commitments to the highest ethical standards. As described below, Brazil decided in 2008 [22] to include in its national research ethics guidelines the requirements established in the 2000 version of the DoH, with the aim of protecting research participants. This could serve as an example for the global community to deliberate how to implement international standards to protect the research participants in their own communities and for further discussions of future revisions of the DoH, as well as seeking for global ethics norms shared equally among all the interested parties.

3 The Essential Role of the Brazilian Unified Health System

In Brazil, the Unified Health System (SUS) has played an essential role in ensuring that the protection and promotion of human rights are translated into health for all.

A Sanitary Reform movement was established as a part of the resistance to the dictatorship (1964–1985) and aimed at strengthening public policies and construction of the bases for a Social Welfare State [23]. It involved civil society, trade unions, health-related/social professionals and organizations, universities, students, and academia, among others, who together considered health as a social and political issue, not just a biological issue dependent exclusively on medical care. The pressure of these movements culminated in 1985 with the end of the 21-year military dictatorship and the restoration of democracy [24].

Soon after, a new constitution was promulgated (1988), defining in its article 196 that health is a universal right and a "duty of the state, and shall be guaranteed by means of social and economic policies aimed at reducing the risk of illness and other hazards, and at universal and equal access to actions and services for the promotion, protection and recovery of health," giving rise to the Unified Health System (SUS).

SUS, a hallmark in the history of public health in Brazil, is based on four fundamental principles: Universality (access to all), Comprehensiveness (comprising prevention, care, and treatment), Equity (the health of all citizens is equally guaranteed by the State), Decentralization of actions from the central government to the states and then to municipalities, and Societal participation (effective participation of the community in the planning, control, and execution of public health services) [25]. The creation of SUS was fundamental to the exemplary confrontation of the AIDS pandemic [25].

4 Establishment of Bioethical Framework in Brazil

4.1 Application of the Principles of Bioethics to Brazilian Context

4.1.1 Intervention Bioethics

The foundational principles of bioethics, which have traditionally guided medical ethics globally, include autonomy, beneficence, non-maleficence, and justice [26]. In Brazil, these principles are generally recognized and applied within medical practice, including in the Brazilian Code of Medical Ethics [27].

The limitations of these traditional bioethical principles in addressing marked social exclusion and inequality have led to the increasing prominence of intervention bioethics [28]. Intervention bioethics, a critical approach that originated in Latin America, recognizes that ethical dilemmas in countries like Brazil often require context-sensitive analysis that considers the broader sociocultural landscape. This approach incorporates additional references such as vulnerability, dignity, solidarity, confidentiality, privacy, responsibility, and prudence to reduce social inequalities, focusing on ethical conflicts and persistent problems from a perspective that goes beyond individualistic concerns. In Brazil, this approach is in accordance with the principles of the Brazilian Unified Health System (SUS).

4.1.2 Emancipation Bioethics

In the struggles for equality and against injustice in research and public health [29], emancipation is used here in the sense described by Paulo Freire, the Brazilian educator and philosopher, who, in *Pedagogy of the Oppressed* [30], argued that emancipation isn't something that is *given* to people by a benevolent leader; rather, it happens when people themselves critically perceive the social, political, and economic contradictions in their lives and act against the oppressive elements of reality. This is exactly how it should be used when discussing citizenship, rights, or confronting the social determinants of health. Emancipation, for Freire, is both critical reflection and transformative action [31].

It is a counterpoint to empowerment, which may be considered no more than a figure of rhetoric, since power is never given and rarely shared. In most cases, the use of the term corresponds to an Illuminist form to marginally providing a bit for the needy, aiming to soothe their claims. It is common for these individuals/countries with economic difficulties to praise their "donors" for helping them to get something that is indeed their right. This attitude may perpetuate disparity and dependency.

Marx saw emancipation—true human freedom—as something that won't be granted from above (not by chance, nor by benevolence from rulers or capitalists), but as something that must be won through active human struggle.

Thus, agreeing with Freire and Marx, justice only becomes reality when the oppressed themselves—those who experience exploitation, alienation, and injustice—become conscious of their situation and emancipate to fight for their rights [32].

4.2 The National Health Council (CNS) and the Role of the Brazilian Research Ethics Commission (CONEP) in Protecting Research Participants

The Brazilian Society of Bioethics was created in 1995 [33], and its members are committed to the strengthening of a legal framework for the protection of research participants as well as for the defense of public health. Followingly, the role of a comprehensive set of Brazilian research ethical guidelines in the promotion and protection of research participants and the Brazilian Unified Health System (SUS)—Brazilian research ethics framework has developed vis-à-vis international guidelines, especially the Declaration of Helsinki (Table 1.1). These guidelines emphasize fundamental bioethical principles, aiming at protecting the rights of research participants, including the recognition and affirmation of dignity, rights to confidentiality, and a due informed consent process, liberty, autonomy, beneficence, non-maleficence, justice, and equity. Furthermore, legal standards related to patient rights are enshrined in Brazilian law and professional codes, providing a legal framework for ethical conduct in health care. There is usually tension between individual autonomy and collective health needs, which requires careful ethical consideration, especially given the Brazilian constitutional emphasis on health as a universal right.

Table 1.1 The process and timeline of establishment of the bioethical framework in Brazil

End of the military dictatorship	1985
Update of the Brazilian code of medical ethics	1988
New Federal Constitution is approved	1988
Creation of the universal unified health system (SUS)	1990
Establishment of the Brazilian Society of Bioethics (SBB)	1995
Creation of the National Health Research Commission (CONEP)/research ethics committees system and of a national research ethics guideline	1996
Establishment of highest standard in the DoH in terms of placebo-controlled trial and post-trial access to avoid exploitation	2000
Incorporation of the limits to the use of placebo and the right to post-trial access in Brazilian guidelines	2008
Downgrade of the above principles in the main text of the DoH (kept until 2024)	2008 (to 2024)
Issuing of the resolution 466/2012, succeeding res 196/96	2012

The National Health Council (CNS), promulgated by Law 8142/1990 as part of SUS [24, 25], is a deliberative and permanent committee linked to the Brazilian Ministry of Health, with the mission to participate in the formulation and control the execution of National Health Policy at a federal level, as well as to systematically monitor public health policies. It comprises 48 members elected to represent public health system users, workers, SUS managers, and health service providers.

In 1996, the CNS established the National Research Ethics Commission (CONEP) and local Research Ethics Committees (CEPs). This framework is responsible for the ethical supervision of all research projects involving humans, which are submitted to more than 800 CEPs distributed throughout the country. This system was established by Resolution CNS 196/96, and succeeded by Resolution 466/2012, with guidelines and regulatory norms for all research involving humans.

The existence of a well-established national system for ethical review of human research through the CEP/CONEP system underscores that a strong infrastructure for ensuring ethical conduct in research is particularly important during pandemics, when rapid research and development of treatments and vaccines are crucial.

4.3 *Brazilian Research Ethics Guidelines vis-à-vis the Declaration of Helsinki: Placebo and Post-trial Access*

Brazil clearly opposed the proposed changes in the 2008 version of the DoH to incorporate the Notes of Clarifications issued in 2002 for the use of placebo and in 2004 downgrading the right to post-trial access. This actually lowered the ethical requirements established in the 2000 version. By sovereign decision, the CNS approved Resolution 404/2008 [34], which ensured that placebo-controlled trials must be limited to the situation when there is no proven intervention; and that at the end of a clinical trial, all participants must be guaranteed access to the best proven prophylactic, therapeutic and diagnostic options identified by the respective clinical trial, both in accordance with the 2000 DoH. Resolution 404/2008 also included that "the discussion on access to health care and products that prove effective for all who need them should be expanded" in public health. This was subsequently expanded and incorporated into Resolution 466/2012 [35] (Table 1.2).

However, in May 2024, Law 14,874 was approved by the Brazilian legislator that "Provides for research with human beings and establishes the National System of Ethics in Research with Human Beings." Of interest in this chapter are the items on limits to the use of placebo and post-trial access. The former is as strict as in Resolution 466/2012 but post-trial access will be for a "determined" period of time, while currently the guidelines specify that it is for an indefinite period, as a responsibility of the sponsor [36]. As of April 2025, the decree regulating its application has not yet been issued, leaving the opportunity to counterargue this setback *in the protection of research participants.*

Responding to such situation, Brazilian Society of Bioethics, with international collaborators, organized a web-based symposium to broaden the horizons of discussion on the issues the global community is confronting concerning the 2024 revision of the DoH [7].

5 Ethical Challenges and Responses During the AIDS Pandemic in Brazil

The AIDS pandemic in Brazil and elsewhere presented a multitude of complex social, sanitary, and ethical challenges, especially in its early stages, when the disease was poorly understood and highly stigmatized. In this context, all the

Table 1.2 Texts of Item III.3 of Resolution 466/2012 by the National Health Council incorporating the principles to justify placebo-controlled trials only when there is no proven intervention, and guaranteeing post-trial access

Item III.3 of resolution 466/2012 (underlined by author)
"All research using experimental methodologies in the biomedical area, involving human beings… b) when using placebo, such use shall be fully justified as to its non-maleficence and methodology requirements, where the benefits, risks, difficulties, and *effectiveness of a new therapeutic method shall be tested, comparing it to the best current prophylactic, diagnostic and therapeutic methods. No use of placebo or any other treatment is permitted in studies where there are no proven methods of prophylaxis, diagnosis, or treatment*; … and d) *guarantee to all participants, at the end of the study and for an unlimited time, free access to the best prophylactic, diagnostic and therapeutic methods with proven efficacy*; d.1) access will also be guaranteed in the interval between the end of the individual's participation and the end of the study, in which case this guarantee could be granted using extension studies, according to a duly justified analysis from the physician assisting the research participant".

Note: d) means continuation after the end of the study period, for as long as needed; d.1) means continuation between the end of individual participation and the end of the study period, using extension studies

bioethical issues were discussed in depth, and deliberations and struggles in the society led to the successful implementation of principles into practice.

Stigma and discrimination against people living with AIDS were widespread, significantly hindering their access to prevention, testing, and treatment, and social inclusion. This disproportionately affected vulnerable populations such as men who have sex with men, LGBTQIA+ people, sex workers, and people who use drugs, who have faced increasing social marginalization, which has impeded effective public health interventions.

Confidentiality issues also posed significant ethical challenges, as health-care professionals struggled with the tension between the duty to maintain patient confidentiality and the need to prevent HIV transmission to others. Balancing the right to individual privacy with the public health imperative of controlling the epidemic required careful ethical analysis.

In 1982, Brazil reported its first cases of AIDS as the country was emerging from a military dictatorship (1964–1985). In 1986, the Ministry of Health (MoH) established the National STD/AIDS Program, which involved extensive participation from interested parties, including nongovernmental organizations (NGOs), public universities, and state/municipal health secretariats. These initiatives included, but were not limited to, mandatory screening for bloodborne infectious diseases and the prohibition of paid blood donations in 1993 [37], the establishment of nationwide specialized clinical services, and the implementation of a harm reduction program [38]. They also included the financing of NGOs projects and the creation, in 1987, of a National AIDS Commission (CNAIDS) to advise the National AIDS program on the response to the pandemic. Furthermore, national networks were implemented to monitor HIV genetic diversity [39], perform CD4+ T-cell counts, quantify viral load, and monitor HIV drug resistance. In 1990, SUS was created, and in 1996 a federal law provided free access to antiretrovirals (ARV) through SUS [40]. Morbidity and mortality have decreased, and while infection rates have fallen to 0.5% in the general population (2019), high prevalence continued among key populations [41]. The comprehensive response implemented to combat the AIDS epidemic was made possible through funding provided mainly by the Brazilian taxpayers.

The collaboration between ministries was also very clear and effective. For example, the alignment between the Ministry of Health and Foreign Affairs facilitated the prominent position in various international fora, especially WHO/UNAIDS, in defense of human rights, against discrimination, and for a worldwide expansion of access to antiretrovirals.

Brazil's response to the AIDS pandemic has been strongly informed by bioethical principles, with an emphasis on human rights and solidarity. Importantly, strong social mobilization and ensuring access to antiretroviral therapy through SUS as a human right were fundamental in shaping Brazil's successful response to the AIDS pandemic.

Furthermore, Brazil has a proactive stance on local pharmaceutical production, particularly by state-owned laboratories. Its strategic use of negotiating power to obtain lower drug prices and its use of "TRIPS flexibilities" to issue in 2007 a compulsory license. This framework was instrumented for the antiretroviral drug Efavirenz not only to ensure the financial sustainability of its universal access program in Brazil but also for a broader impact on global access to essential medicines, setting a precedent for other developing countries facing similar challenges [42]. "TRIPS flexibilities" were explicitly articulated in 2001 by the Doha Declaration [43] of the World Trade Organization (WTO)'s Agreement on Trade-Related Aspects of Intellectual Property Rights (TRIPS Agreement), clarifying that public health is prioritized over the protection of intellectual property rights. By prioritizing the health needs of its population, Brazil has demonstrated an exemplary ethical commitment to the well-being of its citizens. While legal and policy changes have played a crucial role in ensuring access to care and treatment, addressing the social determinants of health has proven to be a more complex ethical task in dealing not only with legal and policy issues but also with the underlying social and cultural factors that contribute to vulnerability and discrimination.

6 Ethical Challenges and Responses During the COVID-19 Pandemic in Brazil

Nearly four decades later, to effectively combat COVID-19, lessons learned from Brazil's experience with AIDS should have been applied to counter growing difficulties with unprecedented rapidity, e.g., combatting isolationism, strengthening international solidarity and cooperation, investing adequately in science and quality public health services, addressing anti-science rhetoric and anti-vaccine movements, and prioritizing equal access to technological progress [44]. These two epidemics have distinct epidemiological, biological, and clinical profiles. When faced with both infections, various scientific communities responded quickly by identifying and sequencing the causative agents, developing diagnostic tests, characterizing each infectious agent, setting up clinical trials to evaluate potential treatments, and providing clinical facilities to care for infected individuals. Regarding COVID-19, effective vaccines were developed rapidly using innovative technologies. Brazil took part in a number of global placebo-controlled trials of vaccines collaborating with external sponsors. Considering the experience gained during the AIDS pandemic, the broad scope of the SUS, and the presence of a robust national immunization program (PNI) capable of vaccinating millions per day, Brazil had all the necessary tools to respond ethically and effectively to the COVID-19 pandemic.

However, the Brazilian president has denied the severity of COVID-19 [45] and, without scientific evidence for its use, pressed on early treatment with chloroquine, through regulations issued by the Ministry of Health. Government officials also downplayed the use of masks and social distancing, and publicly denied the efficacy of developed vaccines. The situation was aggravated by the position of the Federal Medical Council, which supported the unlimited autonomy of doctors in prescribing repositioned drugs [46]. These positions were both unethical and unscientific.

Additional factors may help explain Brazil's antithetical responses to the COVID-19 versus to AIDS. Disparities among Brazilians across the country have required coordinated efforts in both cases. As already described on AIDS pandemic response, alignment between the Ministry of Health and the Ministry of Foreign Affairs enabled Brazil to occupy a prominent position in international forums and to exert leadership in negotiations to expand universal access to antiretrovirals. However, on the contrary, since the beginning of the COVID-19 pandemic, there has been a lack of coordination and inadequate strategies between these ministries regarding the acquisition of diagnostic kits, masks, active pharmaceutical ingredients (API), and vaccines.

It is important to note that the dimension of the pandemic has put enormous pressure on the Brazilian health-care system, worsening the already existing shortages of medical resources such as ICU beds, ventilators, and, initially, vaccines. This scarcity, combined with the aforementioned disarticulation and lack of political will, has raised unprecedent ethical challenges regarding their allocation. Decisions had to be made about which patients would receive life-saving treatments, often based on complex and ethically fraught criteria such as disease severity, age, prognosis, and the presence of comorbidities. These ethical challenges underscored the urgent need for clear, transparent, fair and ethically sound criteria to guide the prioritization of access to critically scarce medical resources.

Confronting such critical challenges, the Brazilian Society of Bioethics developed and issued a set of recommendations (Recommendation 01/2020) [47], with clear objectives to enhance respect for and protection of human dignity and human rights, aiming at ensuring equalitarian access to health care during public health emergencies [48].

Another particularly damaging ethical challenge during the COVID-19 pandemic in Brazil has been the widespread dissemination of misinformation and denial of scientific evidence, often originating from government sources. This "infodemic" [49] has undermined public trust in science and public health institutions, hindering adherence to preventive measures and potentially contributing to higher infection and mortality rates. Brazil, with less than 3% of the world's population, has accounted for 10% of the world's deaths (700,000) from COVID-19.

Additionally, cases of potential exploitation in research and treatment have emerged, involving the use of unproven treatments and allegations of significant ethical violations [50], including lack of informed consent and data manipulation. Also, vaccine hesitancy, fueled in part by misinformation, presented an ethical challenge to achieving widespread immunization coverage.

It is worth noting that Brazil was involved in the early development of vaccines and, despite all the problems mentioned above, the CEP/CONEP system rejected placebo-controlled trials soon after effective and safe vaccines were approved, and Brazil committed to ensuring access to these interventions for its own citizens.

The experiences of other countries and the ethical debates surrounding resource allocation during this and previous health crises offer valuable lessons that could inform the development of robust ethical guidelines for future public health emergencies in Brazil.

7 Comparative Analysis and Lessons Learned

A comparison of the ethical challenges and responses in Brazil during the AIDS and COVID-19 pandemics reveals both similarities and stark differences. Both pandemics brought to the forefront the overarching challenges of stigma and discrimination, albeit manifested in different ways, and the critical importance of ensuring access to care for vulnerable populations. Both also presented ethical dilemmas regarding the balance between implementing necessary public health measures and respecting individual rights and autonomy.

As discussed above, significant differences emerged in the overall approach and effectiveness of responses. The early AIDS response in Brazil was characterized by early government response, by a more unified and rights-based approach, driven by strong social mobilization and a commitment to universal access to care and treatment. In contrast, the response to COVID-19 has often been fragmented, marked by political interference and often characterized by denial of scientific evidence.

The role of civil society has also differed; while AIDS activist groups have played a key role in shaping the AIDS response, their influence was less evident during the early stages of the response to the COVID-19 pandemic. Furthermore, the global context was significantly different. The AIDS pandemic saw more international solidarity, particularly on the issue of access to medicines, while the COVID-19 pandemic has often been characterized by nationalistic tendencies and fierce competition for essential resources such as vaccines [51] and personal protective equipment (PPE).

8 Discussion: Recommendations for Strengthening Bioethical Preparedness and Response

8.1 Protecting Dignity, Rights, and Promote Equity During Pandemic

Key lessons emerge from this comparative analysis regarding the application of bioethics in preventing exploitation and achieving health for all during public health crises. A strong ethical framework, firmly grounded in human rights and social justice, is paramount. The active participation and advocacy of civil society organizations are crucial for holding governments accountable and ensuring that the needs of vulnerable populations are addressed. Transparent and equitable mechanisms for allocating scarce resources are essential to maintain public trust and ensure fairness during times of crisis. Addressing the underlying social determinants of health is not only a matter of social justice but also a critical component of effective pandemic preparedness and response. The dangers of political interference and the spread of misinformation in public health responses cannot be overstated, as they can undermine effective interventions and erode public trust. Finally, international collaboration and solidarity are vital for ensuring equitable access to essential resources and for sharing knowledge and best practices.

The experiences of the AIDS pandemic in Brazil offered valuable lessons that could have informed the ethical and practical aspects of the response to COVID-19. The strong ethical commitment of the Brazilian response to AIDS in ensuring universal access to care and treatment, combating stigma through public health campaigns, and engaging civil society can serve as a blueprint for future health emergencies.

The significant rise of misinformation and science denialism during the COVID-19 pandemic presented a major ethical challenge that was less prominent during the AIDS pandemic. While stigma and some level of misinformation were present during the AIDS epidemic, the scale and political support for science denialism during COVID-19 were unprecedented. This underscores the evolving nature of public health threats and the critical need for robust strategies to proactively combat "infodemics" and promote accurate health information to ensure ethical and effective pandemic responses.

8.2 The Fundamental Role of a Unified Universal Health System (SUS) as an Example: Reinforce the Role of SUS by Better Financing

Learning from the past experiences of global societies, specific and targeted strategies must be strengthened and adequately financed to prevent the exploitation and stigmatization of vulnerable populations during pandemics, emphasizing the need for tailored interventions and robust protections for marginalized communities, who are often disproportionately affected by health crises. Ultimately, it is crucial to integrate bioethical considerations into all aspects of public health policy and practice, not just during emergencies, to build a more resilient and equitable public health-care system in Brazil and elsewhere. Therefore, the existence of a truly universal public health system is essential to reaffirm and strengthen the commitment to ensure that reducing social inequalities is a central priority in health policies and pandemic preparedness plans, based on the concept that health products and interventions are not commercial commodities but common public goods [52–54].

In this direction, a proposal is underway to establish a Brazilian national institution for the prevention and control of health emergencies [55], linked to the Ministry of Health, in line with the principles and guidelines of the Brazilian Unified Health System (SUS).

8.3 The Role of a Research Ethics Network System by CNS and CONEP as an Example: Reinforce Strengthened Independency and Capacity for Protecting Research Participants

It is essential to strengthen the role and independence of a network of national and regional research ethics frameworks such as the Brazilian National Health Council (CNS) and the CEP/CONEP system to ensure robust ethical oversight of research and public health measures, particularly during emergencies when rapid decisions are required. Developing clear and ethically sound guidelines for the allocation of scarce medical resources during public health emergencies is paramount, ensuring transparency, fairness, and explicit consideration of vulnerability.

Such a system could expand the capacity of each country, community, or institution to establish specific rules on research ethics that are more rigorous and more protective than guidelines such as CIOMS [6] and DoH [1], which leads to strengthened, independent research ethics committees (REC) to ensure the rights of research participants and to avoid exploitation, maintaining scientific integrity.

Integrating comprehensive bioethical training and education into the curriculum for health-care professionals across all disciplines and for public health policymakers is essential to build a workforce that is equipped to navigate the complex ethical challenges of pandemic response.

8.4 The Participation of All Interested Parties

It is required to invest in public health education and communication strategies to promote health literacy, effectively combat misinformation, and build public trust in scientific evidence and public health institutions. A good example is the engagement of Japanese research participants in the interpretation of the DoH from their point of view [56]. This process will strengthen and stimulate the participation of participants in all stages of research, as the newly added paragraph of the 2024 DoH states. It will also foster stronger and more sustained collaborations between government agencies, civil society organizations, researchers, and health-care professionals in pandemic preparedness and response planning is crucial for a more coordinated and ethically informed approach.

It is also prerequisite to establish specific and targeted strategies to prevent the exploitation of already-vulnerable populations during pandemics, including those at risk of human trafficking, emphasizing the need for tailored interventions and robust protections for marginalized communities, who are often disproportionately affected by health crises [57].

Specific strategies, with effective participation of civil society with the inclusion of research participants, must be developed and implemented to address stigma and discrimination related to infectious diseases, ensuring equitable access to prevention, testing, treatment, and care for all populations. This must include a thorough revision of the TRIPS agreement to ensure access to drugs and vaccines for all who need them globally.

9 Conclusion

The experiences of Brazil in confronting the AIDS and COVID-19 pandemics underscore the critical importance of bioethics in guiding responses to public health crises. Brazil's response to the AIDS pandemic serves as a notable example of ethics- and human rights-based successful intervention. On the other hand, the COVID-19 pandemic has revealed significant shortcomings in the application of these principles, particularly in regard to social equity, international solidarity, and the role of scientific evidence.

Lessons learned from both pandemics highlight the enduring need for a robust bioethical framework and a universal public health system that prioritizes the promotion and protection of human rights for all individuals, and prevents the exploitation of vulnerable populations, while striving to achieve health for all. Strengthening international collaborations and actively supporting global efforts for equitable access to vaccines, treatments, and other essential resources during pandemics is both an ethical imperative and a practical necessity. It is crucial to integrate bioethical considerations into all aspects of public health policy and practice, not just during emergencies, to build a more resilient and equitable health-care system. By strengthening ethical oversight mechanisms, investing in public health education, fostering collaboration across sectors, and with the effective participation of civil society countries can better prepare for and respond to future pandemics in a manner that upholds the highest ethical standards and ensures more just and equitable outcomes for all its citizens.

The DoH, although prepared by physicians and directed to physicians, has become, on its own merits, the most referred research ethics document worldwide and this achievement is undoubtedly worth commemorating on its 60th anniversary in 2024. The DoH is a living document, thus discussions about its content will stimulate ample debates among people from different parts of the world, aiming at achieving the highest standards equally applied to everyone, everywhere.

Acknowledgments The author was officially engaged in discussions for the revisions of the Declaration of Helsinki from 2000 to 2008 as a member of the Brazilian Medical Association and Brazilian Medical Council teams; he also participated in the CIOMS 2016 guidelines as a member of the working group; he was invited to the regional meeting in Sao Paulo by the WMA to discuss the ethics of placebo-controlled trials; and he participated, with partial funding from CIOMS, in the 2024 General Assembly to adopt the 2024 revision of the DoH.

Conflict of Interest The author declares no competing interests relevant to the content of this chapter.

References

1. The World Medical Association. WMA Declaration of Helsinki – Ethical Principles for Medical Research Involving Human Participants. First adopted in 1964, last amended in 2024. https://www.wma.net/policies-post/wma-declaration-of-helsinki/.
2. World Medical Association. Declaration of Helsinki: ethical principles for medical research involving human subjects. 2000. Rescinded. https://www.wma.net/wp-content/uploads/2018/07/DoH-Oct2000.pdf.
3. Potter VR. Bioethics, the science of survival. Perspect Biol Med. 1970;14(1):127–53. Project MUSE. https://doi.org/10.1353/pbm.1970.0015.
4. Helsinki Statement Stakeholders. Helsinki statement. Clin Eval. 2024;52(3). http://cont.o.oo7.jp/52pop/HelsinkiStatement_18Oct24_final.pdf.
5. Kurihara C, Matsuyama K, Baroutsou V. World medical association's declaration of Helsinki, 2024 revision: celebrating the 60th anniversary, at Helsinki. Clin Eval. 2025;52(3):W1–W27. http://cont.o.oo7.jp/52pop/52pop_contents_e.html.
6. Council for International Organizations of Medical Sciences. International ethical guidelines for health-related research involving humans. 2016. https://cioms.ch/publications/product/international-ethical-guidelines-for-health-related-research-involving-humans/.
7. Co-organized by The Brazilian Society of Bioethics (SBB); International Federation of Associations of Pharmaceutical Physicians and Pharmaceutical Medicine (IFAPP); Clinical Evaluation Inc.; The 129th Pharmaceutical Study Group. Supported by The Japan Association for Bioethics. Proceedings: The 2024 Declaration of Helsinki: Taking Forward Bioethics and Human Rights. Clin Eval. 2025;52(3): 375–452. http://cont.o.oo7.jp/52pop/52pop_contents_e.html.
8. Haidar H, Shamsi Gooshki E. Trial-centrism in the declaration of Helsinki: a challenge for post-study access and preventing data colonialism in AI-driven Health Research. Can J Bioeth / Revue canadienne de bioéthique. 2024;7(4):138–9. https://doi.org/10.7202/1114971ar.
9. The World Medical Association. The Declaration of Taipei on ethical considerations regarding Health Databases and Biobanks. 2002. Last revised 2016. https://www.wma.net/policies-post/wma-declaration-of-taipei-on-ethical-considerations-regarding-health-databases-and-biobanks/
10. Sindicato Médico del Uruguay. World Medical Association Submission of a Motion to Amend (on Paragraph 33, 34, submitted to) 227th Council Session and General Assembly, Helsinki 2024, October 2024. 8, 10 Oct 2024.
11. CONFEMEL. Intervención de la Dra. Zaida Arteta en la Asamblea General de la AMM. 16 octubre, 2024. https://www.confemel.com/intervencion-de-la-dra-zaida-arteta-en-la-asamblea-general-de-la-amm/.
12. Sindicato Medico del Uruguay. SMU propone cambios en la Declaración de Helsinki. October 16, 2024. https://www.smu.org.uy/el-smu-en-la-asamblea-de-la-asociacion-medica-mundial-propone-cambios-en-la-declaracion-de-helsinki/.
13. Angell M. Ethical imperialism? Ethics in international collaborative clinical research. N Engl J Med. 1988;319(16):1081–3. https://doi.org/10.1056/NEJM198810203191608. PMID: 3173435.
14. Lurie P, Wolfe SM. Unethical trials of interventions to reduce perinatal transmission of the human immunodeficiency virus in developing countries. N Engl J Med. 1997;337(12):853–6.
15. Angell M. The ethics of clinical research in the third world. N Engl J Med. 1997;337(12):847–9. https://doi.org/10.1056/NEJM199709183371209.
16. Macklin R. Against relativism: cultural diversity and the search for ethical universals in medicine. Oxford University Press; 1999.
17. Pérez AC, Smith RN. The revised declaration of Helsinki: interpreting and implementing ethical principles in biomedical research. Int J Pharm Med. 2001;15:131–43. https://doi.org/10.2165/00124363-200106000-00006.

18. Lurie P, Greco DB. US exceptionalism comes to research ethics. Lancet. 2005;365(9465):1117–9.
19. Trials on trial. Nature. 2008;453(7194):427–8. https://doi.org/10.1038/453427b.
20. Kurihara C, Greco D, Dhai A, Saio T, Tsubaki H. Ethics of placebo-controlled trials: historical analysis including experiences during the COVID-19 pandemic. In: Kurihara C, Greco D, Dhai A, editors. Ethical innovation for global health: pandemic, democracy and ethics in research. Springer; 2023. p. 195–224.
21. Kurihara C, Greco D, Dhai A. Post-trial access: historical analysis considering the experience of COVID-19 pandemic. In: Kurihara C, Greco D, Dhai A, editors. Ethical innovation for global health: pandemic, democracy and ethics in research. Springer; 2023. p. 225–41.
22. Conselho Nacional de Saúde. Resolução n° 404, de 1° de agosto de. 2008. https://www.gov.br/conselho-nacional-de-saude/pt-br/acesso-a-informacao/atos-normativos/resolucoes/2008/resolucao-cns-no-404.pdf/view. Accessed 29 Apr 2025.
23. Fleury S. Brazilian sanitary reform: dilemmas between the instituing and the institutionalized. Ciênc Saúde Colet. 2009;14(3):743. https://doi.org/10.1590/S1413-81232009000300010.
24. Paim J, Travassos C, Almeida C, Bahia L, Macinko J. The Brazilian health system: history, advances, and challenges. The Lancet [Internet]. 2011;377(9779):1778–97. https://linkinghub.elsevier.com/retrieve/pii/S0140673611600548.
25. Greco DB, Galvão-Castro B. Democracy restoration in Brazil, *the constitutional guarantee of health as a right for all, giving rise to a universal health system* (SUS) and of a National Research Ethics Commission. In: Kurihara C, Dhai A, Greco D, editors. Ethical innovation for global health: pandemic, democracy and research ethics. Springer; 2023.
26. Beauchamp TL, Childress JF. Principles of biomedical ethics. Oxford University Press; 1979.
27. Ferreira S, Porto D. New code of medical ethics, bioethics and hope code of medical ethics. Rev Bioét. 2018;26(4) Brasília https://doi.org/10.1590/1983-80422018264000.
28. Garrafa V, Porto D. Intervention bioethics: a proposal for peripheral countries in a context of power and injustice. Bioethics. 2003;17(5-6):399–416. https://doi.org/10.1111/1467-8519.00356.
29. Greco D. Emancipation in the struggle for equality in research involving human volunteers. Rev bioét (Impr.). 2013;21(1):19–30. https://revistabioetica.cfm.org.br/revista_bioetica/article/view/786/875.
30. Freire P. Pedagogy of the Oppressed, 50th Anniversary Edition. Bloomsbury Academic; 2018. p. 229.
31. Streck DR, Redin E, Rowman ZJJ, editors. Paulo Freire Encyclopedia. Littlefield Publishers; 2012. p. 490. 978-1-4422-1656-3
32. Marx, K. 'On the Jewish Question', in Karl Marx, Early Writings, trans. Rodney Livingstone and Gregor Benton, intro. Lucio Colletti (Penguin Books, Harmondsworth. 1975. pp. 228–234. Originally written in 1843. https://www.degruyterbrill.com/document/doi/10.7312/blau17412-061/html?lang=en.
33. Parizi-Carvalho RR, Martins GZ, Greco DB. The Brazilian Society for Bioethics: a bioethics of commitments. Rev Bioét. (Impr.). 2017;25(2):218–23. https://doi.org/10.1590/1983-80422017252181.
34. Conselho Nacional de Saúde. Resolução n° 404, de 1° de agosto de 2008. https://www.gov.br/conselho-nacional-de-saude/pt-br/acesso-a-informacao/atos-normativos/resolucoes/2008/resolucao-cns-no-404.pdf/view.
35. Conselho Nacional de Saúde. Resolução nº 466, de 12 de dezembro de 2012. https://www.gov.br/conselho-nacional-de-saude/pt-br/acesso-a-informacao/atos-normativos/resolucoes/2012/resolucao-no-466.pdf/view.
36. Santos CAA, Rodrigues RG. Does Law 14874, A New Ethical Regulation of research with human beings in Brazil, Violate Human Rights? Rev Iberoamericana Bioética. 2024;26:1–14) [ISSN 2529-9573]. https://doi.org/10.14422/rib.i26.y2024.008.
37. Fernandes MS, Silla L, Goldim JR, Martins-Costa J. Brazilian legal and bioethical approach about donation for research and patents of human body parts. J Community Genet. 2017;8(3):199–208. https://doi.org/10.1007/s12687-017-0303-y. Epub 2017 May 29.

38. Fonseca EM, Ribeiro JM, Bertoni N. Bastos, FI Syringe exchange programs in Brazil: preliminary assessment of 45 programs. Cad. Saúde Pública, Rio de Janeiro. 2006;22(4):761–70. https://www.scielo.br/j/csp/a/ZsNnjWY4XBBvdnzVhH9KLgz/?format=pdf&lang=en.
39. Galvão-Castro B, Couto-Fernandez JC, Mello MA, Linhares-de-Carvalho MI, Castello-Branco LR, Bongertz V, et al. A nationwide effort to sistematically monitor HIV-1 diversity in Brazil: preliminary results. Mem Inst Oswaldo Cruz. 1996;91(3):335–8.
40. Galvão J. Access to antiretroviral drugs in Brazil. Lancet. 2002;360(9348):1862–5. https://doi.org/10.1016/S0140-6736(02)11775-2.
41. MS – Ministério da Saúde. Boletim epidemiológico de HIV/ Aids. 2024. https://www.gov.br/aids/pt-br/central-de-conteudo/boletins-epidemiologicos/2024.
42. Greco DB, Simão M. Brazilian policy of universal access to AIDS treatment: sustainability challenges and perspectives. AIDS. 2007;21(Suppl 4):S37–45. https://doi.org/10.1097/01.aids.0000279705.24428.a3.
43. World Trade Organization. Doha Declaration on the TRIPS Agreement and Public Health. Adopted by the WTO Ministerial Conference of 2001 in Doha on November 14, 2001. https://www.wto.org/english/thewto_e/minist_e/min01_e/mindecl_trips_e.htm.
44. From Galvão-Castro B, Grassi MFR, Castilho EA, Greco D. Inst Oswaldo Cruz. Rio de Janeiro. 2021;116:e210071. https://www.scielo.br/j/mioc/a/zLgC44cLh44wkKBDCghFfyw/?lang=en&format=pdf.
45. The Lancet. COVID-19 in Brazil: "So what?". Lancet. 2020;395(10235):1461.
46. Sociedade Brasileira de Bioética Nota Pública: SBB solicita revogação imediata da orientação do Ministério da Saúde sobre uso da cloroquina em pacientes com covid-19 (Public Note: SBB requests immediate revocation of the Ministry of Health's guidance on the use of chloroquine in patients with covid-19) pg 173 in Greco, DB, Martins, GZ, Oliveira, RA, Diniz, NM Ações da Sociedade Brasileira de Bioética Editora Rede Unida 419 pg, 2024. DOI 9786554621199. Available https://doi.org/10.18310/9786554621182. https://editora.redeunida.org.br/wp-content/uploads/2024/06/Livro-ACOES_DA_SOCIEDADE_BR_DE_BIOETICA.pdf.
47. Sociedade Brasileira de Bioética. Recomendação SBB nº 1/2020. 2020. https://drive.google.com/file/d/1-Iy-xGRe-3g8HfdJUG2cMj7Pc_AIpRkL/view.
48. Greco D, Kurihara C, Saio T, Nakazawa E, Kimura R, Reid L, Criscuolo D, Kerpel-Fronius S, Imamura K.COVID-19 and bioethics: Part1 Ethical Challenges and COVID-19: The Recommendation No. 01/2020 of the Brazilian Society of Bioethics (SBB). Clin Eval 2020; 48(3): 661-684.
49. Islam MS, Sarkar T, Khan SH, Mostofa Kamal AH, Hasan SMM, Kabir A, Yeasmin D, Islam MA, Amin Chowdhury KI, Anwar KS, Chughtai AA, Seale H. COVID-19-Related Infodemic and Its Impact on Public Health: A Global Social Media Analysis. Am J Trop Med Hyg. 2020;103(4):1621–9. https://doi.org/10.4269/ajtmh.20-0812.
50. The Skeptic Reason with Compassion. Pasternak N, Orsi C. Prevent Senior' in Brazil accused of conducting unethical medical experiments on Covid patients. 1 October 2021. https://www.skeptic.org.uk/2021/10/prevent-senior-in-brazil-accused-of-conducting-unethical-medical-experiments-on-covid-patients/. Accessed 29 Apr 2025.
51. New MSF report: High-income countries must stop hoarding 870 million excess COVID-19 vaccines doses and redistribute them to save lives. Published: Oct 18, 2021, Updated: May 25, 2023. https://www.doctorswithoutborders.ca/new-msf-report-high-income-countries-must-stop-hoarding-870-million-excess-covid-19-vaccines-doses-and-redistribute-them-to-s-ave-lives/
52. Corporate author: UNESCO International Bioethics Committee, World Commission on the Ethics of Scientific Knowledge and Technology. Statement on COVID-19: ethical considerations from a global perspective. April 2020. https://unesdoc.unesco.org/ark:/48223/pf0000373115.
53. Corporate author: International Bioethics Committee, World Commission on the Ethics of Scientific Knowledge and Technology. UNESCO's ethics commissions' call for global vaccines equity and solidarity: joint Statement by the UNESCO International Bioethics Committee (IBC) and the UNESCO World Commission on the Ethics of Scientific Knowledge and Technology (COMEST). February 2021. https://unesdoc.unesco.org/ark:/48223/pf0000375608.

54. Corporate author: International Bioethics Committee, World Commission on the Ethics of Scientific Knowledge and Technology. Joint statement of UNESCO's ethics commissions on ensuring equal access for all to vaccines and therapeutics developed to confront COVID-19: joint statement of the UNESCO International Bioethics Committee (IBC) and UNESCO World Commission for the Ethics of Science and Technology (COMEST). September 2021. https://unesdoc.unesco.org/ark:/48223/pf0000379042.
55. Centro de Prevenção e Controle de Doenças do Brasil (CPCD). https://cpcdbrasil.org.br/. Accessed 29 Apr 2025.
56. Kurihara C, Inoue K, Kai H, Suzuki K, Saeki H, Funabashi Y, Kishi N, Kuge A, Murakami T, Saito Y, Uchida E, Tsutsumi N, Imamura K. Our “WMA Declaration of Helsinki”: opinions and proposals from patient and public for research ethics. In: Kurihara C, Greco D, Dhai A, editors. Ethical innovation for global health: pandemic, democracy and ethics in research. Springer; 2023. p. 243–69.
57. Todres J, Diaz A. COVID-19 and Human Trafficking—the Amplified Impact on Vulnerable Populations. JAMA Pediatr. 2021;175(2):123–4. https://doi.org/10.1001/jamapediatrics.2020.3610.

Chapter 2
Ubuntu Principles: Solidarity to Include Vulnerable Communities for Equitable Access

Ames Dhai and Chieko Kurihara

Abstract The Ubuntu Principle underscoreing "*We are, therefore I am. I am, therefore we are*" is an African idea reflecting indigenous values. It reinforced African collaborations as the COVID-19 pandemic was confronted. It was pivotal in promoting meaningful community engagement and is incorporated in the 2024 revision of the Declaration of Helsinki (DoH) of the World Medical Association (WMA).

The vulnerability has also been extensively discussed during the consultations for the revisions for the 2024 DoH. The DoH 2024 advances research targeting vulnerable populations, with strengthened safeguard and benefit sharing. Prior to this, UNESCO's Declaration on Bioethics and Human Rights and UNESCO's statements regarding COVID-19 as a global public good have facilitated access to health for all. In addition, the CIOMS ethical guidelines for health-related research involving humans and their reports on research in resource-limited settings have promoted the protection of vulnerable people, prevention of exploitation, and directed for access to research results.

In South Africa, constitutional provisions and a health system were established to overcome the ravages of apartheid and protect the rights of research participants. From a global perspective, international cooperation to promote bioethics and the right to equal health is essential to eliminate health inequalities and to achieve the sustainable development goals (SGDs) as promulgated by the United Nations.

Keywords Ubuntu principle · Community engagement · Resource-limited settings Benefit sharing · Health equity · Sustainable development goals (SGDs)

This chapter is constructed by consolidating previous publications of lecture proceedings of the reference No. 4, 17, 21, under permission of the publisher of these proceedings, with some new consideration/analysis and updated information.

A. Dhai
School of Clinical Medicine, University of the Witwatersrand, Johannesburg, Gauteng, South Africa

C. Kurihara (✉)
Kanagawa Dental University, Yokosuka, Kanagawa, Japan
e-mail: chieko.kurihara@nifty.ne.jp

C. Kurihara et al. (eds.), *The 2024 Declaration of Helsinki*,
https://doi.org/10.1007/978-981-96-9294-1_2

1 Ubuntu Principle and Meaningful Community Engagement

During the 2 years process for the 2024 revision of the Declaration of Helsinki (DoH), eight regional meetings were held worldwide. These were organised by the World Medical Association (WMA), in collaboration with American Medical Association (AMA), to which the chairperson of the Working Group for this revision belongs. There were also collaboration with national associations from each region. The sixth regional meeting was held in Johannesburg, South Africa, on 18 and 19 February 2024. It was co-organised by WMA, AMA, and the South African Medical Association (SAMA). The main topic of this meeting was "vulnerability" and the importance of "community engagement" was discussed extensively in the context of protecting vulnerable research participants in resource-limited settings. At the onset of the meeting, the "**Ubuntu Principle**" (Ubuntuology) reflecting African indigenous values was discussed.

This Principle defines a human being in the context of others and the community. "*We are, therefore I am. I am, therefore we are*" was described by John Samuel Mbiti (1931–2019), a scholar who contextualised such African indigenous theology within an academic discipline [1, 2]. This articulation of the African communitarian world views can be contrasted with the Western individualistic views, as Descartes "*I think, therefore I am*", as well as the Kantian principle defining human beings as autonomous individuals, never to be used as a mere means to an end, i.e., the principle of respect for autonomy.

During the meeting, several examples of experiences and practices of community engagement were discussed. Vulnerability, community engagement, as well as post-trial access in the contexts of research in resource-limited settings were explored in depth during the consultations and were not limited to the Johannesburg meeting only. These issues were considered in Vatican City (January 2024), Munich (May 2024), and whenever other opportunities arose. Consequently, important new texts were incorporated in paragraph 6 of the 2024 DoH.

> *Since* ***medical research takes place in the context of various structural inequities****, researchers should carefully consider how the benefits, risks, and burdens are distributed.*
>
> ***Meaningful engagement with potential and enrolled participants and their communities should occur before, during, and following medical research****. Researchers should enable potential and enrolled participants and their communities to share their priorities and values; to* ***participate in research design, implementation, and other relevant activities; and to engage in understanding and disseminating results.***

Table 2.1 Ubuntu principle in African community likened to meaningful community engagement

"I am because we are, we are therefore I am." (Mbiti J): The **importance of the community** in African indigenous value systems, and what it means to be a human person.
*"**Umuntu Ngumuntu Ngabantu**"* (Nguni saying): a human being is a human being because of other human beings.
Interdependence and interrelatedness: Greater good for all while protecting vulnerable individuals and groups (for vaccine access and hesitancy).
African tradition of ***lekgotla*** where all gather under a tree to discuss and exhaust all the options
» likened to **meaningful community engagement** in the vaccines discussions.

Source: Dhai A. Ethics in vaccine allocation in developing countries. Clinical Evaluation. 2021; 49(Sup 38): 63–71 [4]

2 COVID-19 Response in Africa and the Multilateral Approach

During the COVID-19 pandemic, African communities experienced tremendous disruptions in health, economic and social systems, where huge numbers of vulnerable people were affected by expanding deleterious effects. Early on, the African Union embarked on a multilateral approach [3–5], and as early as February 2020, it compiled the Joint Continental Strategy. This was organised by the Africa CDC together with the African Task Force for the Vaccine. The Africa CDC together with the African Union, launched a Values Framework for the Fair, Equitable, and Timely Allocation of COVID-19 Vaccines in Africa. This framework was basically a collaboration between the South African Medical Research Council and the Africa CDC, and was a result of widespread consultation on the African continent, together with various capacity development activities [4, 5].

African indigenous values and Ubuntu (Umuntu) principles, stressing the interdependence and interrelatedness of humans (Table 2.1) [4] applied both for vaccine access and vaccine hesitancy. The African tradition of lekgotla, where all gathered under a tree to discuss and exhaust all options was highlighted and was likened to "meaningful community engagement". Furthermore, COVID-19 vaccines were to be considered as a "Global Public Good".

3 Vaccine Equity, Solidarity, and Benefit Sharing

3.1 Understanding Global Public Good

During the COVID-19 pandemic, there were a number of international documents that called for global vaccine equity and solidarity [6, 7], and promoting the idea of regarding COVID-19 vaccines as Global Public Goods [8–10]. Based on such an understanding, the "TRIPS Waiver" was proposed jointly by South Africa and India to the World Trade Organisation for waiver from the application of the TRIPS

Agreement (Agreement on Trade-Related Aspects of Intellectual Property Rights) to COVID-19-related intellectual property rights [11].

Two criteria determine a public good [12]. It has to be non-rival in consumption. Consumption by one person must not interfere with the goods being available equally to others. It has to be non-excludable. Suppliers cannot deny it to those who are unable to pay its market price [4].

3.2 Benefit Sharing in the UNESCO Declaration on Bioethics and CIOMS Documents

There are some supporting documents for benefit sharing in the context of bioethics and research ethics. The UNESCO Universal Declaration on Bioethics and Human Rights [13], 2005, emphasises that human vulnerability must be considered in applying and advancing scientific knowledge, medical practice, and associated technologies. Article 13 encourages solidarity and cooperation. Article 15 promotes the sharing of benefits as an issue of justice.

The CIOMS 2016 guidelines for health-related research [14] also focus on benefit sharing, avoiding exploitative research, especially in partnerships between high-income and low-income countries. A study from a high-resource country, conducted in a low-resource country might fail to establish plans to make products and services resulting from research available locally– this must be safeguarded against. Several controversial cases including cases of "ethics dumping" have been presented in the CIOMS 2021 Report on research in resource-limited settings [15].

4 CIOMS Approach for Benefit Sharing in Low-Resource Settings

4.1 Access and Interrelated Health Equity with Research Equity

These Guidelines [14] and the Report [15] by CIOMS articulate that equity in health and research are closely related. Health equity means everyone has equitable opportunities and resources to access health. This aligns with the World Health Organisation (WHO)'s definition of "health equity" [16]:

> *the absence of unfair, avoidable or remediable differences among groups of people, whether those groups are defined socially, economically, demographically, or geographically or by other dimensions of inequality* (e.g. *sex, gender, ethnicity, disability, or sexual orientation).*

This notion of health equity is based on the concept of social justice, which underscores that having differences in access to health is ethically immoral and unjustifiable.

How is this translated to research equity? [17, 18] Drawing from the definition of health equity, the definition of research equity can be considered to be those affected by research or who can benefit from its outcomes having equal opportunities to contribute to research and benefit from research. Research equity includes those who have been absent or silent in research and those who have been deliberately left out. Where research includes all who gain to benefit from its results, research equity is achieved. Hence research equity is intricately linked to health equity and is an important means of addressing health equity.

Guideline 3 of the 2016 CIOMS [14] recommends equitable distribution of benefits and burdens in selecting individuals and groups of research participants. Equitable distribution would allow participants to be drawn from qualified populations in geographical areas of study where the results can be applied. There should be no unfair discrimination regarding inclusion and exclusion criteria, based on various issues like privilege, disadvantage, race and sex. Health disparities will be perpetuated where there is underrepresentation in research.

Research benefits must address the diverse needs of different classes or groups of people. It is unjust to selectively include disadvantaged individuals or groups. They already have risks and burdens. They can be overused in research. Moreover, they are most likely to be excluded from research benefits or have difficulty in access, especially when there is no commitment to post-trial access. Broader inclusion of different social groups will help ensure broader access with strengthened safeguards both research and healthcarc for diverse individuals and groups.

4.2 Social Value

Guideline 1 of the 2016 CIOMS [14] describes scientific and social values and respect for individual rights. During the current revision process, social value was proposed in the first draft of the DoH but not included in the second draft [19] and the final adopted version. However, social value is an essential consideration. The information the study produces must be able to promote individual or public health. The research would have social value if the endpoints are related to clinical decision-making and if clinicians, policymakers, and others are likely to alter their practice based on the study findings. It is only if the research has social value that the associated risks, costs, and burdens of the research can be justified.

4.3 Research in Low-Resource Settings

Guideline 2 of the 2016 CIOMS [14] is specific to research conducted in low-resource settings. The consideration in this guideline was expanded to the CIOMS 2021 Report on research in resource-limited settings [15]. These documents highlight that where resources are decreased, there is vulnerability to exploitation.

Where research is conducted in low-resource settings, local social values must be respected. There must be a critical benefit. The guidelines also highlight that the responsiveness of research to health needs and priorities is important to provide social value to communities and populations. There has got to be a shared responsibility in terms of providing social value [18]. This shared responsibility would be shared by sponsors, funders, researchers, governments, etc.

Ultimately, post-trial access is a prerequisite. It is not only on interventions developed but also in terms of the knowledge generated by the research to be disseminated, regardless of whether it's a low-income setting or a high-income setting. This knowledge needs to be generated equitably as well. It is best to engage with communities to ascertain how best to get this done adequately. In this manner, the notion of "ethics dumping", as articulated in the CIOMS 2021 Report can be countered [15].

5 Sustainable Development Goals (SDGs) and Clinical Research

In terms of the sustainable development goals (SDGs) [20], designated by the United Nations (UN), healthy lives need to be ensured and well-being for all promoted, with universal access to needed medicines and vaccines being essential. However, to attain the realization of the SDGs, good quality research to identify and address unmet needs is necessary, and therefore we need to have access to this type of research for all.

Clinical research drives healthcare advancement. Without clinical research in low-resource settings, entire populations miss out on these advancements.

6 Vulnerability and Avoiding Exploitation

Trust is essential in clinical research in resource-limited settings. Research is crucial for the improvement of health and development of society. However, local populations often do not understand the aim and nature of research. Some see research as exploitative, with researchers from high-income countries taking advantage of low-cost and underregulated environments in low- and middle-income countries (LMICs). When entering these sites and building collaborations, we must start off with a foundation of trust right at the onset.

It is important to include special and vulnerable populations. In the past, certain physiologically special populations with medical and pathological problems were excluded from research. Recently, we have seen a movement from exclusion to inclusion. But we need to look beyond physiological differences only, as poverty and socio-economic situations in low-income settings can also render research participants vulnerable.

While unnecessary research with vulnerable populations should be avoided, it is essential that vulnerable persons, like any other societal groups, are included in the research to ensure that they can be treated with the medicines safely and effectively. It would be noteworthy that the 2024 DoH included the following principle for vulnerability:

> *Some individuals, groups, and communities are in a situation of more vulnerability as research participants due to factors that may be fixed or contextual and dynamic, and thus are at greater risk of being wronged or incurring harm.* ***When such individuals, groups, and communities have distinctive health needs, their exclusion from medical research can potentially perpetuate or exacerbate their disparities. Therefore, the harms of exclusion must be considered and weighed against the harms of inclusion.*** *In order to be fairly and responsibly included in research, they should receive* ***specifically considered support and protections****. (para. 19).*

> *Medical research with individuals, groups, or communities in situations of particular vulnerability is* ***only justified*** *if it is* ***responsive to their health needs and priorities*** *and the individual, group, or community* ***stands to benefit from the resulting knowledge, practices, or interventions****. (para. 20).*

7 Steve Biko, South African Constitution, and Research Ethics

Ethics intends to safeguard human dignity and to promote justice, equality, truth, and trust. This is so pertinent in the South African context, especially considering the hundreds of years of apartheid that was eventually abolished in 1994. What is ethics in research? It is simply about how we can ensure that people who are vulnerable are protected from exploitation and other forms of harm. We also have seen an evolving language moving from a passive subject to an active participant.

It was in 2007 that the Steve Biko Centre for Bioethics was established at the University of the Witwatersrand, commemorating Steven Bantu Biko (1946–1977), who was the leader of the black consciousness movement and was killed by the apartheid regime [21]. While a student at a medical school at a university in South Africa, he greatly influenced the lives and thinking of the relevant communities on human rights, equality, and medical ethics. At that time, the apartheid government did not allow teaching of bioethics at the black medical school as this might have raised consciousness of human rights. Biko was detained in a jail where he was subjected to harsh brutal treatment by the South African police during his interrogation. South African doctors were called in to assess his medical condition. Unfortunately, they colluded with the regime of the time and *inter alia*, wrote false medical information subordinating the welfare of their patient to the interest of the security policy.

Just after the abolishment of the apartheid regime, the South African Constitution was established in 1996 [22]. Section 12 states that "Everyone has the right to bodily and psychological integrity, which includes the right ... not to be subjected to medical or scientific experiments without their informed consent." This is very similar wording to the International Covenant on Civil and Political Rights [23].

Based on this Constitution, the National Health Act of 2003 [24] defines clinical trials and sets out the framework of the National Health Research Ethics Council (NHREC) and guidelines for Health Research Ethics Committees. Clinical trials of medicines and other products are governed under the Good Clinical Practice regulations. This framework, defined by the NHA, covers broad types of research involving humans, including social science research where there is an implication on health as well. SAGCP guidelines are closely aligned to ICH-GCP, which also more broadly reflect years of controversial ethical issues in clinical trials involving vulnerable people in developing countries.

8 Conclusion

"*We are, therefore I am. I am, therefore we are*", as promulgated by Ubuntu, has supported the African COVID-19 response as well as promoted meaningful community engagement, which is newly recommended in the 2024 DoH.

Constitutional provisions and the associated health system in South Africa post-apartheid are stringent governance documents for protecting the rights of research participants, especially those that are vulnerable, and for promoting benefit sharing.

To overcome structural iniquity and health disparity worldwide, international cooperation to promote bioethics and the right to equal health is essential for the future achievement of SDGs, thereby assuring fundamental human rights, both in research and healthcare, for everyone in the world.

Acknowledgement CK appreciates the International Federation of Associations of Pharmaceutical Physicians and Pharmaceutical Medicine (IFAPP) for granting her travel expenditure for her in-person participation in the WMA Regional Meeting in Tel Aviv (2022 December); and the WMA to grant her travel expenditure for her in-person participation and presentation as a panellist in WMA's Regional Meeting in Washington D.C. (2024 August) and waiver of the registration fee from WMA as a journalist to attend the General Assembly in Helsinki (2024 October). She participated in all the other WMA Regional Meetings at her own expense or online.

Conflict of Interest AD was a Board member of the South African Medical Association, which was a member of the Workgroup of the World Medical Association for the 2024 revision of the Declaration of Helsinki. The authors have no other competing interests to declare that are relevant to the content of this chapter.

References

1. Mbiti JS. African religions and philosophy. Johannesburg: Heinemann; 1969.
2. Gathogo J. John Mbiti's Ubuntu theology: was it rooted in his African heritage? Studia Hist Ecc. 2022;48(2) https://doi.org/10.25159/2412-4265/10292.
3. African Union; Africa CDC. Framework for fair, equitable and timely allocation of COVID-19 vaccines in Africa. 31 Jan 2021. https://africacdc.org/download/framework-for-fair-equitable-and-timely-allocation-of-covid-19-vaccines-in-africa/. Assessed 15 May 2025. Accessed 3 May 2025.
4. Dhai A. Ethics in vaccine allocation in developing countries. Clin Eval. 2021;49(Suppl 38):63–71.
5. Dhai A. Equitable access to COVID-19 vaccines, vaccine research, and vaccine apartheid on African continent: challenges and recommendations. In: Kurihara C, Greco D, Dhai A, editors. Ethical innovation for global health: pandemic, democracy and ethics in research. Singapore: Springer; 2023. p. 39–55.
6. The Lancet COVID-19 Commission. Enhancing global cooperation to end the COVID-19 pandemic. February 2021.
7. Corporate author: UNESCO International Bioethics Committee, World Commission on the Ethics of Scientific Knowledge and Technology. Statement on COVID-19: ethical considerations from a global perspective. April 2020. https://unesdoc.unesco.org/ark:/48223/pf0000373115.
8. Corporate author: International Bioethics Committee, World Commission on the Ethics of Scientific Knowledge and Technology. UNESCO's ethics commissions' call for global vaccines equity and solidarity: joint Statement by the UNESCO International Bioethics Committee (IBC) and the UNESCO World Commission on the Ethics of Scientific Knowledge and Technology (COMEST). February 2021. https://unesdoc.unesco.org/ark:/48223/pf0000375608.
9. Corporate author: International Bioethics Committee, World Commission on the Ethics of Scientific Knowledge and Technology. Joint statement of UNESCO's ethics commissions on ensuring equal access for all to vaccines and therapeutics developed to confront COVID-19: joint statement of the UNESCO International Bioethics Committee (IBC) and UNESCO World Commission for the Ethics of Science and Technology (COMEST). September 2021. https://unesdoc.unesco.org/ark:/48223/pf0000379042.
10. Dhai A. Access to COVID-19 vaccines as a global public good: a co-ordinated global response based on equality, justice and solidarity is key. S Afr J Bioethics Law. 2021;14(1):2–3. https://doi.org/10.7196/SAJBL.2021.v14i1.768.
11. Council for Trade-Related Aspects of Intellectual Property Rights. Waiver from certain provisions of the TRIPS Agreement for the prevention, containment and treatment of COVID-19: Communication from India and South Africa. 2 Oct 2020; IP/C/ W/669. https://docs.wto.org/dol2fe/Pages/SS/directdoc.aspx?filename=q:/IP/C/W669.pdf&Open=True
12. van den Berg RD. Evaluation of the funding of global public goods-Note for the OECD/DAC Evaluation Network. 2015.
13. United Nations Educational, Scientific and Cultural Organization. Universal declaration on bioethics and human rights. 19 October 2005. https://www.unesco.org/en/legal-affairs/universal-declaration-bioethics-and-human-rights?hub=66535.
14. Council for International Organizations of Medical Sciences. International ethical guidelines for health-related research involving humans. 2016. https://cioms.ch/publications/product/international-ethical-guidelines-for-health-related-research-involving-humans/.
15. Council for International Organizations of Medical Sciences. Clinical research in resource-limited settings. 2021. https://cioms.ch/publications/product/clinical-research-in-low-resource-settings/#description.
16. World Health Organization. Health equity. https://www.who.int/health-topics/health-equity#tab=tab_1. Assessed 15 May 2025.
17. Dhai A. Access and equity in research: Justice, vulnerability and low-resource settings. In: Co-organized by The Brazilian Society of Bioethics (SBB); International Federation of

Associations of Pharmaceutical Physicians and Pharmaceutical Medicine (IFAPP); Clinical Evaluation Inc.; The 129th Pharmaceutical Study Group. Supported by The Japan Association for Bioethics. Proceedings: The 2024 Declaration of Helsinki: Taking Forward Bioethics and Human Rights. Clin Eval. 2025; 52(3): 428–432. http://cont.o.oo7.jp/52pop/52pop_contents_e.html.
18. ENGAGE. Equity in research. Equity in research | Engage Assessed 15 May 2025.
19. Kurihara C, Greco D, Dhai A, Matsuyama K, Baroutsou V. Vulnerability, social value and the equitable sharing of benefits from research: beyond the placebo and access debates. Front Med. 2024;11:1432267. https://doi.org/10.3389/fmed.2024.1432267.
20. United Nations. Resolution adopted by the General Assembly on 25 September 2015—Transforming our world: the 2030 Agenda for sustainable development. 21 October 2015. https://www.un.org/en/development/desa/population/migration/generalassembly/docs/globalcompact/A_RES_70_1_E.pdf.
21. Dhai A. Ethics and regulations of clinical research in South Africa. Clin Eval. 2017;45(3):681–701. http://cont.o.oo7.jp/45_3/w15-w35.pdf
22. The Constitution of the Republic of South Africa. As adopted on 8 May 1996 and amended on 11 October 1996 by the Constitutional Assembly.
23. United Nations. International Covenant on Civil and Political Rights. Adopted by General Assembly resolution 2200A (XXI) of 16 December 1966, entry into force 23 March 1976. https://www.ohchr.org/en/instruments-mechanisms/instruments/international-covenant-civil-and-political-rights.
24. No. 61 of 2003: National Health Act, 2004. Government Gazette. Vol. 469 Cape Town 23 July 2004 No. 26595.

Chapter 3
Strengthening Ethical Safeguards Protecting Vulnerable Populations in Egyptian: Role of Research Ethics Committee

Hany Sleem

Abstract The ability of the Egyptian Network of Research Ethics Committees (ENREC) to foster consensus on the signing of the Helsinki Statement, which evaluates 10 items of the improvement of the 2024 revision of the Declaration of Helsinki (DoH) and clarifies five remaining challenges, within a short time frame reflects its commitment to ethical research practices in Egypt, ensuring the rights and protection of research participants. ENREC believes that promoting ethical medical research in Egypt requires adherence to fundamental principles that guarantee participant rights, enhance the social value of research, and ensure transparency and fairness in the dissemination of and access to the results. Therefore, the majority of ENREC members endorsed the principles of the Helsinki Statement, recognizing it as a valuable addition to the latest revision of the DoH in 2024 and believes the Statement facilitates protecting research participants and enhances their rights, especially in low- and middle-income countries.

This chapter introduces the background of the research ethics framework in Egypt and the initiative of the ENREC, particularly its consensus development process on the Helsinki Statement, and emphasizes the protection of vulnerability, which is one of the significant revisions of the 2024 DoH.

Keywords Vulnerability · Research ethics committee · Low- and middle-income countries · Capacity development · Social value

---

H. Sleem (✉)
National Hepatology and Tropical Research Institute, Cairo, Egypt

Egyptian Network of Research Ethics Committees (ENREC), Cairo, Egypt

C. Kurihara et al. (eds.), *The 2024 Declaration of Helsinki*,
https://doi.org/10.1007/978-981-96-9294-1_3

1 Introduction: Consensus Development of the ENREC to the Helsinki Statement

The Egyptian Network of Research Ethics Committees (ENREC) is an organization operating under the auspices of the Egyptian Society for Healthcare Development (ESHD), a nongovernmental organization (NGO). It was created in 2008 to promote harmonization among Research Ethics Committees (RECs). Its mission is to strengthen ethical oversight through collaboration and capacity building, among the 85 or more registered RECs nationwide.

The Helsinki Statement [1] was disseminated on October 18, 2024, one day before the adoption of the World Medical Association (WMA)‘s Declaration of Helsinki (DoH) [2]. The Statement was developed by voluntary collaborating individuals independent of the WMA, who had been extensively discussing ways for the better revision of the DoH. It celebrates 10 items of improvements in the 2024 revision of the DoH. Simultaneously they pointed out five items of remaining challenges (Table 3.1) [3].

Responding to their call for signatories to endorse the Statement, the ENREC quickly prepared an Arabic translation [4] and discussed the Statement and eventually approved to endorse it. The names of 32 individual members and ENREC were published as individual and institutional signatories being included in the Helsinki Statement.

To the Arabic version, a note was added to state the change in the DoH of the word “subject” to “participant” is not relevant in Arabic, as the same meaning word as “participant” was already used in the Arabic translation of the DoH. This raised

Table 3.1 Summary of the contents of the Helsinki statement

Celebrate the DoH 2024 for the revisions: (1) Change “research subjects” to “research participants”[a]; (2) Scope to include nonphysicians, teams, and organizations; (3) Recognition of various structural inequities in research; (4) Meaningful engagement with participants and their communities at all the stages of research; (5) DoH must be adhered to public health crises; (6) Inclusion of vulnerable people with adequate protection, recognizing contextual vulnerability; (7) Strengthened research ethics committee reflecting community values; (8) The preferences and values of incapable research participants shall be considered; (9) Data and research specimens shall be handled per the Declaration of Taipei on health databases and biobanks; (10) Clinical use of unproven intervention must never be undertaken to circumvent the protection set forth in the DoH.
Remaining challenges: (1) Plain language understandable for patient and the public; (2) Promoting consideration of “social value”; (3) Risk minimization in placebo-controlled trials; (3) Posttrial access for participants who still need it, as well as for those in need in the host community and globally.

Source: This table is slightly modified from the table in the article, reference No. 3, reproduced under the permission of the IFAPP and authors

[a] The following note was added in the Arabic translation to item No. 1: “This issue was not consistently present in the Arabic translation. This is because the English terms “research subjects” and “research participants” convey the same meaning in Arabic, aligning with the intended spirit of the English term “research participants”

the interest of an organizer of the Helsinki Statement and it was explained that "*Since the literal Arabic translation of the word "subject" (شيء) is typically reserved for inanimate objects, we opt for a more suitable term from the outset. Consequently, we use the word "مشارك," which most accurately captures the essence of the English word 'participant'.*" This explanation was included in a publication [5] by three of the organizers of the Statement. This publication is a report to introduce the process of the 2024 revision of the DoH.

This chapter introduces the background of the research ethics framework in Egypt and the initiative of the ENREC with the objective of clarifying how such consensus development was possible, especially focusing on the issue of protecting vulnerability, which is one of the significant revisions of the 2024 DoH.

2 Egyptian Society and Need for Protecting Vulnerability

2.1 Social Background

Egypt has a population of 110 million, the largest of Arabic countries, with a young and growing population, with a significant portion residing in urban centers. There are many challenges despite efforts to improve access to health care. While the country has made strides in improvement, challenges remain, particularly regarding equitable distribution and quality of health services. Egypt faces economic challenges, including inflation. Egyptian society has a rich cultural heritage and strong family values. However, significant social inequalities exist, with disparities in access to education, and access to health care. Religious background is predominantly Muslim, with a significant Coptic Christian minority. Religious identity plays a significant role in Egyptian society and culture.

2.2 Protecting Vulnerability

Vulnerable populations are defined to be those with diminished autonomy or increased susceptibility to harm, e.g., children, elderly, individuals with disabilities, and those facing socioeconomic hardship, which is particularly prevalent in developing countries like Egypt. Limited access to advanced health-care and socioeconomic disparities heighten vulnerability, obviously which requires specialized ethical considerations. Above all, we have to care about economically disadvantaged people in Egypt, including ensuring fair access to research benefits, preventing exploitation due to limited health-care options, and addressing potential power imbalances.

For this reason, ethics committees are crucial for safeguarding participant rights and well-being. Independent review is essential especially for protecting vulnerable populations.

3 Legal Farmwork of Clinical Trial in Egypt

3.1 Egyptian Constitution

The old Egyptian Constitution in 1971 (Article 43) defined that "no medical or scientific experiment may be performed on any human being without their free consent." This is based on the United Nations International Covenant on Human Rights [6]. This constitution has been strengthened, and the current Egyptian Constitution 2014 (Article 60) defines that "…No medical or scientific experiment may be performed on any human being without their free, authenticated consent, in accordance with established principles in the field of science…" [7].

3.2 Egyptian Clinical Trials Law No. 214/2020

Under this constitutional definition, Egyptian Clinical Trials Law No. 214/2020 includes the following items:

- The state maintains a strong commitment to the highest ethical standards in research, with ongoing efforts dedicated to safeguarding vulnerable populations.
- The law mandates research ethics committee (REC) approval for all types of clinical trials, not only drugs or devices but also surgical or procedural trials.
- The law promotes key protections:
  - Informed consent in Arabic, with literacy/cultural adaptations.
  - Explicit safeguards for children, pregnant women, and refugees.
  - Penalties for noncompliance (fines, trial suspension).

3.3 Vulnerability: Categories Eligible for Additional Protection

Article 3 of the Egyptian Clinical Trials Law mentions the vulnerability:

> *The conduct of medical research may not be restricted to vulnerable group or to groups who deserve additional protection unless the research is necessary and relates to diseases of their own, with the availability of scientific and ethical justifications for their use, and provided that the informed consent is obtained from each of them. If the medical research is to be carried on one of the groups entitled to additional protection, the consent of the parents, or whoever has the guardianship in the event of the death of one or both of the*

parents, or the legal representative must be obtained, in accordance with the controls and procedures specified by the executive regulations of this law. (The texts in italics are direct translation of the law in Arabic.)

Article 3 addresses the categories of group who deserve additional protection. The Arabic language lacks a direct equivalent for "vulnerability." Instead, the phrase "categories requiring additional protection" is employed. This phrase inherently links the concept to necessary action, prompting a logical response. Consequently, the RECs readily acknowledge and fulfill their duty to provide these groups with comprehensive protection. Importantly, the Law requires clinical trials involving vulnerability for:

- Equitable access: Requires trials to benefit underserved regions;
- Compensation: Mandates free care for trial-related injuries;
- Transparency: Public registration of trials in a publicly open database to prevent exploitation;
- Community engagement: RECs must include civil society representatives.

4 Roles of the Egyptian Network of Research Ethics Committees (ENREC)

4.1 Capacity Development: Communication and Accreditation

For the effective implementation of the above legal framework, since its establishment in 2008, ENREC has been engaged in various activities to strengthen the capacity of RECs in Egypt. It includes harmonization of local practices in line with international standards, e.g., the Council of International Organizations of Medical Sciences (CIOMS), the World Health Organization (WHO), providing training materials and opportunities along with networking practices. The impact of the ENREC on each REC's review practice could be described as follows:

- Facilitates information sharing and best practices.
- Promotes harmonization of ethical review processes.
- Enhances the quality and consistency of ethical oversight across Egypt.
- Provides training and resources for ethics committee members.
- Strengthens the ethical infrastructure in Egypt.

ENREC has communication Tools as below (Fig. 3.1):

- WhatsApp group for REC Chairmans and consultants (157 Members).
- Facebook group for REC members (200 members).
- Facebook Page (1700 Followers) for Community Outreach.
- Website with distance learning services:

www.enrec.org
Email: enrec.egypt@gmail.com

Fig. 3.1 ENREC Facebook page for community outreach

This screen capture is used under the authorization of the designated coordinator of the ENREC, Hany Sleem (author of this chapter)

It is also important that in Egypt there is a national system of REC accreditation based on the following international and domestic references.

- **International references:**
  - WHO Standards and operational guidance for ethics review of health-related research with human participants.
  - Research Ethics Committee (REC) Self-Assessment Tool.
- **National standard:**
  - Egyptian Clinical Trials Law 2020 and its Executive Regulations 2022.
  - Egyptian Drug Authority (EDA) Rules of Procedure.
  - National Authority for Quality Assurance and Accreditation of Education.
  - Central Administration for Research and Development. Ministry of Health.

4.2 Capacity Development: Training Programs

There is also a training curriculum for research ethics (for human participants and experimental animals). A vulnerability module is included in all training. The training is divided into three levels, including the following items:

- **One-day training:** Introductory to all areas of research ethics.
  - Research Ethics History
  - Research Ethics Principles
  - Risk and Benefits Assessment

- REC Function
- Informed Consent
- Vulnerability

- **Two- or three-day training:** Standard training in the field of human experimentation ethics. This is a standard practice for individuals engaged in conducting research involving human participants. The goal is to ensure that researchers, and potentially community representatives, understand the ethical principles and regulations designed to protect the rights, safety, and well-being of participants.
  - Historical ethical context: Reviewing significant historical events and ethical failures in research (like the Nuremberg trials, the Tuskegee Syphilis Study).
  - Ethical principles: Introduction to fundamental ethical principles, the process of obtaining informed consent.
  - Vulnerable populations: Specific considerations and enhanced protections required when conducting research on those groups.
  - Confidentiality and privacy: Understanding the importance of protecting participants' personal information and how to manage research data ethically.
  - Research Ethics Committees (REC): Explaining the role and function of these committees.
  - National and International Regulations and Guidelines: Overview of relevant international ethical guidelines (like the Declaration of Helsinki, CIOMS guidelines) and Good Clinical Practice (GCP).
- **Advanced training:** In the field of human experimentation ethics and GCP. This will be online. The training curriculum is accredited by the University of Maryland, USA.

4.3 Protecting Vulnerable Groups: ENREC'S Strategy

ENREC focuses on protecting vulnerable groups through these steps:

- Established Clear Guiding for All RECs: ENREC developed simple, consistent rules for ethics committees across Egypt. This ensured all RECs followed the same steps to protect vulnerable people fairly.
- Organized Training for Ethical Decision-Making: ENREC ran workshops to help REC members tackle tough ethical questions in research. These programs are targeted areas with fewer resources, where experts and tools were scarce.
- Respected Local Values: ENREC blended global ethics rules with Egypt's culture. They talked to communities to understand their concerns, ensuring research to respect local traditions.
- By prioritizing fairness, education, ongoing care, and cultural respect, ENREC strengthened protections for vulnerable groups while supporting ethical research in Egypt.

Two case examples are provided here to clarify what ENREC-trained RECs are ensured:

- **Case Example 1:** A Hepatitis C Treatment trial in Egypt (low literacy, high poverty)
- Consent processes using visual aids and local norms.
- Community members training to address mistrust issues:
  - Historical factors, socioeconomic disparities, and previous health-care experiences can contribute to mistrust of medical researchers or external organizations, particularly in vulnerable communities. Given the high burden of Hepatitis C in Egypt, communities may have complex feelings related to past public health campaigns or access to treatment. To build trust and facilitate recruitment and retention.
- Posttrial access to health care for participants:
  - In Egypt, where many joined hepatitis C treatment trials hoping to obtain life-saving direct-acting antibiotics (DAAs) at a time when they were unable to afford them, research ethics committees should have ensured that trials provide aftercare for ethical considerations.
  - Even as the Egyptian government expands its efforts to improve health services, poverty leaves gaps in care, and participants should not lose effective treatment simply because the trial ends. Research sponsors should ensure they continue to receive it (or similar care) until national programs make it widely available.
  - This is consistent with global ethical guidelines that require sponsors to provide lasting benefits. This care should not be limited to research participants alone, but also to the communities they represent and their peers.
  - Success in this depends on collaboration. Sponsors should partner with the Egyptian Ministry of Health to integrate proven treatments into public programs—negotiating prices, distribution, or provision of interim medications. The goal? To ensure that research achievements continue after the study ends, sustainably.
- **Case Example 2:** Pediatric Oncology Trials (high vulnerability due to parental desperation and lack of alternatives). In this case, ENREC provided recommendations:
  - Required standalone child assent processes (age-appropriate).
  - Ensured free palliative care for nonresponsive cases.
  - Monitored compensation for families (travel costs, lost wages of the parents).

5 Critical Role of the Declaration of Helsinki and Consensus on the Helsinki Statement

5.1 Declaration of Helsinki (October 2024 Edition): Guiding Principles

For Egypt, the DoH is a foundational document for ethical principles in medical research. It emphasizes informed consent, confidentiality, risk-benefit assessment, etc. It provides specific guidance for protecting vulnerable populations focusing on:

Highlights the Need for Special Protections and Considerations Emphasizes the importance of equitable access to research benefits.

- Requires careful evaluation of potential risks and benefits.

5.2 ENREC Consensus on the Helsinki Statement 2024

ENREC's ability to foster consensus on the signing of the Helsinki Statement within a short time frame reflects its commitment to ethical research practices in Egypt, ensuring the rights and protection of research participants. ENREC believes that promoting ethical research in Egypt requires adherence to fundamental principles that guarantee participant rights, enhance the social value of research, and ensure transparency and fairness in the dissemination of and access to results. Therefore, the majority of ENREC members endorsed the principles of the Helsinki Statement, recognizing it as a valuable addition to the latest version of the DoH in 2024. In addition, ENREC believes that the Helsinki Statement enhances the protection of research participants and most importantly the Statement endorses the rights of participants, especially vulnerable people in low- and middle-income countries (LMICs).

Representatives of the ENREC attended the online consultative meetings held by the organizers of the Helsinki Statement, two times (October 16 and 20) during the WMA's General Assembly, where ENREC members' inquiries were raised and discussed. Following these web meetings, online discussions were conducted with the remaining network members on the Statement, after its translation into Arabic for ease of access and alignment with Egyptian social values. The network's WhatsApp group for committee chairs and the Facebook group for committee members were utilized to achieve consensus among the network members. The statement received member approval, and it was agreed that the network coordinator would sign the statement as a representative of the network, while also encouraging other members to sign on behalf of their respective institutions, provided they obtained official approvals. ENREC benefited from its status as a nonprofit, nongovernmental organization, unburdened by the bureaucratic obligations that may hinder governmental institutions, as we represent our members' views rather than an official stance.

5.3 ENREC'S Commitment to Ethical Research: The Helsinki Statement 2024

ENREC's quick adoption of the Helsinki Statement shows its strong dedication to ethical research, especially concerning the protection of vulnerable groups.

- **Participant Autonomy and Informed Consent:**
  - Emphasis on rigorous informed consent to protect individuals, especially those in LMICs or with limited support.
- **The Responsibility of Posttrial Benefits:**
  - Where sponsors ensure continued access to beneficial interventions for participants and broader strategies to address health disparities.
- **Addressing Vulnerability: ENREC's Focus on Equity:**
  - Vulnerability Focus: ENREC's application of the Helsinki Statement directly addresses systemic risks faced by LMIC participants.
  - Acknowledges that socioeconomic, educational, and political factors amplify vulnerability. Prioritizes equity, aiming to reduce coercion and unfair burdens on vulnerable groups.
  - Outcome: ENREC's approach aims to ensure ethical rigor in research involving vulnerable populations, promoting fairness and respect.

6 Challenges and Strategies in Implementation

It is important to navigate diverse cultural, socioeconomic, and health-care access disparities, which are significant in a developing country like Egypt. It is important to ensure adequate resources and training for ethics committees, especially in regions with limited infrastructure. It is necessary to maintain ongoing monitoring and evaluation across diverse settings. It is especially important to promote South-South collaboration for context-specific guidelines.

Strengthening public awareness and engagement, particularly in underserved communities should be promoted. It is necessary to invest in continuous education and training programs, with a focus on addressing the unique challenges of conducting ethical research in a developing country, and to promote interdisciplinary collaboration, including partnerships with community leaders and organizations that understand the local context.

Special care should be provided in adapting to emerging technologies and research methodologies. We have to enhance international collaboration and knowledge sharing, continuing to prioritize the protection of vulnerable populations.

7 Conclusion

By aligning its work with Egyptian policies, ENREC led the implementation of a strong framework for protecting research participants, especially vulnerable groups. Sustained investment in RECs is vital for ethical research in Egypt's healthcare landscape. A call for continued domestic and international collaboration and commitment to the highest ethical standard ant its implementation is a prerequisite for the achievement of global health, without leaving no one behind.

Acknowledgments The author expresses special appreciation to the vulnerable communities participating in ethical research, the Egyptian Ministry of Health, and the Supreme Council for Reviewing the Ethics of Clinical Medical Research.

Conflict of Interests There is no conflict of interest related to this manuscript.

Note This manuscript is based on the author's presentation on April 9, 2025, during the International Conference on Pharmaceutical Medicine (ICPM) 2025, which is the annual meeting of the International Federation of Associations of Pharmaceutical Physicians and Pharmaceutical Medicine (IFAPP), at the session titled: "Vulnerability, social value and community engagement: voices from patients and Global South." Author's Presentation title: "Role of Ethics Committee for protecting vulnerable people in Egypt."

References

1. Helsinki Statement Stakeholders. Helsinki statement 2024. Clin Eval. 2025;52(3) http://cont.o.oo7.jp/52pop/HelsinkiStatement_18Oct24_final.pdf.
2. The World Medical Association. WMA Declaration of Helsinki—Ethical principles for medical research involving human participants. First Adopted in 1964, Last Amended in 2024. https://www.wma.net/policies-post/wma-declaration-of-helsinki/.
3. Kurihara C, Matsuyama K, Baroutsou V. Collaboration with the WMA and key contribution to the 2024 revision of the declaration of Helsinki. IFAPP Today. 2024;49:8–12. https://ifapp.org/wp-content/uploads/2024/11/IFAPP-TODAY-49-NovDec-2024.pdf.
4. Arabic Translation of Helsinki Statement Clin Eval 2025; 52(3): W60–W63. Online Publication on Oct 18, and Updated on Oct 21, 2024. https://Cont.O.oo7.Jp/52pop/HelsinkiStatement_ar.pdf.
5. Kurihara C, Matsuyama K, Baroutsou V. World medical association's declaration of Helsinki, 2024 Revision: celebrating the 60th Anniversary, at Helsinki. Clin Eval. 2025;52(3):W1–W27. Http://Cont.O.oo7.Jp/52pop/52pop_contents_e.Html.
6. United Nations. International covenant on civil and political rights. Adopted by General Assembly Resolution 2200A (XXI) of 16 December 1966, Entry into Force 23 March 1976. https://www.ohchr.org/en/instruments-mechanisms/instruments/international-covenant-civil-and-political-rights.
7. Constitution of the Arab Republic of Egypt. English Translation. https://www.sis.gov.eg/up/dustor/dustor-english002.pdf.

Chapter 4
The DoH, the DoT, and the Duty to Participate in Data Research: A Case Reflection from Taiwan

Daniel Fu-Chang Tsai and Yu-Chen Juang

Abstract Research ethics is a dynamic and evolving domain, continually shaped by developments in scientific practice and technological innovation. The adoption of the Declaration of Taipei (DoT) and recent revisions to the Declaration of Helsinki (DoH) reflect the growing need to engage with the ethical challenges posed by data-intensive biomedical research. This article examines a decade-long legal dispute in Taiwan concerning the secondary use of National Health Insurance (NHI) data for research purposes, illustrating the ethical tension between the protection of individual data privacy and the promotion of socially valuable research. A central concern addressed here is whether there exists a moral obligation to participate in research, particularly in contexts involving minimal risk and substantial public benefit. While the protection of human participants remains a foundational principle of contemporary research ethics, the authors argue that an uncritical application of protective norms may hinder the advancement of health-related knowledge. In response, we propose a recalibration of the ethical framework—one that recognizes not only the rights of individuals but also their responsibilities as members of a solidaristic community. In this view, participation in low-risk research, including studies using health data and biospecimens, may constitute a civic duty grounded in reciprocity and public interest.

Keywords The Declaration of Taipei · The Declaration of Helsinki · The Taiwan National Health Insurance · Health information · Health database

D. F.-C. Tsai (✉)
Research Institute & Department of Medical Education & Bioethics, National Taiwan University College of Medicine, Taipei, Taiwan
e-mail: fctsai@ntu.edu.tw

Y.-C. Juang
Institute of Health Policy and Management, National Taiwan University, Taipei, Taiwan

C. Kurihara et al. (eds.), *The 2024 Declaration of Helsinki*,
https://doi.org/10.1007/978-981-96-9294-1_4

1 From the Declaration of Taipei to the Declaration of Helsinki

The 67th World Medical Association (WMA) Congress, held in 2016, announced the Declaration of Taipei on Ethical Considerations Regarding Health Databases and Biobanks [1] (hereafter referred to as the Declaration of Taipei, DoT). This Declaration seeks to provide a comprehensive ethical framework for the collection, storage, and utilization of identifiable health data and biological materials beyond the context of individualized patient care. It introduces additional ethical principles specifically designed for the governance of health databases and biobanks (Article 3). The DoT defines "health information" as data systematically recorded by health-care providers during the course of medical treatment, which is used to inform ongoing patient care (Article 2). Furthermore, a health database is described as an organized system for collecting, organizing, and storing health information (Article 4). According to the DoT, medical records and examination reports generated by physicians and health-care teams in the course of treatment are encompassed within the definition of a health database.

Building upon the foundational principles outlined in the Declaration of Helsinki (DoH) [2], the DoT establishes rigorous ethical standards aimed at safeguarding human dignity, autonomy, privacy, and confidentiality in the realm of medical research. Central to these principles is the requirement to obtain informed consent prior to the utilization of identifiable personal health data (Article 1). For data stored in health databases that are intended for unspecified future research purposes—referred to as secondary use—the DoT stipulates that broad consent, which is a form of general consent for future uses, should be obtained (Article 12). In cases where data and biological specimens pertain to individuals unable to provide consent—such as minors or those who have lost their capacity to consent due to medical conditions—the DoT mandates that reasonable efforts must be made to acquire consent once these individuals are capable of giving it, thus allowing continued use of their data for research purposes (Article 13).

In practice, the collection of health data and specimens often occurs initially for the purpose of patient treatment, not for research. It is only later that researchers may seek to retrospectively analyze medical records for research purposes, a trend that has grown increasingly significant, particularly with the integration of artificial intelligence (AI) and machine learning in health care. The revision of the Declaration of Helsinki in October 2024 acknowledges this evolving reality, reflecting the growing importance of retrospective research [3].

The revision, particularly the rewording of Article 32, introduces a critical shift in how informed consent is approached for secondary uses of health data. The revised paragraph mandates that informed consent must be obtained prior to the collection, processing, and storage of data for foreseeable secondary use. However, during discussions within the working group, a debate arose regarding the feasibility of applying the DoT's provisions in contexts where obtaining consent proves difficult. Some members proposed directly implementing the DoT's requirements for obtaining consent from individuals capable of giving it, while others, including the author, argued that while informed consent remains a cornerstone of ethical research, it is not the only means of ensuring participant protection. In practice, waiving consent for low-risk research is an established practice, and the advancement of medical knowledge, which is essential for improving patient health, necessitates the continuation of research without overly burdensome consent requirements.

Given the prevalence of retrospective research in clinical practice, mandating informed consent in every case is neither practical nor feasible. Imposing such a requirement may render valuable data inaccessible for research, ultimately hindering the progress of medical science. This would be counterproductive to the overall goal of advancing patient health. Consequently, the revised DoH explicitly includes an exception in Article 32, allowing the secondary use of preexisting data and specimens "where consent is impossible or impracticable to obtain," provided that such use undergoes review and approval by a research ethics committee (REC). This modification strikes a delicate balance between facilitating medical advancement and upholding the ethical obligation to protect the health and well-being of research participants.

2 Lessons from Taiwan's National Health Insurance Database Litigation: Striking a Balance Between Health Data Research and the Right to Information Privacy

Taiwan stands as a prominent example of a nation with a robust national health insurance system since 1995, having achieved an impressive coverage rate of 99% by 1998. In 2000, the Taiwanese government established the National Health Insurance Research Database (NHIRD), thereby facilitating external access to health insurance data for research purposes. This initiative was further expanded in 2008, when the NHIRD began integrating health insurance data with other vital health-related information, such as cancer registries and mortality statistics [4]. Despite the potential for significant medical advancements through such a comprehensive database, the Taiwanese Association for Human Rights (TAHR), a civil society organization, raised concerns regarding the potential privacy risks associated with the database. In 2012, TAHR sought the removal of personal data from the database on behalf of individuals, a request that was denied by the National Health Insurance Administration. Subsequently, TAHR pursued legal recourse by filing an administrative lawsuit, which was ultimately dismissed by the Supreme Administrative Court in 2017 [5]. The organization then sought constitutional interpretation, leading to a ruling in 2022 that brought the case to a close [6].

This protracted legal dispute underscores critical ethical dilemmas surrounding the intersection of personal data privacy and the advancement of medical research. The decisions rendered by both the Supreme Court and the Constitutional Court reflect an ongoing dialogue about how to balance these often competing interests.

The case centered on three fundamental issues: first, whether the government's use of data initially collected for health insurance purposes to create a health research database infringes upon individuals' right to informational privacy; second, the standards for data de-identification when it is used for research purposes; and third, whether individuals retain the right to opt out of the database—commonly referred to as the right to withdraw.

In 2005, Taiwan's grand justices (now the Constitutional Court) issued Interpretation No. 603, which affirmed that individuals' right to informational privacy, understood as the right to control one's personal data, is constitutionally protected. However, the ruling also clarified that this right is not absolute. When the government acts in pursuit of significant public interests, it may collect and use personal data, provided that such actions are explicitly regulated by law and that appropriate safeguards are in place to ensure the accuracy and security of the data. In the present case, the Supreme Court recognized the substantial value of using National Health Insurance (NHI) data for academic research, particularly for national health policy formulation, disease prevention, and treatment development, thereby aligning the use of such data with significant public interests. Nonetheless, the Constitutional Court emphasized that the legal framework for using NHI data remains insufficiently clear. Although the use of NHI data for research purposes was deemed legitimate, the Court underscored the need for clear legislative regulations that govern the use and oversight of such data, ensuring that individuals' informational privacy rights are adequately protected.

Concerning the use of NHI data for research, the Supreme Court held that data, once processed (e.g., through encryption or coding) in such a manner that it is no longer easily traceable to specific individuals, loses its personal attributes and may thus be exempt from the provisions of Taiwan's Personal Data Protection Act (PDPA). However, the Constitutional Court further clarified that if there remains any objective possibility, however remote, of reidentifying individuals from the data, it must still be regarded as personal data and subject to the PDPA. This stipulation is especially pertinent for NHI data, which constitutes sensitive and unique categories of information. The Court emphasized that even the potential for reidentification through extreme means must be taken into account. Consequently, the use of processed NHI data for research should still adhere to the principles of personal data protection, and mechanisms must be established to safeguard the accuracy and security of this information throughout the research process. Some advocates have proposed complete de-identification of NHI data; however, such an approach would undermine the data's utility for meaningful comparative medical analysis, thereby detracting from its potential to serve the public interest. As such, a more measured approach—where data is processed to prevent reidentification through reasonable means and its use is restricted to official or academic research purposes—represents the least invasive method. Nevertheless, authorities must remain vigilant in updating data protection protocols in line with technological advancements to ensure the continued protection of individuals' privacy.

Finally, regarding the right to withdraw from the database, the Supreme Court ruled that the government may lawfully refuse individuals' requests to withdraw their data. The Court reasoned that excluding individuals' data from the comprehensive NHI Research Database, based solely on privacy concerns, could significantly hinder the realization of the public health benefits associated with such a database. The Constitutional Court adopted a more nuanced approach, acknowledging the absence of clear legal regulations governing individuals' right to withdraw their data from research use. The Court mandated that, within 3 years, the relevant authorities must enact or amend legislation to explicitly define the terms under which withdrawal is permitted, as well as any exceptions to this right. The Constitutional Court affirmed that the ability to withdraw from the database—or conversely, the limitation of such a right—falls within the constitutional protection of informational privacy, provided such measures are carefully defined and legally regulated.

3 A Moral Duty to Participate in Research

Empirical studies conducted in Taiwan have demonstrated that, following a process of public deliberation, the proportion of individuals willing to permit the reuse of their biological samples without requiring re-consent increases. In one such study, 80% of patients and 72% of the public affirmed that, provided the research is reviewed and approved by an ethics committee, residual samples could be utilized for other research purposes without necessitating re-consent. The study suggests that when research institutions establish robust ethical review processes, coupled with privacy protection regulations, and promote greater public engagement in governance mechanisms, the financial and logistical burdens associated with re-consent can be substantially mitigated [7].

These findings invite a reexamination of the ethical questions surrounding research participation. Contemporary research ethics emerged as a response to the ethical transgressions witnessed in human subject research during and after World War II, thereby placing a strong emphasis on protecting research participants. However, when such protective measures become overly rigid, resulting in complex and multilayered safeguards, there is a concern that they may hinder the progress of biomedical research, particularly in low-risk studies. In this context, some scholars have proposed that participation in medical research can be understood as a moral obligation. Furthermore, there has been renewed scrutiny of the justification for the right to withdraw from research.

Do individuals have a moral obligation to participate in medical research? Harris (2005) contends that, in accordance with the principle of non-maleficence, we possess a moral duty not to inflict harm upon others. Extending this reasoning through the principle of beneficence, one derives the "rule of rescue," which asserts that we have an obligation to prevent significant harm where it is reasonable to do so. Harris argues that medical research is an essential vehicle for alleviating the suffering caused by disease, and as such, individuals have a moral duty to engage in research efforts that may lead to advancements in medical knowledge. Additionally, grounded in the principle of fairness, individuals have an ethical responsibility to avoid being "free riders"—those who benefit from the efforts of others without contributing themselves. In situations that yield collective benefits, such as social welfare initiatives, individuals are expected to share in the sacrifices required to sustain such endeavors [8].

If medical research is recognized as a matter of public interest, there may be circumstances in which mandatory participation could be morally justified. In such cases, the following considerations should be borne in mind [8]: (1) Individuals should not be presumed unwilling to act for the public good, particularly under conditions of minimal burden and risk; (2) People are generally willing to participate in activities that offer benefits both to themselves and to society; (3) When a general obligation to act in the public interest is recognized, the challenges associated with obtaining consent are alleviated, and participation in research is less likely to be perceived as exploitation.

Schaefer et al. [9], advancing the public goods argument, maintain that biomedical knowledge constitutes a public good, accessible to all, including those who have not contributed to its development. Participation in medical research, they argue, is vital to sustaining this public good, and therefore everyone has a moral obligation to participate. Schaefer and colleagues further assert that individuals should be permitted to refuse participation only if they present a compelling reason for doing so [9]. Similarly, Rhodes [10, 11] posits that everyone is a stakeholder in the outcomes of biomedical research and should contribute in proportion to the benefits they receive. She advocates for a revision of traditional research ethics, shifting away from the view of participants solely as potential victims of exploitation [10]. Given that health care needs are universal, and that medical progress hinges on societal cooperation, which cannot be achieved without active participation, Rhodes emphasizes the necessity of public education to foster awareness about the social value and responsibility of contributing to biomedical research, especially in light of the persistent challenge of inadequate participant recruitment in many studies [11].

Forsberg et al. [12] further highlight that access to effective health care cannot be taken for granted. Health care relies on ongoing medical research, and individuals cannot predict the medical needs they may face in the future. Given the shared societal need for medical progress, the obligation to participate in research should be framed within the context of a "social contract." Forsberg and colleagues argue that individuals have a duty to engage in research that meets the following criteria: (1) it involves minimal risks; (2) it imposes no undue burdens on participants; and (3) it is likely to yield beneficial outcomes. Research utilizing stored tissue samples or health data meets these conditions and therefore falls within the scope of this moral obligation [12]. Accordingly, from the perspective of the social contract, public participation in such research becomes an ethical duty.

In contrast to current regulations that mandate individual consent for the use of health data in research, except in cases where obtaining consent is impractical, Ballantyne and Schaefer [13] challenge this framework. They argue that the moral obligation to participate in health data research can justify the secondary use of identifiable personal health information from the public sector without requiring consent. Ballantyne and colleagues suggest that Institutional Review Boards (IRBs) or Research Ethics Committees (RECs) should shift their focus away from the practicality of obtaining consent and instead prioritize the potential public benefits of the research, as well as ensuring public engagement and transparency in the process [13].

In summary, participation in medical research is not merely an altruistic act but a moral obligation grounded in the principle of reciprocity, as medical research generates public benefits that all citizens stand to gain. This obligation becomes particularly compelling when the burdens on participants are minimal, as is the case in studies involving medical records, health data, or low-risk observational research, while the potential benefits of the research are significant.

3.1 Limiting the Right to Withdraw

In clinical trials and human subject research, the right to withdraw from a study at any time without consequence is widely recognized. However, Schaefer and Wertheimer seek to refine the justification for this right, positing that it arises from four key aspects: (1) information imbalance; (2) an inability to mitigate risk; (3) inherent uncertainty; and (4) involuntary bodily invasion [14]. From this standpoint, in studies with risks significantly lower than those found in clinical trials, or where risks are virtually nonexistent, it may be reasonable to limit the right to withdraw.

Holm [15] agrees with this reasoning, asserting that the right to withdraw should not be absolute. While this right should extend to biobank research, Holm argues that the scope and content of the right need not mirror those in clinical research settings, where physical or psychological risks are higher [15]. Furthermore, while it may not serve as the justification for the right to withdraw, this right plays a crucial role in fostering public trust in research [15]. Helgesson and Johnson [16] argue that safeguarding the right to withdraw is essential for maintaining public confidence, as it influences individuals' willingness to participate in research, including their consent to provide samples for biobanks [16].

While traditional clinical research upholds the right to withdraw as fundamental, there are ethical grounds for limiting this right in low-risk research contexts, such as biobanks and health data studies. Grounded in the principles of the rule of rescue, avoidance of free-riding, reciprocity, fairness, and the promotion of public good, the contribution of one's medical records and health data to low-risk, high-social-value research possesses strong moral legitimacy. In such contexts, the moral obligation of citizens to participate in research is not only ethically defensible but, arguably, ethically imperative. This obligation is particularly compelling in jurisdictions such as Taiwan (and countries alike), where the establishment of a universal National Health Insurance (NHI) system ensures that the population's basic health care needs are met through a collectively funded and solidaristic model. The ethical weight of public benefit, when coupled with minimal individual burden and risk, justifies a reassessment of individual rights in the context of secondary data use. Specifically, in the case of health data research conducted under ethical oversight and robust data protection mechanisms, it may be morally permissible to place limits on the right of individuals to withdraw. Such a position does not negate respect for autonomy, but rather situates it within a broader framework of civic responsibility and solidarity that underpins the legitimacy of publicly funded health-care systems.

4 Conclusion: Reframing Research Ethics in the Age of Data Science

Research ethics is a dynamic and continually evolving field, shaped by the epistemic and technological transformations that redefine both the possibilities and the boundaries of biomedical inquiry. The development of the World Medical Association's Declaration of Taipei (DoT), alongside recent revisions to the Declaration of Helsinki (DoH), signals the enduring need for ethical frameworks that can respond to emerging dilemmas in data-driven and population-level research. As scientific paradigms shift—from interventional trials to algorithmic modelling, and from individual consent to systemic oversight—the normative task becomes one of balancing the imperative to advance medical knowledge with the obligation to safeguard the rights and interests of research participants.

This article has examined a pivotal case study: a decade-long legal dispute in Taiwan concerning the secondary use of National Health Insurance (NHI) data for research purposes. The rulings of both the Supreme Court and the Constitutional Court reflect a growing recognition that existing legal and ethical frameworks must evolve to accommodate the realities of contemporary health data governance. These decisions carry significant implications for the regulation of big data and artificial intelligence in health care, not only in Taiwan but as a reference point for broader global debates.

Taiwan's longstanding investment in ethical infrastructure—through the widespread establishment of Institutional Review Boards (IRBs), the pursuit of international accreditation (such as FERCAP [17] and AAHRPP [18]), and a regulatory regime that at times exceeds Western standards—demonstrates its commitment to participant protection and scientific integrity. Yet this institutional robustness also invites a reevaluation of prevailing assumptions. While the principle of protecting research participants remains foundational, we argue that it must now be situated within a broader ethical discourse that includes the moral obligation of citizens to contribute to the collective enterprise of medical research.

The integration of electronic health records with biological specimens, genomic data, and even nonclinical information reflects a profound shift toward data-intensive biomedical science. Such research infrastructures support critical innovations in fields including artificial intelligence, public health surveillance, precision medicine, and pandemic preparedness. The ethical legitimacy of these activities rests not only on safeguarding privacy and ensuring data security but also on their alignment with the public good. As beneficiaries of publicly funded health systems, individuals arguably hold civic responsibilities alongside their rights.

Realizing the promise of data-driven health care requires the sustained and cooperative engagement of multiple actors: patients, researchers, institutions, and regulatory bodies. The public good cannot be pursued without public participation. In this light, we contend that the ethical discourse on research participation must move beyond individual risk mitigation toward an account that affirms solidarity, reciprocity, and shared responsibility. Particularly in contexts of minimal risk and high social value, participation in health data research should be understood not merely as a matter of personal choice but as a civic duty rooted in our mutual dependence on a just health system.

Acknowledgments DFT is a Board member of Medical Ethics Committee at the Taiwan Medical Association, and has been participating in several Workgroups of the WMA, including the 2024 revision of the Declaration of Helsinki.

Conflict of Interest The authors have no competing interests to declare that are relevant to the content of this chapter.

References

1. World Medical Association. WMA Declaration of Taipei on Ethical Considerations regarding Health Databases and Biobanks. First adopted October 2002, revised October 2016. https://www.wma.net/policies-post/wma-declaration-of-taipei-on-ethical-considerations-regarding-health-databases-and-biobanks/.
2. World Medical Association. Declaration of Helsinki: ethical principles for medical research involving human subjects. 2000. Rescinded. https://www.wma.net/wp-content/uploads/2018/07/DoH-Oct2000.pdf.
3. Resneck JS Jr. Revisions to the declaration of Helsinki on its 60th anniversary: a modernized set of ethical principles to promote and ensure respect for participants in a rapidly innovating medical research ecosystem. JAMA. 2025;333(1):15–7.
4. Tsai DFC, Juang YC, Chen YH, Yeh HY. The ethical and legal implication of using medical record and health data in research. Formosan J Med. 2020;24(5):471–89. https://doi.org/10.6320/FJM.202009_24(5).0001.
5. Taiwan Supreme Administrative Court Judgment 106 Pan-Zi No. 54. 2017.
6. TCC Judgment 111-Hsien-Pan-13. 2022. https://cons.judicial.gov.tw/en/docdata.aspx?fid=5534&id=352245.
7. Lin KM, Rei W. In: DFC T, editor. ISBN: 9789869490542 Citizen Deliberation of Human Subject Research: Report of National Research Program for Biopharmaceuticals ELSI office. Taipei: Health Culture; 2017.
8. Harris J. Scientific research is a moral duty. J Med Ethics. 2005;31(4):242–8.
9. Schaefer GO, Emanuel EJ, Wertheimer A. The obligation to participate in biomedical research. JAMA. 2009;302(1):67–72.
10. Rhodes R. Rethinking research ethics. Am J Bioeth. 2005;5(1):7–28.
11. Rhodes R. When is participation in research a moral duty? J Law Med Ethics. 2017;45:318–26.

12. Forsberg JS, Hansson MG, Eriksson S. Why participating in (certain) scientific research is a moral duty. J Med Ethics. 2014;40(5):325–8.
13. Ballantyne A, Schaefer GO. Consent and the ethical duty to participate in health data research. J Med Ethics. 2018;44(6):392–6.
14. Schaefer GO, Wertheimer A. The right to withdraw from research. Kennedy Inst Ethics J. 2010;20(4):329–52.
15. Holm S. Withdrawing from research: a rethink in the context of research biobanks. Health Care Anal. 2011;19(3):269–81.
16. Helgesson G, Johnsson L. The right to withdraw consent to research on biobank samples. Med Health Care Philos. 2005;8(3):315–21.
17. SIDCER. The SIDCER-FERCAP Foundation. https://www.sidcer-fercap.org/pages/home.php. Accessed 4 May 2025.
18. Association for the Accreditation of Human Research Protection Program. https://www.aahrpp.org/. Accessed 4 May 2025.

Chapter 5
Ethical, Legal, and Social Issues of Biobanks, Including Brain Banks: Experiences in South Korea, Taiwan, and Japan, and Future Perspectives for Global Principles

Young-Joon Ryu, Daniel Fu-Chang Tsai, Takeo Saio, and Chieko Kurihara

Abstract In the 2024 revision of the Declaration of Helsinki (DoH) by the World Medical Association (WMA) for medical research involving human participants, the Declaration of Taipei (DoT) for health databases and biobanks was referred to for the first time. This will have a significant impact on the secondary use of data and biological materials obtained from research participants. The increased awareness of the DoT will not only impact research but also have an even greater impact on the management and governance of health databases and biobanks worldwide. A further revision process of the DoT started in April 2025.

The authors had an opportunity to conduct joint research on the ethical, legal, and social issues of biobanks, especially focusing on brain banks, in South Korea, Taiwan, and Japan. This chapter describes our findings and proposals for further consideration. Although this is a collection of anecdotal descriptions of each situation in the three countries, it is desired that our findings will contribute to further development of global ethical principles for health databases and biobanks, through the revision process of the DoT, which is now a requisite item of the DoH.

Y.-J. Ryu
Department of Medical humanities and Medical Ethics, Department of Pathology, College of Medicine, Kangwon National University, Chun-Cheon, South Korea

D. F.-C. Tsai
Research Institute & Department of Medical Education & Bioethics, National Taiwan University College of Medicine, Taipei, Taiwan

T. Saio
Department of Internal Medicine and Psychiatry, Fuji Toranomon Orthopedic Hospital, Shizuoka, Japan

C. Kurihara (✉)
Kanagawa Dental University, Yokosuka, Kanagawa, Japan
e-mail: chieko.kurihara@nifty.ne.jp

C. Kurihara et al. (eds.), *The 2024 Declaration of Helsinki*,
https://doi.org/10.1007/978-981-96-9294-1_5

Table 5.1 Governance framework recommended in the DoT [1]

Paragraph 12 for items to be informed for "valid consent"
The purpose of the Health Database or Biobank; *The risks and burdens associated with collection, storage, and use of data and material;* *The nature of the data or material to be collected;* *The* ***procedures for the return of results including incidental findings****;* *The* ***rules of access*** *to the Health Database or Biobank;* *How privacy is protected;* ***The governance arrangements as stipulated in paragraph 21;*** *In case the data and material are made non-identifiable the individual may not be able to know what is done with their data/material and that they will not have the option of withdrawing their consent;* *Their fundamental rights and safeguards established in this Declaration; and* *When applicable,* ***commercial use and benefit sharing, intellectual property issues and the transfer of data or material to other institutions or third countries****.*
Paragraph 21 for governance arrangement
The purpose of the Health Database or Biobank; *The nature of health data and biological material that will be contained in the Health Database or Biobank* ***Length of time for which the data, or material will be stored****;* *Regulations of the disposal and destruction of data or material;* *How the data and material will be documented and* ***traceable in accordance with the consent of the concerned persons****;* *how the data and material* ***will be dealt with in the event of change of ownership or closure****;* *Arrangement for obtaining appropriate consent or other legal basis for data or material collection;* *Arrangements for protecting dignity, autonomy, privacy and* ***preventing discrimination****;* ***Criteria and procedures concerning the access to and the sharing of the health data or biological material*** *including the systematic use of* ***Material Transfer Agreement (MTA)*** *when necessary;* *The person or persons who are responsible for the governance;* *The security measures to prevent unauthorized access or inappropriate sharing;* *The* ***procedures for re-contacting participants*** *where relevant;* *The procedures for receiving and addressing inquiries and complaints.*

Source: Extracted from the WMA Declaration of Taipei [1]
Bold, underline are by authors, for the texts which are notable items beyond the items ordinarily written in the informed consent form

The European Union (EU)'s General Data Protection Regulation (GDPR) in 2016 [12] promulgated a strict policy requiring consent to the objective of usage of personal data. The system in this framework, in which the EU grants "adequacy decision" to the countries with sufficient personal data protection framework, has had a significant global impact, as it eases restrictions on the transfer of personal data to the countries that have established EU-equivalent data protection systems. Later, the European Health Data Space [13] has opened the possibility of data use without explicit consent, but provided the opportunity of "opt-out" to data subjects to reject the use of their data upon seeing the disclosed information, on condition that a data governance mechanism has been duly established. Unreasonable executions of the opt-out right have been rejected in some court decisions [14]; thus, it would be important to promote social consensus on the moral obligation of the public to support and collaborate for public health.

The DoT [1] would take the role of an ethical basis for these frameworks, just as the DoH gives a basis for research regulations, and in all the cases, ethics committee approval is needed for such exceptional data use without explicit consent.

In addition to such consideration, the brain bank needs tissue donation from deceased donors, and their willingness to participate in research during their life time. Additionally, consent (or permission) from the family after the death of this person is required in many jurisdictions.

Here we introduce our experiences as case studies regarding ethical, legal, and social issues of brain banks and biobanks, discussed in each of the countries of South Korea, Taiwan, and Japan, with the desire that such information would contribute to the further development of global ethical principles for health databases and biobanks, as explored in the process of the revision of the DoT [1].

3 Brain Banks in South Korea, Taiwan, and Japan

3.1 South Korea

3.1.1 Historical Development and Contemporary Status of Brain Bank in South Korea

It has been difficult to obtain human brain tissue for research in South Korea. Traditionally, human brain tissue research could have been conducted under limited opportunities for specific purpose (Table 5.2). Until a few years ago, only anatomists, pathologists, and forensics who were interested in some neurological studies had been able to collect brain tissue from the dead for personal research needs. Most

Table 5.2 Limited opportunities for traditional brain autopsy research in South Korea

Institutions	Purposes	Autopsy subjects	Researchers
Medical school	Anatomy education	Donor to a medical school	Anatomists
University affiliated hospital	Pathological cause of death identification	Body of a person who died in a hospital	Pathologists
Crime autopsy institution	Law-enforcement related crime cases	Someone who died in a criminal case	Forensics

autopsies are carried out with the consent of the family, except in criminal cases, where an autopsy could be executed according to court order without any consent.

The brain tissue collection for research on the brain nervous system was started in 2014 by the Korean government. In 1998, the Ministry of Science and ICT enacted the *Brain Research Promotion Act* [15], in which Article 17 allowed the establishment of a brain research institute. Based on this article, in 2011, "Korea Brain Research Center" was established, followed by a facility construction at Daegu in 2014. Focusing on the needs of brain research, the second phase plan of the "Basic Brain Research Promotion Plan," which was started in 2008 and planned to reestablish every 4 years, was implemented in 2013. This development plan for brain science has led to an increasing demand for research-purpose human brain tissue and related clinical information. Brain research scientists always need to conduct research using human brain tissue in order to verify their academic findings from animal experiments; however, they could not easily obtain human tissue for research. Researchers applied for brain tissue at high cost from brain banks in the United States or Europe, where human brains have been collected and stored for a long time. But this does not meet their needs. Researchers needed not only human brain tissue but also related clinical information or blood, etc., but foreign brain banks did not provide such information for various reasons. Thus, Korean researchers wanted to be provided with human brain resources for brain research in Korea and assigned the newly established Korea Brain Research Institute to play that role.

The Korea Brain Research Institute established the "Korea Brain Bank (KBN)" in the same building in 2014. The KBN was formed in cooperation with the Sao Paulo Medical School in Brazil, the world's largest brain bank, the Brain Research Institute of the Netherlands, and the Brain Bank of Niigata University in Japan. In 2015, for the first time, two medical hospitals (Seoul National University and Korea University) were designated to pilot a brain tissue collection project. Afterwards, in 2017, Asan Medical Center, Kyungbook National University Chilgok Hospital, Chonnam National University Hospital, Kangwon National University Hospital, and in 2018, Inje University Busan Paik Hospital were added to operate a regional network. This system is called "Korea Brain Bank Network (KBBN)."

Another government-led brain bank project is being conducted by the Ministry of Health and Welfare (MOHW). The scope of these brain banks is "dementia." Under the MHW, The National Health Center is conducting these plans, and the Department of Chronic Diseases in the Korean National Institute of Health (NIH) funded two brain banks at two university-affiliated hospitals (Seoul National University Hospital in 2017 and Busan University Hospital in 2018).

In total, eight brain banks have been established under the KBBN Project: an alliance between the central brain bank within the Korea Brain Research Institute, the five regional-based brain banks, and two dementia brain banks supported by the MHW. In 2019, the KBBN announced that it had collected 100 human whole brains during the 5 years since its implementation.

3.1.2 Legal and Regulatory Framework of Brain Bank in South Korea

The so-called "brain research resources" include not only brain tissue, cerebrospinal fluid (CSF), and blood, but also clinical information, such as electroencephalogram (EEG), psychological tests, and imaging data. These resources can be collected from both live and dead bodies, and researchers want to get integrated information. However, Korea's legal system has caused difficulties because cases of living bodies are under the *Bioethics and Safety Act* [16], while the cases of dead bodies are under the *Act on the Dissection and Preservation of the Dead* [17]. The *Korean Civil Act* [18] interprets that human rights and duties are given at birth and end at death. In this respect, the human tissue of the living and the human tissue of the dead are not considered equal. Therefore, the collection of human tissues from living bodies for research should be distinguished from the collection of human tissues from dead bodies.

However, Korean law does not yet provide comprehensible regulations for brain banks collecting brain tissue from the dead and integrating it with information obtained while life time. The existing legislation on research using dead bodies, the *Act on the Dissection and Preservation of the Dead* [17], was enacted in 1962 and is criticized for not reflecting the current status of brain research. For example, brain tissue obtained by autopsy can only be studied at medical schools and affiliated hospitals, but it is prohibited by law to transfer it to other researchers. This regulation can prevent the implementation of a multi-organizational cooperation system necessary for modern brain research. Another example is the technical limitation when conducting studies of disease, which requires a mutual comparison and convergence of information obtained by research during both donors' life time and after their death.

In the Korean government, the *Bioethics and Safety Act* [16] and the *Act on the Dissection and Preservation of the Dead* [17] are governed by the MHW, while the *Brain Research Promotion Act* [15] is governed by the Ministry of Science and ICT. The Ministry of Science and ICT wants to get research-purposed brain bank and to enhance brain tissue collaborative research, while the Ministry of Health and Welfare feels no need for revision of the legal system. One of the causes can be considered the conflict between the development-oriented policy of the Ministry of Science and ICT and the ethical and safety-oriented policy of the MHW. The two ministries have discussed overlapping issues in many cases, but they have proved to be different in many ways. The two ministries disagreed on the legal basis of the Brain Bank and have been through a coordination process by the Prime Minister's Office during 2018 and 2019. Although some members of the Parliament submitted legislation on this matter to Parliament to resolve the issue, it is in an unclear state.

3.1.3 Social and Ethical Dimensions of Brain Bank in South Korea

In South Korea, the autopsy of the body and the culture of deceased body donation are not yet common to people. These have been regarded as taboo or extraordinary things. During the 500 years of the Choseon Dynasty (1393–1897), Confucian culture and funeral culture viewed the autopsy of the dead as damaging filial piety. In addition, traditional medicine, which has been practiced for a long time since the Middle Ages, has not paid much attention to autopsy or the brain. Only starting from the seventeenth century did scientific interest in the brain begin. But even in the late nineteenth century, when modern Western medicine was being introduced in the Korean peninsula, autopsy was considered a heterogeneous medical practice only useful for death cause identification. The medical and scientific society of Korea has been modernized rapidly since the introduction of Western medicine and has made academic advances as much as those of the West. However, in contrast, the core ideas of Korean society and its members are still influenced by traditional values.

Recent poll results, which were obtained in the research project [19], support this claim. Among the 1028 South Koreans asked if they were "willing to donate their brains to a brain bank," 49.4% replied they were willing to donate while 51.5% replied they would not. These results indicate that strong contrasting views on brain donation still exist. It enables the hypothesis that people consider modern and premodern thinking to coexist, or that the idea of brain donation is influenced by the current social atmosphere and system. Regardless of the high rate of willingness in the result of the polls, it is very difficult to receive consent to donation in South Korea. It is relatively familiar to Koreans that most of the purpose of the donation of corpses after death is solely for the education of medical students. Actually, more than 120 years of modern medical education in South Korea have been supported by deceased body donations for educational purposes, but the number of donations still does not meet the educational needs. Such a limitation could be overcome by raising public awareness of the value of brain banks, while respecting the culture of funerals and the feelings of bereaved families.

Second, the establishment and operation of brain banks in South Korea have not been accompanied by significant social conflicts or controversies. On the contrary, there has been a steady increase in the number of individuals who, after encountering brain bank advertisements in public spaces such as city buses and subways, proactively reach out for consultations. An analysis of cases in which these consultations have led to actual postmortem examinations and donations has revealed a noteworthy ethical issue. An analysis of donors and their motivations indicates that some donors lack direct descendants to care for them and are instead supported by relatives or third parties. Among the benefits offered by brain banks to such prospective donors are medical support for patients with degenerative brain diseases, funeral assistance, and post-funeral memorial services. While these incentives are practically necessary for the operation of brain banks, concerns have been raised that they may be perceived as compensation for donation, thereby giving rise to ethical concerns.

A third issue pertains to intra-family disagreements. Under current legal provisions, an individual's consent to brain donation, given prior to death, is recognized as legally binding. However, in practice, cases frequently arise in which the donor's expressed will is disregarded due to strong opposition from bereaved family members. This tendency appears to stem from the brain banks' reluctance to engage in physical or legal disputes with the bereaved families. Nevertheless, this situation has been criticized as an infringement on individual autonomy, as it allows the donor's will to be overridden by familial opposition.

3.2 *Taiwan*

3.2.1 Historical Development and Contemporary Status of Brain Bank

Despite the global recognition of brain banks as indispensable infrastructures for advancing research on neurodegenerative and psychiatric disorders—with over 100 such repositories established worldwide—Taiwan did not inaugurate a national brain bank until 2023. The main obstacle was the *Human Autopsy Act* of 1984 [20], which prohibited physicians from conducting postmortem examinations within 6 h of death. Such a delay would preclude the timely preservation of neural tissues, particularly the proteins and nucleic acids essential for high-fidelity neuropathological analyses.

Further compounding the challenge was the enactment of the *Human Biobank Management Act* [21] in 2012, a legislative framework designed with rigor, yet one that imposed procedural and bureaucratic thresholds difficult for individual researchers to navigate without institutional backing. The result was a paradox: while the moral and scientific imperative for brain banking grew ever more urgent, the regulatory environment remained inhospitable to its realization.

Recognizing this impasse, in 2017, a coalition of neurologists, patient advocates, and legal scholars established the Taiwan Brain Bank Task Force and subsequently, the Taiwan Brain Bank Association—two entities that served as mediators between scientific aspiration and regulatory prudence. Through sustained dialogue with the Ministry of Health and Welfare (MOHW) and legislative authorities, consensus gradually emerged regarding the lawful parameters under which brain tissues could be retrieved and archived.

In 2019, the MOHW issued an administrative clarification stating that the collection of postmortem brain tissue for research purposes could proceed under the auspices of the *Human Biobank Management Act* [21], thereby exempting such activities from the restrictive provisions of the 1984 *Human Autopsy Act*. This interpretive shift paved the way for the formal establishment of the Taiwan Brain Bank, which received governmental approval in July 2023. Its official inauguration was held on November 5, 2023, at the College of Medicine, National Taiwan University.

The Bank operates with financial support from the MOHW and is subject to ethical oversight by a dedicated review committee, in compliance with the national biobank framework. Within the first year of its existence, by August 2024, 38 individuals had consented to posthumous brain donation, and eight had fulfilled this commitment. These figures, modest though they may seem, signify a profound shift in Taiwan's cultural, ethical, and scientific landscape.

3.2.2 Legal and Regulatory Framework of Brain Bank in Taiwan

In the early development of brain banking in Taiwan, physicians often refrained from retrieving brain tissue for research purposes, even when donors had provided prior consent. This hesitancy stemmed primarily from the legal ambiguity surrounding the *Human Autopsy Act* of 1984 [20], a law originally designed to govern the anatomical dissection of cadavers donated to medical schools. While well-intentioned, the Act imposed a mandatory 6-hour waiting period following death before any autopsy or embalming could proceed—a buffer period ostensibly aimed at allowing prosecutors to review the death report and rule out foul play. From a neuroscientific perspective, however, this delay proved detrimental: key biomolecules such as RNA and proteins in neural tissue degrade rapidly postmortem, rendering delayed specimens scientifically compromised and impeding meaningful neuropathological research.

Crucially, the scope of the *Human Autopsy Act* [20]—focused on physician conduct during gross anatomical dissections—did not foresee nor accommodate the long-term storage, governance, or research use of postmortem tissues in a systematic repository such as a biobank. In essence, the ambitions of neuroscience had outpaced the legal scaffolding meant to regulate its practice.

In response to these limitations, the Ministry of Health and Welfare (MOHW) convened a panel of legal and medical experts in September 2019. This culminated in the issuance of a formal administrative clarification on November 11 of that year. The new interpretation decisively narrowed the applicability of the *Human Autopsy Act* [20] to anatomical education within medical schools, distinguishing it from the postmortem procurement of biological materials for research under the *Human Biobank Management Act* (2012) [21]. Furthermore, it affirmed that forensic review by prosecutors is required only in cases of nonnatural deaths—consistent with organ donation protocols—and clarified that the 6-hour restriction does not apply to physicians retrieving brain tissue from donors who died of natural causes.

Importantly, this reinterpretation positioned brain banking squarely within the jurisdiction of the *Human Biobank Management Act* [21], thereby establishing a legal foundation for Taiwan's first national brain bank. Yet, this framework also surfaced new ambiguities. Brain donation, unlike other biobank specimen collections, can only commence after death, aligning it more closely with cadaveric organ donation in timing and procedural urgency. However, while organ donation seeks to save lives, brain tissue donation serves the future-oriented aim of advancing research—a distinct though equally altruistic enterprise.

One of the unresolved issues concerns consent, namely, is it allowed for relatives to agree to donate brain tissue after a person's death? Taiwanese scholars Shih and Ho have noted that the *Human Subjects Research Act* [22]—which defines cadaveric tissues as human specimens—offers a more expansive view. Article 13 of that Act permits relatives to provide consent for research use of a corpse, provided it does not contradict the known wishes of the deceased, or where prior consent can be certified by two physicians and family members raise no objection [23]. Some scholars advocate for this law to complement existing biobank regulations, while others call for targeted amendments to the *Human Biobank Management Act* itself [24]. As of this writing, MOHW has not issued a definitive stance. In the absence of regulatory clarity, the Bank has chosen a conservative course: to accept only antemortem consent, thereby minimizing legal risk while potentially limiting donor participation.

3.2.3 Social and Ethical Dimensions of Brain Bank in Taiwan

As a participant in the expert consultations surrounding the legal architecture of the Taiwan Brain Bank, the author observed a shared conviction among stakeholders: that postmortem brain donation is ethically analogous to cadaveric organ donation. Both involve interventions initiated only after death—upholding the "dead donor rule"—and both require swift execution to preserve biological viability. More deeply, both are acts of altruism, undertaken by individuals (or their families) who, in the face of mortality, choose to affirm life: in one case, by extending it through transplantation; in the other, by advancing the scientific understanding of human suffering and disease.

Yet despite these moral parallels, the regulatory treatment of brain and organ donation remains bifurcated. Brain donation is governed under biobank statutes and organ donation under transplantation law. This division, while administratively convenient, imposes an unnecessary procedural burden on potential donors and their families.

The Taiwan Brain Bank did not commence operations until 2023—a testament to both the complexity of its legal pathway and the caution with which it was pursued. The conservatism guiding its early implementation is understandable; trust, once lost, is difficult to regain, particularly in the ethically sensitive realm of posthumous donation. Nevertheless, if the Bank is to fulfill its scientific and public health mission, its policies must evolve to support broader participation. Brain banks, by their nature, depend not on the heroic act of a few but on the sustained generosity of many. This requires not only ethical transparency but also regulatory frameworks that are inclusive, coherent, and responsive to the deeply human act of giving—after death—for the benefit of the living.

3.3 Japan

3.3.1 Historical Development and Contemporary Status of Brain Bank

In Japan, some institutes have been storing brain specimens for a long time, but it was only after 2000 that brain bank activity of the contemporary type began collecting brain tissues based on registration of donors who gave informed consent during their life time and additional consent from the family after the death of this donor. In the early 2000s, the Ministry of Health, Labour and Welfare (MHLW) started to grant support to this kind of activity. Later, a newly established funding agency, the Japan Agency for Medical Research and Development (AMED), succeeded in supporting the project, and then in 2016, the total amount of this grant during the 5 years until 2020 was defined to be 250 million Japanese yen (2.3 million USD). As a result of continuous governmental funding, a collaborative network of five core sites has been organized, named Japan Brain Bank Network (JBBN). There are some other research institutes or hospitals that have stores of brain tissues and can provide them for other researchers.

At the time of the authors' collaborative meeting in 2019, the following information was confirmed: The JBBN holds 1820 frozen specimens, of which 155 are collected from patients with psychiatric disorders and 699 are from those with neurological disorders. The Head Office of JBBN is placed at the National Center of Neurology and Psychiatry (NCNP), Tokyo. The NCNP brain bank has had 257 registrants for 10 years, of which 30 cases were autopsied later. On the other hand, other members of JBBN have a greater number of specimens; e.g., the Brain Research Institute of Niigata University claims to own 30,000 frozen specimens; Tokyo Metropolitan Geriatric Hospital holds over 2300 frozen specimens of brain. This would be partially because collection at JBBN, as a network of contemporary brain banks, is based on the consents of donors during their life time, although there is no specific legal basis for this consent. Meanwhile, most of the traditional collections are presumably based on the consent of the bereaved, although details have not been disclosed to the public. Such traditional consent of the bereaved is based on the *Act on the Dissection and Preservation of Corpses* ("*Autopsy Act*" hereafter) [25], enacted in 1949. Japanese *Civil Code* [26], enacted in 1896, gives human rights only to living persons, and there is no specific legal rule to cover research participants who provide their brain tissue for research.

Among these government-funded projects, there have been some studies to find an association between molecular imaging during life time and pathological data after death. Followingly, some new type of large-size prospective patient registry study, integrating medical information during life time and brain bank network, is going to be initiated.

3.3.2 Legal and Regulatory Framework of Brain Bank in Japan

The biggest challenge for brain banks in Japan is the legitimate handling of brain tissues under the *Autopsy Act* [25]. It has not been amended to fit recent research usage of human tissues derived from the deceased to be shared among the wider research community. Legal bases for obtaining human brain tissues are mainly for medical education, pathological cause of death identification, or law-enforcement forensic or administrative autopsy, allowing the possibility of storage for research. The *Autopsy Act* [25] specifies requirements of lawful autopsy and corpse preservation with dismissal of the crime of destruction of a corpse defined in the *Penal Code* of 1907 [27]. It defines the conditions for the preservation of parts of corpse; however, it does not define conditions on the transportation, storage of tissues, and supply repeatedly to other researchers. Also, it requires the consent of the bereaved but does not define legally valid informed consent of a donor during life time.

For this reason, academic societies and research institutions of Japan, involving scholars of laws or ethics, have collaboratively developed the Ethical Guidelines for Brain Bank ("Brain Bank Guidelines," hereafter), taking the opportunity of the above-mentioned grant projects. During the 2 years of 2015 and 2016, ten societies consisting of the fields of neurology, psychiatry, and pathology agreed to the Brain Bank Guidelines. As the premise, they positioned the ethical implication of the brain bank as the "Gift of Hope" to the next generation, following the policy expressed by brain banks in the world. The dignity of the deceased, courtesy to the bereaved, as well as social common wisdom, are identified as fundamental values we have to respect. Most importantly, they stated that harvest and preservation of the tissue of the deceased must be performed according to the *Autopsy Act* [25]; therefore, tissue acquisition must be after the autopsy with the purpose of providing the definitive pathological diagnosis as the" final service" to the patient and caregivers. They also defined detailed procedures of informed consent from a basically capable donor during life time, as well as consent from the bereaved family at the time of the death of this donor. The *Autopsy Guidelines* under the Act suggest detailed procedures and that only small parts of the deceased body could be utilized in various ways, including research use, supplying outside.

In such a situation, guidelines for human research without a legal basis have been developed by governments since the end of the 1990s. The latest version, named *Ethical Guidelines for Medical and Biological Research Involving Human Subjects* ("*Ethical Guidelines*," hereafter) [28], covers research involving living and deceased humans including individually identifiable data and biological samples. The Ethical Guidelines [28] defined items to be described in the protocol of biobank activity, separately from the protocol of research activity, and items to be described in the informed consent form, common between the case of providing data or samples to health databases or biobanks and the case of participating in research. These items required in the protocol and ICF are insufficient compared to the DoT [1] but brain tissue donations for research have to be conducted under the Ethical Guidelines [28]. The set of the above-described laws and guidelines is summarized in Table 5.3.

Table 5.3 Laws and guidelines related to brain bank activity in Japan

Names of laws and guidelines	Scope and applicability
Postmortem examination and corpse preservation Act (Act no. 204, 1949)	Autopsy and preservation of corps for investigation of death; medical research and education No definition related to brain bank activity but the acquisition of tissue from the deceased must be in accordance with this Act
Autopsy guidelines (Notification and directive No. 693, MHW, 1988)	Detailed governmental guidelines to the above Act and for lawful conduct of autopsy
Ethical guidelines for medical and biological research involving human subjects (MEXT, MHLW, and METI, 2021)	Medical and life science research and biobank/health database activities involving human participants, including individual-identifiable human material or personal information; and the deceased
Ethical Guidelines for Brain Bank (by 10 Academic Societies, 2015–2016)	Brain bank activity

3.3.3 Social and Ethical Dimensions of Brain Bank in Japan

According to an ancient history book describing the beginning of Japan, in A.D. 459, when the emperor's wife died because of an imputation that she carried a child of a servant, and the emperor directed an autopsy and then proved innocent of his wife. This was the first dissection in Japanese history. Later in 701, the first law system was established with reference to the Chinese dynasty Tang's decree, and this disallowed dissection. Then it was only from the mid-1700s that dissections came to be conducted, introducing Dutch medicine, which was the beginning of the introduction of modern Western medicine.

In contemporary days, brain bank activities in Japan have not provoked social debate, compared to the national debate on "brain death" at the time of the establishment of the *Organ Transplantation Act* in 1997. Brain death has raised nationwide debates over "Is brain death a death of a human?" People argued that it is not "natural" that scientific theory provides a "new" definition of death; death is a process that takes time for the bereaved to understand/accept a family member's end of life. On the other hand, brain tissue donation proceeds after the traditionally defined physical death. If informed consent was given by a donor during the life time, it would be justifiable for the bereaved to follow the will, although they may have some sentiment unwilling to permit harmful procedures to the body of the deceased. A shared decision-making process could be gradually developed among the attending physician and a patient during the life time and a caregiver family. This is at limited institutions, and the majority of Japanese hospitals are not engaged in such activities. Some specialized research institutes have developed patient registry systems, preparing leaflets and donor cards. Sometimes decision-making should be done with a coordinator independent from the attending physician, to avoid undue influence. It seems that the Japanese have not given a "special" moral status to brain tissue, which is not consistent with the situation that they are not willing to accept "brain death."

This would be because people can accept "brain donation" after the physical death, if its value is sufficiently explained, with some trustworthy relationship.

The majority of Japanese culture and religious belief is a mixed marriage of Buddhism, Shinto, and Confucianism, and cremation has become popular in a modernized society. In some rural areas, people keep the dead body for several days, and bereaved families tend to reject autopsy. One questionnaire survey showed agreements with future autopsy/donation to brain bank: 92/98% (102/94 respondents) in patients and their families who are the participants of an educational program on parkinsonism; 73/64% (88/64 respondents) in the participants of a public educational program; 78/64% in university students of German language (166/129 respondents); 79/76% (101/80 respondents) in medical students. (More than half of agreement with autopsy are conditional, requiring bereaved family's agreement.) This does not represent the general public; however, education could promote public awareness and consensus on donations to brain banks. It should also be noted that under the *Act on Body Donation for Medical and Dental Education* [29] in 1983, for corpse donations limited only to medical education, nearly 300,000 registrations and more than 130,000 donations had been achieved, as of 2019.

The society-initiated Brain Bank Guidelines have been carefully developed, avoiding public debate. It is related to the notorious "lobotomy" during the 1970s. One prominent professor's research use of a brain sample after psychosurgery, without the patient's consent to the research use of this sample, was accused by the psychiatrists' community. This was a part of the university student movement to protest against authority. This and other issues raised fundamental mistrust in biological psychiatry in the Japanese community of psychiatrists. Considering such bitter experiences, brain bank scientists have developed Brain Bank Guidelines carefully, considering not to provoke public debate.

4 Biobank in South Korea, Taiwan, and Japan

4.1 South Korea

4.1.1 Historical Development and Contemporary Status of Biobanking

South Korea's "human material bank," body-derivatives biobank, started in 2005. Human material banks can be divided into government institutional banks, government-supported banks, and private banks. As of 2019, about 50 human bioderivative banks are registered.

Government-funded institutions such as the ones funded by the MHW operate four centers (Central Human Resource Bank, Masan Tuberculosis Bank, National Cancer Center, and Korea Brain Bank). The Ministry of Science and ICT also funds four centers. In addition, there are 17 university hospitals supported by the Korea Biobank Network (KBN), which has been undertaken since 2008 by the Korea Centers for Disease Control and Prevention.

The Korea National Research Resource Center (KNRRC) Project, which began as a project of the Ministry of Science and Technology in 1995, is currently operating as a research funding project of the Ministry of Science and ICT/Korea Research Foundation. This project collects and manages specialized research materials not only human-derived but also from various fields such as animals, plants, microorganisms, fusion materials, etc. Among them, banks that store human-derived materials include the liver cancer specimen bank, pathogenic progenitor material bank, prostate bank, Korea leukemia bank, Korea women's cancer bank, Korea cell line bank, overseas infectious disease human resources bank, serum sample bank, research lung tissue bank (2012 closure), and Korean eye tissue and blindness-related gene bank (2008 closure).

However, there are no brain neuron tissue banks. The project functions to deposit, store, and distribute research materials that are difficult for individual researchers to obtain.

4.1.2 Legal and Regulatory Framework of Biobank in South Korea

Under the *Bioethics and Safety Act* [16], the human material bank regulates all matters concerning operations within human biobanks, such as permitting, operating, collecting, storing, and distributing based on the law. The minister's certification or registration of the bank is needed, along with audit or supervision. The departments that manage and supervise operations are the MHW and the Centers for Disease Control. Human-derived material banks keep relevant clinical information along with human-derived materials. Thus, the issue of anonymizing information that includes personal information or information that can identify an individual is always dealt with caution. The Act defines the set of basic information to be explained to a person who gives consent to provide the material derived from this person to a human material bank, which is as described in Table 5.4. More details are defined in the Ministerial Decree.

Table 5.4 Necessary information to be informed to a person giving consent to provide human material to a human material bank, defined by the Bioethics and Safety Act

1. The objectives of the human materials research (only applicable where the human material Bank directly conducts the human materials research)
2. Protection and disposal of personal information
3. The scope of researchers and institutions that are provided with the human material, etc.
4. Storage, management, and disposal of the human material, etc.
5. Withdrawal of consent, disposal of the human material, etc. when consent is withdrawn, rights of the donor of the human material, and other matters specified by ministerial decree of health and welfare

Source: Bioethics and Safety Act. Wholly Amended by Act No. 11250, Feb. 1, 2012; amended by Act No. 16372, Apr. 23, 2019
https://elaw.klri.re.kr/eng_mobile/viewer.do?hseq=52559&type=part&key=36#:~:text=The%20purpose%20of%20this%20Act,human%20materials%2C%20etc.%20or%20of

Nowadays, starting from October 2019, the law allows the reuse in hospitals of specimens generated from surgery for the purpose of diagnosis and treatment. In a situation where it is difficult to obtain additional consent, the law stipulates that if there is already an explanation for reuse in the process of obtaining consent, and if samples are donated to the Human Resources Bank, they can be used for research after deliberation by the Institutional Ethics Review Committee.

4.1.3 Social and Ethical Dimensions of Biobank in South Korea

The human material bank is a system that allows individual researchers to formally obtain human derivatives for research. The donor's consent is the most basic requirement for this. The *Bioethics and Safety Act* [16] provides a legal form for consent for donation to individual researchers and for donation to a human material bank. Modification to the legal form needs the approval of the institutional ethics review committee. The problem is legal consent forms that include some sentences that a donor gives up all of the rights related to the economic benefit of the donor. This seems to be "exculpatory language," which is deemed to be inappropriate, which is common in the world. There are also concerns that the human-derived information of numerous people managed by the state through the human-derived material bank could be combined with criminal information in the future to be used for state control of personal information. There would be concerns because the law mentions the exemption of consent for the government's use.

4.2 Taiwan

4.2.1 Historical Development and Contemporary Status of Biobanking in Taiwan

The evolution of biobanking in Taiwan has been neither linear nor smooth; rather, it has proceeded along a path shaped by regulatory uncertainty, ethical awakening, and persistent scientific aspiration. Although the practice of retaining surplus human biological specimens from clinical encounters for future research has long been tacitly embedded within medical culture, it remained largely unregulated until the promulgation of *The Precautions for the Collection and Utilization of Human Samples for Research Purposes* in 2002. This foundational guideline was the first to articulate the ethical prerequisites for biobanking in Taiwan, mandating informed consent and ethics review as conditions for the lawful collection, storage, and use of human samples for research.

Crucially, the guideline addressed the retrospective use of already collected samples, permitting their continued use only if the samples were anonymized, could not reasonably be traced back to donors for re-consent, and had entered the public domain prior to the regulation's issuance. These provisions reflect an early ethical attempt to reconcile the value of accumulated research materials with the emerging imperative to respect individual autonomy and data privacy.

A more comprehensive legal framework emerged with the enactment of the *Human Biobank Management Act* [21] in 2010, which codified the operational, ethical, and institutional standards required for the establishment and governance of biobanks. This legislation laid the groundwork for a national biobanking system, culminating in the formal launch of the Taiwan Biobank in 2012. Its mission is both ambitious and explicitly contextualized: to conduct large-scale cohort and case-control studies focused on diseases prevalent within the Taiwanese population. Specifically, the biobank aims to enroll 200,000 community-based volunteers and 100,000 hospital-based patients with one of the 10–15 most common conditions in Taiwan.

By 2019, the Taiwan Biobank had accrued 125,221 community-based participants and 4357 hospital-based cases. Beyond the national repository, Taiwan's biobanking landscape has grown to include 31 licensed biobanks as of 2020. These encompass institutional, municipal, and private-sector initiatives. The National Health Research Institutes Biobank, for example, prioritizes oncology, with sample collections focused on the liver (9600 cases) and lung cancers (3500 cases). The Tainan City Biobank is notable for its breadth, having archived approximately 950,000 specimens from citizens who underwent health screenings. Meanwhile, the MJ Biobank—a private entity operating through a membership-based health examination model—has accumulated over 100,000 blood samples since 2002. Most other biobanks are embedded within academic medical centers, forming an archipelago of research capacity across the nation.

According to data from the Ministry of Health and Welfare (MOHW), as of 2025, Taiwan's biobanking infrastructure has collectively engaged over 460,000 volunteers and stored more than 4.5 million biological samples. However, this proliferation has not been without growing pains. Fragmentation, inefficiencies in application procedures, and lack of interoperability among repositories have impeded the realization of biobanking's full potential.

To address these systemic limitations, the MOHW launched the *National Human Biobank Integration Platform* in 2019—an ambitious consortium initiative designed to unify Taiwan's disparate biobanks under a shared governance and operational framework. By late 2019, 21 biobanks had signed formal agreements to participate, a number that remained stable through March 2025. As of early 2025, the consortium publicly affirmed its intent to consolidate data from 31 participating repositories, encompassing 460,000 individuals and 4.5 million specimen-linked examinations, with integration pathways open to key government databases such as the National Health Insurance system, cancer registries, and rare disease databases.

The establishment of a "single-window" application mechanism now allows researchers to access coded, medically annotated samples across biobanks through a unified portal. Each participating institution remains responsible for providing ethics review and scientific evaluation of proposed studies, ensuring both local accountability and national coordination. While the platform remains in its formative stages, it signals a profound shift toward a more integrated, ethically robust, and future-oriented biobank ecosystem—one capable of supporting the aspirations of precision medicine, regenerative therapies, and biomedical artificial intelligence in Taiwan.

4.2.2 Legal and Regulatory Framework of Biobank in Taiwan

Legal regulations concerning biobanks in Taiwan include the following laws and directives, as shown in Table 5.5.

Among Taiwan's regulatory instruments governing biomedical research, the *Human Biobank Management Act* [21] (HBMA) occupies a central and foundational position. Enacted to safeguard both the dignity of research participants and the integrity of biospecimen-based inquiry, the HBMA articulates a comprehensive legal architecture for the ethical and scientific stewardship of human biological materials.

The statute delineates essential components of biobank governance, encompassing definitional clarity, procedural standards for establishment, the lawful collection of biospecimens, robust participant protections, institutional responsibilities for ongoing management, the conditions under which stored materials may be accessed for research, and supplementary oversight provisions. In effect, the HBMA seeks to balance two moral imperatives: the advancement of biomedical knowledge and the preservation of individual autonomy and privacy.

Institutional efforts to establish a biobank must undergo rigorous review by the Ministry of Health and Welfare (MOHW), culminating in a formal licensing

Table 5.5 Laws and regulations to cover biobanks in Taiwan

Laws
Human Biobank Management Act (2019.6.12)
https://law.moj.gov.tw/LawClass/LawAll.aspx?PCode=L0020164
Administrative orders
Administrative Regulations on the Establishment of Human Biobank (2011.1.31)
https://law.moj.gov.tw/LawClass/LawAll.aspx?PCode=L0020173
Administrative regulations for benefit feedback from commercial use of biobank (2010.9.8)
https://law.moj.gov.tw/LawClass/LawAll.aspx?PCode=L0020170
Standard review fee of biobank (2011.2.15)
https://law.moj.gov.tw/LawClass/LawAll.aspx?PCode=L0020174
Administrative rules
Rule for information safety of biobank (2010.7.2)
Standard for international transfer and biological sample export of biobank (2017.9.12)
Administrative directives
Notice for biobank application (2014.7.29)
https://dep.mohw.gov.tw/DOMA/cp-3133-12826-106.html
Stand form of informed consent for biobank participation (2013.6.28)
https://dep.mohw.gov.tw/DOMA/cp-3133-12827-106.html

Table 5.6 Items included in the informed consent form for participants designated by the Human Biobank Management Act

1. The legal authorities and their contents governing the establishment of the biobank
2. The identity of the biobank operator
3. Information regarding the identity and the service unit of the biological specimen collectors
4. The reasons why a particular participant was selected
5. The rights and direct benefits to which the participant is entitled pursuant to the human biobank Act
6. The purposes of collection and the range and duration of the use of the collected biological specimens; collection methods; types and quantities of specimens to be collected; and regions where specimens are collected
7. Any complications and hazards that might possibly occur during and as a result of a collection(s)
8. Any possible impacts of the genetic information derived from the biological specimens on the participant, and their relatives or an ethnic group
9. Any reasonable risks or inconvenience which the participant may anticipate
10. The rights which are excluded by these human biobank Act
11. The mechanism designed to safeguard personal privacy and other rights and benefits of the participant
12. The operator's organizational structure and operating principles
13. Specific type of health information of the participant that is expected to be linked in the future
14. Relevant regulations governing the applications of the biobank
15. Anticipated commercial applications
16. The participant may choose whether upon their death or incapacity, their biological specimens and related data and information will continue to be stored and used
17. Other important matters related to the biobank

process. This is not a mere administrative formality but a substantive evaluative mechanism, intended to ensure that each biobank meets the ethical, procedural, and infrastructural criteria set forth in law. At the heart of this regulatory framework lies the doctrine of informed consent, which the HBMA treats not as a perfunctory gesture but as a morally dense and legally binding covenant between researcher and participant.

To that end, the HBMA mandates that the informed consent process include 17 discrete and explicitly enumerated elements—ranging from the scope and purpose of sample use to data protection measures, to the rights of participants to withdraw consent (Table 5.6). This level of specificity reflects a deep normative commitment: that individuals who contribute to the common good through biomedical research do so with full understanding, voluntary assent, and institutional respect. In this way, the HBMA does not merely regulate the mechanics of biobanking—it defines the ethical contours of what it means to treat persons as partners in the scientific endeavor.

4.2.3 Social and Ethical Dimensions of Biobank Governance in Taiwan

The development of biobank regulation in Taiwan presents a paradigmatic case of the moral and institutional tensions that accompany the governance of biomedical innovation. When the Ministry of Health and Welfare (MOHW) initiated the national biobank initiative, it simultaneously undertook the task of codifying its ethical and legal underpinnings. Legal scholars were commissioned to draft what would become the *Human Biobank Management Act* [21] (HBMA), enacted in 2010. Notably, several lawmakers took a personal interest in this legislation, reflecting a broader societal concern over the responsible use of human biological materials in research.

The HBMA, in its Article 3, introduced foundational definitions that would shape the scope and ambition of biobank regulation in Taiwan. A "biological specimen" was defined as any human-derived cell, tissue, organ, or fluid—whether collected or produced through research—that could potentially reveal biometric identifiers. "Biomedical research" was construed primarily as research involving basic biometric data, such as genetic information. A "biobank" was understood as a repository not only for biological specimens but also for personally linked data, maintained for the purposes of population- or group-based research, and, crucially, designed for future research applications without data de-identification.

This broad definitional reach effectively brought most contemporary biomedical research under the purview of the Act. As a result, institutions engaged in such research were required to establish a licensed biobank and obtain MOHW approval before initiating any studies involving human biospecimens. The implications were far-reaching: hospitals and research institutes across Taiwan began independently setting up their own biobanks to comply with legal requirements. By 2020, this had culminated in the existence of 31 officially sanctioned biobanks within a nation of 23 million—a phenomenon that, while technically compliant, raised questions about efficiency, redundancy, and regulatory proportionality.

The approval process itself proved burdensome. Researchers reported waiting periods of 2–3 years for biobank licensing—a delay that, understandably, drew criticism for impeding scientific progress. But the more contentious issues lay not in procedural inefficiency, but in the HBMA's handling of preexisting biospecimens. Once the HBMA took effect, it became legally ambiguous whether previously collected clinical or research samples—gathered without the now-required specific consent—could be lawfully retained or utilized. Should they be incorporated into a licensed biobank? Was re-consent ethically necessary? And if re-consent proved impossible, did this preclude their continued use altogether?

In 2011, amendments to Articles 29 and 30 of the HBMA sought to resolve these questions—but only deepened the controversy. These amendments extended the Act's jurisdiction retroactively to all previously collected human samples intended for biomedical research and mandated that unconsented specimens be destroyed unless re-consent was secured by February 5, 2012. The announcement sent shockwaves through Taiwan's scientific community. Fearing the loss of millions of valuable samples amassed over decades, prominent researchers publicly appealed for a regulatory reprieve. The biomedical sector viewed the policy as a catastrophic blow to national research capacity—what some deemed the most damaging setback in Taiwan's biotechnology history.

In response, the Executive Yuan negotiated with the MOHW to issue an interpretive letter in late 2012, permitting the continued use of such specimens under specific conditions, including institutional review board (IRB) approval. Yet this compromise was met with resistance from certain human rights groups, who argued that it undermined individual privacy and consent rights.

The legal ambiguity was eventually addressed through the enactment of the *Human Subjects Research Act* in December 2011. This law provided a more general framework for human research ethics, including provisions for the use of human biological samples. In August 2012, recognizing the overlapping jurisdiction between the two Acts, legislators repealed Article 29 of the HBMA. This revision clarified that the HBMA would govern licensed biobank research, whereas the *Human Subjects Research Act* would apply to other forms of research involving human materials. In doing so, Taiwan took an important step toward disentangling two regulatory paradigms and establishing a more coherent division of ethical oversight.

Nevertheless, the regulatory turbulence of those years revealed a deeper ethical challenge. The HBMA's initial insistence on re-consent for legacy samples, while rooted in a strong commitment to autonomy, arguably failed to accommodate the evolving realities of biomedical research. Longitudinal studies, large-scale cohort analyses, and data-intensive approaches such as artificial intelligence necessitate rethinking consent not as a fixed transaction but as part of a dynamic governance relationship.

While respect for human dignity and individual rights remains paramount, regulatory frameworks must also recognize values such as solidarity, beneficence, and scientific justice. An excessive emphasis on privacy and autonomy, if unbalanced, risks undermining the collective good that research seeks to promote. When individuals willingly contribute their biological materials to further medical science, their actions reflect not only personal autonomy but also a sense of moral duty—a commitment to a shared future. Failure to recognize this altruistic impulse can foster a culture of entitlement in which society reaps the benefits of research without participating in its burdens.

Therefore, future policymaking must be grounded not only in legal precision but also in moral pluralism. Genuine public engagement—including patients, researchers, healthcare professionals, policymakers, and civil society—must become a routine part of the deliberative process. It is only through such inclusive dialogue that regulatory systems can evolve to reflect both the ethical complexity and the scientific promise of modern biomedicine.

4.3 Japan

4.3.1 Historical Development and Contemporary Status of Biobanking in Japan

In Japan, many medical institutions have collected pathological specimens for a long time, but in recent decades, biobank projects involving large-scale genome cohorts have been developed in response to the US "Human Genome Project" and its successor project, "Precision Medicine Initiative." [30] Among them, the following have been called "three major biobanks (As of 2019)." These are biobank projects of the contemporary type where participants give "broad informed consent" [10] to future unknown research use of their biological materials along with their medical records, including genetic analysis information.

- BioBank Japan
  Started in 2003, 270,000 registrants, and 51 diseases
  It started as a project with the grant by the MEXT and since 2015 by the AMED. Samples provided by ten partner medical institutions nationwide have been stored in the biobank equipped at the Institute of Medical Science, the University of Tokyo
- Tohoku Medical Megabank (ToMMo)
  Started in 2012, 146,400 registrants, and ten categories of diseases
  Established at Tohoku University as a part of a reconstruction policy in response to the Great East Japan Earthquake in 2011, with the aim of building a community-based cohort for disaster-stricken areas and building a base for future-oriented medicine
- National Center Biobank Network (NCBN)
  Started networking activities in 2012, 28,512 registrants, six major disease areas related to the six national research centers (cancer; cardiovascular diseases; neurology and psychiatry; infectious diseases; pediatrics and gynecology; geriatrics)

Not limited to the above, many medical research institutions have the same kind of biobanks, with various objectives. In Japan, there is no governmental accreditation system of biobank, thus, establishing a biobank is not so difficult, except for following the cumbersome and bureaucratic procedures of the above-mentioned "*Ethical Guidelines*" [28] and each institution, both of which are not legally binding. In addition, most of the biobanks have their own procedural rules of self-regulation. Such variety and ambiguity of procedural steps also cause administrative burdens for researchers.

Table 5.7 Items included in the protocol of biobank and/or health database designated by the Ethical Guidelines for Medical and Biological Research Involving Human Subjects [28]

1. Framework of organization including names of organization and researcher of HDB/BB
2. Purpose and social value of the HDB/BB
3. Method and time period
4. Types of data and/or biological materials
5. Procedures of informed consent according to the guidelines
6. Handling of personal data or other related information
7. Risk, burden, and benefit, as well as overall risk/benefit assessment and risk minimization measures
8. Method or storing and quality control
9. Handling of data and biological material at the time of closing HDB/BB
10. Funding and conflict of interest
11. Response to and consultation with participants and other individual concerned
12. Financial expenditure on or remuneration for participants, if there is
13. Handling of results from research
14. Future possibility of utilization of data and/or materials and the way the participants could confirm in future how their data and/or materials are utilized and/provided outside of the HDB/BB

4.3.2 Legal and Regulatory Framework of Biobank in Japan

In Japan, although there is no specific regulatory framework for health databases and/or biobanks (HDB/BBs), the "*Ethical Guidelines*" [28] has a section listing items that have to be included in the protocol of HDB/BBs (Table 5.7), which must be approved by Research Ethics Committee (REC). This means that a small portion (items of protocol description) of the scope of the WMA DoT [1] constitute part of the scope of the DoH [2]; thus, all the other parts of rules for research are also applied to HDB/BBs, as far as the items are applicable.

The required items for HDB/BBs in the protocol are also common to research and HDB/BBs. One characteristic item for the HDB/BBs is "the way of quality control" of the HDB/BBs (No. 8 of Table 5.6). However, this item does not attract much attention from the Japanese research community. The *Ethical Guidelines* [28] are not legally binding; therefore, many of the HDB/BBs do not include this item in the protocol and REC does not so much care about it.

The *Ethical Guidelines* [28] do not explicitly define the conditions for "broad informed consent," and the required items of the informed consent form (ICF) are also common to research and HDB/BBs. For this reason, consent of an individual to provide data or material could be called "broad informed consent," which is not mere blanket consent, although it does not meet the items defined by the DoT [1]. In general cases, research using data or biological material provided by HDB/BBs is required to obtain REC approval and provide information disclosure; and in most cases "opt-out procedure" is required. This means that individuals could reject the use of their data/materials upon seeing the disclosed information, and the items to be disclosed for this opt-out procedure are defined in the *Ethical Guidelines*.

In recent years, the *Ethical Guidelines* [28] have been revised several times aiming for the achievement of the "adequacy decision" granted by the EU under GDPR [12]. "Adequacy decision" was already granted to the overall legal framework, which means that the Japanese regulatory framework for personal data protection is equivalent to the EU, thus personal data transfer from the EU to Japan has been eased. However, the exceptional framework for research and public health activities (including HDB/BBs) has not achieved it. Aiming for "adequacy decision" on the frameworks of research and public health, every detailed part of the *Ethical Guidelines* related to data protection has been finely revised. While details are fine, the fundamental construction of the framework is simple: the use of personal data without consent can be approved by a REC, confirming the appropriateness of information disclosure and the opt-out procedure. In addition, the *Next Generation Medical Infrastructure Act* [31] was enacted in 2017 and revised in 2023, to expand the framework for utilizing a large amount of anonymized or pseudanonymized data. Under this act, both entities engaged in personal data processing for anonymizations and entities engaged in data use are certified. This framework was designed to follow the framework of European Health Data Space [13] to facilitate multiple uses of personal data in the framework of GDPR, in the field of health.

There is one prominent addition in the *Ethical Guidelines* [28] of the items of the ICF that does not attract much public attention. Item No. 14 of Table 5.7 (items of protocol) "*Future possibility of utilization of data and/or materials and the way the participants could confirm in future how their data and/or materials are provided/utilised outside of the HDB/BB.*" should be also included in ICF, if relevant. This is an important element of "dynamic consent" [11]. Dynamic consent means a type of broad informed consent, where the participants can follow future utilizations of their data and are provided opportunities for continuous decisions on whether or not to maintain their consent, considering the progress of and/or results generated from secondary uses. This enables participatory collaboration between researcher and participants, while well-designed data management procedures are required. There are several projects to apply this kind of patient involvement policy; however, it has not become a standard practice in Japan.

4.3.3 Social and Ethical Dimensions of Biobank in Japan

Various ethical issues have been discussed regarding large-scale biobank projects. The representative example that raised ethical and social problems is the case of Tohoku Medical Megabank. People living in the earthquake-affected areas, with nearly 20,000 deaths, were defined as the target population of this project. The project includes a reconstruction plan and medical support for local residents. This plan was criticized, as such kind of support could serve as incentive for the residents to participate in the project, which may have undue influence on vulnerable populations. Some local groups argued that this target population is a "vulnerable group" as defined by the DoH and asked for the opinion of the Governor of Miyagi

Prefecture. The Governor replied and denied this argument. Many other criticisms have been raised, but they do not have a significant impact on the implementation of the project, and the project has proceeded in accordance with governmental guidelines.

5 Discussion: Recommended Points to Be Considered

The evolving landscape of research dealing with large-scale data and materials derived from human participants has revealed the need for the evolution of research ethics, not limited to the obligation of individual physicians to protect the dignity and rights of individual participants but also to their dedication to public health. This situation led to the incorporation of the DoT [1] in the DoH [2], as well as the revision of the DoT to explore global consensus on the governance framework of health databases and biobanks. There is also a need to develop an international consensus on a new type of broad informed consent that maintains the fundamental rights of individuals to their own data and biomaterials, while allowing their multipurpose uses based on altruism and solidarity of each participant. In addition to this, consensus is needed on what the circumstances are in which historical data and biological materials are used without obtaining consent as it is not reasonable, through opt-out procedures, or in which risk reduction renders even opt-out procedures unnecessary.

In addition to such changes in the principles of consent in an evolving research environment, special considerations are needed to promote the utilization of brain banks. Brain bank is one type of biobank; therefore, international standards defined in the DoT [1] should be the basis of its governance framework. Furthermore, as it inevitably requires brain tissue donation from a deceased donor, some additional principles and conditions are needed as follows:

- A legal framework for the justification for obtaining, storing, and supplying of brain tissue, defined in the style of contemporary regulations for large-scale usage of human data or materials, in accordance with globally recognized traditional laws and regulations concerning autopsy, along with the dismissal of the crime of destruction of a corpse.
- Informed consent from both the donor during the life time and the bereaved family after the death of the donor should be defined, along with the development of a patient registry system during their life time. In-depth exploration of the principles and procedures is needed to avoid the family's decision overriding the will of the diseased while ensuring that the rights of inevitably vulnerable donors are not exploited.
- Most of the candidate donors are mentally disordered or cognitive impaired patients; therefore, at the occasion of registration during their life time, capability of consent should be carefully continuously assessed. Shared decision-making among the candidate donor, family, and attending physician supported by a

coordinator are recommended, or otherwise, decision-making should be with a coordinator independent from the attending physician, to avoid undue influence.
- The family's sentiment unwilling to permit harmful procedures to the body of the deceased should be also carefully considered.
- The dignity of the deceased should be especially respected, sharing the ethical value of "Gift of Hope" with the public.

6 Conclusion

The inclusion of the DoT on health databases and biobanks in the 2024 DoH for research involving human participants, as well as the further revision of the DoT, would be a driving force to promote the ethical conduct of data-driven research, including the secondary use of data and biological materials obtained in research. Further deliberations are needed by the relevant communities for implementation regarding broad informed consent, dynamic consent, and opt-out procedures, which are different from traditional informed consent. Furthermore, among biobanks, brain banks require an international consensus on an ethical and legal framework in which the wishes of both the deceased donor and the bereaved family are respected and are consistent with relevant traditional legislation, e.g., laws related to autopsy. This situation would help the maturation of people's altruism and moral duty, with solidarity, to share both the efforts of participating in research and the benefits resulting from research in the society at large.

Acknowledgments Some part of this collaborative research was granted by the research project led by YJR, the Bio & Medical Technology Development Program of the National Research Foundation (NRF), funded by the Korean government (MSIT) (NO. 2019M3E5D2A02064496), including travel expenditures to invite TS and CK for meetings to discuss related topics in Seoul in 2019 and in Chuncheon in 2023.

The authors also appreciate the Japan Association for Bioethics (JAB) for granting travel expenditure to invite YJR and DFT to participate in the Annual Meeting of the JAB in 2018 in Kyoto to discuss related topics.

DFT is a board member of the Medical Ethics Committee at the Taiwan Medical Association and has participated in several Workgroups of the WMA, including the 2024 revision of the Declaration of Helsinki.

CK appreciates the International Federation of Associations of Pharmaceutical Physicians and Pharmaceutical Medicine (IFAPP) for granting her travel expenditure for her in-person participation in the WMA Regional Meeting in Tel Aviv (2022 December), and the WMA for granting her travel expenditure for her in-person participation and presentation as a panelist in WMA Regional Meeting in Washington DC (2024 August) and waiver of registration fee from WMA as a journalist to attend the General Assembly in Helsinki (2024 October). She participated in all the other WMA Regional Meetings at her own expense or online.

Conflict of Interest The authors declare no competing interests relevant to the content of this chapter.

References

1. World Medical Association. WMA Declaration of Taipei on Ethical Considerations regarding Health Databases and Biobanks. First adopted October 2002, revised October 2016. https://www.wma.net/policies-post/wma-declaration-of-taipei-on-ethical-considerations-regarding-health-databases-and-biobanks/.
2. World Medical Association. WMA Declaration of Helsinki—Ethical Principles for Medical Research Involving Human Participants. First adopted in 1964, last amended in 2024. https://www.wma.net/policies-post/wma-declaration-of-helsinki/.
3. World Medical Association. Rescinded: WMA Declaration on Ethical Considerations regarding health databases. Adopted October 2002, revised October 2016. https://www.wma.net/wp-content/uploads/2016/10/DoE-Oct2002.pdf.
4. Kurihara C, Baroutsou V, Becker S, Brun J, Franke-Bray B, Carlesi R, Chan A, Collia LF, Kleist P, Laranjeira LF, Matsuyama K, Naseem S, Schenk J, Silva H, Kerpel-Fronius S. Linking the declarations of Helsinki and of Taipei: critical challenges of future- oriented research ethics. Front Pharmacol. 2020;11:579714. https://doi.org/10.3389/fphar.2020.579714.
5. Dhai A. The WMA declaration of Taipei: human database and biobanks for the common good. S Afr J Bioeth Law. 2016;9:50–1.
6. Green ED, Watson JD, Collins FS. Human genome project: twenty-five years of big biology. Nature. 2015;526(7571):29–31. https://doi.org/10.1038/526029a.
7. Snaedal J. Interviewed by Kurihara C, Saio T. Interview with Dr. Jon Snaedal, Chair of the World Medical Association workgroup for the Taipei declaration; representative of the Icelandic Medical Association: health databases, biobank and fundamental human rights. Rinsho Hyoka (Clinical Evaluation). 2018;46(1):165–70.
8. Chen I. Challenges in regulating biomedical research: the human subjects research act and the human biobank management act in Taiwan. Clin Eval. 2014;42:459–66.
9. Council of Europe. Recommendation CM/Rec(2016)6 of the Committee of Ministers to member States on research on biological materials of human origin. Adopted by the Committee of Ministers on 11 May 2016 at the 1256 th meeting of the Ministers' Deputies.
10. Council for International Organizations of Medical Sciences. International Ethical Guidelines for Health-related Research Involving Humans. 2016.
11. Council for International Organizations of Medical Sciences (CIOMS). Patient involvement in the development, regulation and safe use of medicines. CIOMS Working Group report. Geneva, Switzerland: 2022. https://doi.org/10.56759/iiew8982.
12. Regulation (EU) 2016/679 of the European Parliament and of the Council of 27 April 2016 on the protection of natural persons with regard to the processing of personal data and on the free movement of such data, and repealing Directive 95/46/EC (General Data Protection Regulation) (2016). Official Journal of the European Union. 4.5.2016: L 119/1-88.
13. Regulation (EU) 2025/327 of the European Parliament and of the Council of 11 February 2025 on the European Health Data Space and amending Directive 2011/24/EU and Regulation (EU) 2024/2847.
14. Tsai DFC. The ethical challenges of implementing declaration of Taipei in the era of big data research. Japanese Association for Bioethics. December 8, 2018; Kyoto, Japan.
15. Brain Research Promotion Act. 1998.
16. Bioethics and Safety Act. December 29, 2003. Act No. 11250.
17. Act on the Dissection and Preservation of the Dead. 1962.
18. Civil Act. 1960.
19. Jeong SJ, Lee IY, Jun BO, Ryu YJ, Sohn JW, Kim SP, Woo CW, Koo JW, Cho IJ, Oh U, Kim K, Suh PG. Korea brain initiative: emerging issues and institutionalization of neuroethics. Neuron. 2019;101(3):390–3. https://doi.org/10.1016/j.neuron.2019.01.042.
20. The Human Autopsy Act. December 21, 1948, amended June 16, 1984. https://law.moj.gov.tw/ENG/LawClass/LawAll.aspx?pcode=L0020015.

21. Human Biobank Management Act. February 3, 2010, amended January 20, 2021. https://law.moj.gov.tw/Eng/LawClass/LawAll.aspx?PCode=L0020164.
22. Human Subjects Research Act. December 28, 2011, amended January 2, 2019. https://law.moj.gov.tw/ENG/LawClass/LawAll.aspx?pcode=L0020176.
23. Shih TS, Ho JJ. Analysis of legal and ethical issues in the establishment and operation of Taiwan's brain banks. Law Life Sci. 2020;9(2):57–86.
24. Li YQ. Legal Analysis on the Establishment and Operation of Brain Banks. 10th Session Topic Analysis and Special Research [Report made by Legislative Yuan]. https://www.ly.gov.tw/Pages/ashx/File.ashx?FilePath=~/File/Attach/235834/File_19796591.pdf.
25. Act on the dissection and preservation of corpses. June 10, 1949. Act No. 204. http://arjamalev2012.blog.fc2.com/blog-entry-723.html.
26. Civil Code. April 27, 1896. Act No. 89 of https://www.japaneselawtranslation.go.jp/ja/laws/view/4314.
27. Penal Code. April 24, 1907. Act No. 45. https://www.japaneselawtranslation.go.jp/ja/laws/view/3581.
28. Ministry of Education, Culture, Sports, Science and Technology, Ministry of Health, Labour and Welfare, Ministry of Economy, Trade and Industry Ethical Guidelines for Medical and Biological Research Involving Human Subjects 2021 March 23. https://www.mext.go.jp/content/20250325-mxt_life-000035486-01.pdf.
29. Act on Body Donation for Medical and Dental Education. 1983. Act No 56. https://www.japaneselawtranslation.go.jp/ja/laws/view/1114.
30. Kurihara C, Inoue T. Chapter 6. Ethics, regulations and clinical development of precision medicine: Activating with molecular imaging. In: Inoue T, David Y, Huan G, editors. Personalized pathway-activated systems imaging in oncology: principal and instrument. Springer; 2017. p. 105–26.
31. Act on anonymized medical data that are meant to contribute to research and development in the medical field. Act No. 28 of May 12, 2017.

Chapter 6
From COVID-19 to Global Standards: Redefining Korean Bioethics and Research Ethics

Ock-Joo Kim

Abstract This chapter examines the impact of the COVID-19 pandemic on South Korea's research ethics framework and its subsequent realignment with evolving global standards, particularly the 2024 revision of the Declaration of Helsinki (DoH). South Korea's robust biomedical research output and significant participation in international clinical trials underscore the critical need for a strong ethical foundation. The chapter analyzes South Korea's rapid response to the pandemic's immediate challenges, focusing on streamlined ethical review, adapted informed consent procedures, and modifications to clinical trial conduct. A thematic analysis explores key ethical challenges such as prioritizing participant safety, ensuring effective national research governance, and fostering meaningful community engagement. The chapter then investigates the long-term initiatives implemented to strengthen national research governance, enhance institutional capacity for research ethics, and establish a comprehensive framework for research integrity. Finally, it assesses the alignment of the Korean Bioethics and Safety Act with the 2024 DoH, identifying areas needing revision, such as terminology, the treatment of vulnerable populations, and data governance, to ensure continued harmonization with evolving international ethical standards and promote a truly participant-centric research culture. The Korean experience during the pandemic informs recommendations for future improvements in research ethics and governance.

This chapter summarizes and synthesizes the following three papers published by the author. Park YS, Kim OJ. Government initiatives for research ethics during COVID-19 pandemic in Korea. J Korean Med Sci 2024;39(12):e116 (Ref. 25); Kim OJ. The 2024 Revision of the Declaration of Helsinki and the future directions of Korea's Bioethics and Safety Act. J Korean Med Sci. 2025 Feb;40(9):e99 (Ref. 28); Park YS, Doris S, Kim OJ. Research ethics challenges in pandemic Korea and their implications for the revised 2024 Declaration of Helsinki. J Korean Med Sci. Forthcoming 2025 (Ref. 50).

O.-J. Kim (✉)
Department of the History of Medicine and Medical Humanities, Seoul National University College of Medicine, Seoul, South Korea
e-mail: okim9646@snu.ac.kr

C. Kurihara et al. (eds.), *The 2024 Declaration of Helsinki*,
https://doi.org/10.1007/978-981-96-9294-1_6

Keywords Korean bioethics · COVID-19 pandemic · Research ethics governance · Vulnerable populations · Data governance

1 Introduction

The 2024 revision of the World Medical Association's Declaration of Helsinki (DoH) [1] marks a pivotal moment in the ongoing evolution of global bioethics. Adopted initially in 1964 and previously updated in 2013, the DoH has served for decades as the foundational ethical framework guiding medical research worldwide [2]. This latest revision, culminating in its official adoption in October 2024, reflects a concerted effort to adapt the DoH to the complexities of twenty-first-century biomedical research and to address emerging ethical challenges [3–10]. The significant interval since the previous revision underscores the necessity to reevaluate and strengthen ethical principles in light of both rapid advancements in scientific technology and the profound societal shifts that have reshaped the global research landscape.

The global COVID-19 pandemic presented an unprecedented crisis for research ethics worldwide, necessitating a reevaluation of established principles and the development of adaptable frameworks [11–24]. Globally, the pandemic exposed vulnerabilities within existing research ethics frameworks, highlighting the need for agile, adaptable, and ethically sound responses. In South Korea, a nation with a strong history of contributions to global biomedical research and a relatively mature research ethics infrastructure, the pandemic posed unique and significant challenges [25]. Emphasizing the tension between the urgent need for rapid research and the imperative to protect the rights and well-being of research participants, the resulting discussions within the Korean research community and government agencies offer valuable insights into the practical challenges of navigating this complex ethical terrain.

The considerable volume and international impact of South Korea's biomedical research underscores the critical importance of robust ethical oversight. In 2021, South Korea's contributions to the global scientific community were substantial, with the nation ranking 12th worldwide in both SCI-indexed publications (83,680) and total citations (169,443) [26]. The predominance of clinical medicine within this research output (15,173 publications, representing 18% of the total) further highlights the need for strong ethical frameworks. This prominence is also reflected in South Korea's robust participation in international clinical trials, with a global ranking of fourth (2023) in the number of registered trials and Seoul securing the top spot among cities worldwide. The continued high ranking in single-country clinical trials (third in 2023) underscores the necessity for continuous alignment with international ethical guidelines [27]. The ethical conduct of research is therefore paramount, not only to maintain the high quality of scientific output but also to uphold the respect for the rights and safety of human research participants.

The COVID-19 pandemic exposed critical vulnerabilities in global research ethics, demanding a fundamental reevaluation of established principles and the development of more adaptable ethical frameworks. This chapter examines South Korea's response to this crisis, analyzing both its immediate, emergency measures and its longer-term initiatives to enhance preparedness, while simultaneously exploring the major ethical challenges that emerged through a thematic analysis. The analysis ultimately assesses the alignment of the Korean bioethics framework with the 2024 DoH revision and its implications for harmonizing Korean bioethics with evolving global standards. This chapter investigates the challenges in the South Korean context, analyzing how the urgent need for rapid research intersected with the ethical imperative to safeguard participant rights and well-being, prompting significant modifications to existing policy adjustments. This includes the dynamic interplay between the unprecedented ethical challenges posed by the COVID-19 pandemic and the subsequent evolution of South Korea's bioethics and research governance. The analysis situates this evolution within the broader context of South Korea's prominent role in global biomedical research and the existing legal framework established by the 2013 Bioethics and Safety Act [28]. With the initial, rapid response to the pandemic's immediate ethical concerns and the longer-term initiatives designed to enhance research ethics preparedness, this multifaceted analysis considers both the chronological development of the government's responses and thematic exploration of the key ethical challenges, ultimately assessing the implications for aligning Korean bioethics with global standards. Finally, this chapter addresses the interplay between these pandemic-driven challenges in South Korea and the key revisions incorporated into the 2024 DoH. By examining the congruence and divergence between the updated ethical guidelines and existing Korean bioethics legislation, it aims to assess their implications for the future trajectory of Korean bioethics and research governance. This analysis seeks to provide a deeper understanding of how global ethical standards are being negotiated and implemented within a specific national context, highlighting both successes and areas requiring further development.

2 The Korean Context: Research Ethics Before the Pandemic

South Korea's approach to research ethics has undergone significant transformation in recent decades, shaped by internal developments and evolving global standards. While initially lacking a comprehensive regulatory framework, the landscape shifted dramatically following the Hwang Woo-Suk scandal of 2006–2007 [29]. This event, characterized by unprecedented breaches of research ethics and scientific integrity, had a profoundly negative impact on public trust in South Korean biomedical research, particularly stem cell research, and significantly hampered the field's advancement. Paradoxically, the scandal served as a powerful catalyst for rapid advancements in Korean research ethics and governance [30, 31].

2.1 Institutional Bioethics Committees (IBCs) and Government Support

The Hwang Woo-Suk affair exposed critical deficiencies in the oversight of research ethics, particularly highlighting the inadequacy of existing ethical review mechanisms. A rudimentary ethics committee had been established within the government-funded National Research Program for Cell Therapy, launched in 2002, which also funded part of Hwang's stem cell research. However, this committee lacked the authority and institutional capacity to prevent the serious ethical violations that later came to light. The committee's requirement that researchers obtain approval from their institution's Institutional Review Boards (IRBs) underscored a growing awareness of the need for robust ethical review, although the IRB system itself was still nascent at that time. In the Korean context, following the introduction of the Korean Good Clinical Practice (KGCP) guidelines in 2001, IRB review was mandated only for clinical trials involving drugs or medical devices that required regulatory approval from the Korea Food and Drug Administration (KFDA). Surgical or other types of clinical studies were not necessarily subject to IRB review at that time. The widespread lack of experience with nonclinical IRBs and the prevalence of IRBs focusing primarily on clinical trials meant that many researchers were unfamiliar with the process and unable to secure necessary IRB approvals for their projects.

The enactment of the Bioethics and Safety Act in 2005 marked a pivotal moment. While initially focused on regulating embryo and genetic research, the Act mandated the establishment of Institutional Bioethics Committees (IBCs), equivalent of IRBs, within relevant institutions. These IBCs were established under the authority of the Bioethics and Safety Act, in contrast to Institutional Review Boards (IRBs), which are governed by the Korean Good Clinical Practice (KGCP) guidelines under the Pharmaceutical Affairs Act and primarily oversee drug and medical device clinical trials. However, the lack of clear guidelines regarding IRB function and the prevailing ambiguity regarding their role contributed to further ethical lapses—as exemplified by the Hwang Woo-Suk case itself. The subsequent exposure of these IBC deficiencies prompted significant reforms. Amendments to the Bioethics and Safety Act in 2008 introduced provisions for government oversight and support of IBCs, including funding, evaluation, and training for committee members [31]. The government's engagement was further strengthened by formally recognizing and supporting the Korean Association of IRBs (KAIRB), initially established in 2002 [32]. This organization played a crucial role in developing training programs, providing educational resources, and enhancing the capacity of IBCs nationwide. This systematic approach addressed the earlier shortcomings in IRB training and the expertise needed to appropriately oversee a wider range of research projects.

2.2 *Expansion to a Comprehensive Bioethics Framework*

The Bioethics and Safety Act enacted in 2004 initially centered on embryo and genetic research, evolved significantly with its full revision in 2013. The amended act is now considered the foundational Bioethics and Safety Act of South Korea. This significant expansion reflected lessons learned from the Hwang Woo-Suk scandal and a commitment to aligning Korean ethical guidelines with evolving international standards [31]. This comprehensive legislation covered all aspects of human subjects research and the use of human-derived materials, mandating IRB review for all such research activities. The 2013 comprehensive revision of the Bioethics and Safety Act in South Korea differs from the United States' Common Rule in that the Common Rule applies only to research funded by federal government sources. In contrast, the South Korean Bioethics and Safety Act is applicable to all human subjects research and research involving human-derived materials conducted within the country, regardless of the source of research funding. In this regard, all human subjects research and studies involving human-derived materials carried out in South Korea, whether publicly or privately funded, are protected under this legislation in terms of the rights, safety, and welfare of participants or donors. The updated framework also included provisions for institutional ethics education and guidance, post-IBC approval research oversight, and the establishment of a national-level infrastructure for bioethics policy development and research. The Act's revised purpose statement explicitly included the prevention of harm to humans and the protection of human dignity, underscoring a shift toward a more comprehensive and human-centered approach to research ethics. Furthermore, the Act codified six fundamental principles, mirroring the spirit of the UNESCO Universal Declaration on Bioethics and Human Rights (2005) [33], establishing a more robust national bioethics governance structure.

3 Early Pandemic Response in Korea: A Case Study in Rapid Adaptation

South Korea's response to the COVID-19 pandemic was characterized by a rapid and decisive mobilization of resources, reflecting a proactive approach to both public health management and ethical considerations in research [34–49]. The initial phase of the pandemic (January 2020–2021) was dominated by the urgency to develop and deploy diagnostics, therapeutics, and vaccines. This urgency necessitated prompt adaptations to existing ethical guidelines and regulations to accelerate the research process [25].

Several key governmental initiatives during this period highlight the prioritization of both rapid response and ethical conduct:

1. Streamlining Ethical Review: Recognizing the critical time constraints, the government took swift action to expedite the ethical review process for COVID-19-related research. This included the implementation of multiple strategies, such as:
   - Joint operation of Institutional Review Boards (IRBs): This allowed for increased review capacity and efficiency for multicenter studies.
   - Exempt or expedited review by a specially designated review committee: The National Bioethics Policy Institute, a governmental agency, established and operated a special review committee specifically for studies involving COVID-19-related data. This mechanism enabled expedited ethical review of research deemed essential for pandemic control.
   - Designation of a supervising agency: The Korean Academy of Medical Sciences was designated as the overseer for the Clinical Trial Safety Support Institution and the Central IRB, ensuring expert guidance on ethical review. This centralized oversight improved efficiency and consistency in the review process.
2. Facilitating Informed Consent: The pandemic necessitated the rapid adaptation of informed consent procedures. The government's measures included:
   - Permission for non-face-to-face consent methods: This allowed for remote consent acquisition through telephone explanations or digital transmission of consent forms. This mitigated the risks of infection transmission associated with in-person interactions.
   - Guidelines for urgent review scenarios: This clarified the protocols for expediting the ethical review of time-sensitive studies.
3. Adapting Clinical Trial Conduct: The government addressed the logistical challenges of conducting clinical trials during a time of social distancing and lockdowns through guidelines that allowed for:
   - Telephone counseling and prescription: This measure ensured that access to vital clinical interventions was maintained.
   - Proxy prescription: This facilitated participation in clinical trials for patients who were unable to attend in-person appointments due to isolation or quarantine measures.
   - Online ethics training: This enabled the rapid upskilling of clinical trial personnel.

These early interventions demonstrate South Korea's commitment to a swift and responsible approach to managing the ethical implications of pandemic-related research. However, the effectiveness of these measures in balancing the urgency of the situation with the ethical conduct of research warrants further analysis, particularly when considering the long-term implications of these changes to the research ecosystem.

4 Ethical Challenges in Pandemic Research: A Thematic Analysis

The COVID-19 pandemic in South Korea exposed a complex interplay of ethical challenges within the nation's research ecosystem. While the preexisting regulatory framework, established through the 2013 Bioethics and Safety Act, provided a foundation for ethical review, the pandemic's unique characteristics necessitated rapid and flexible adaptations. This section presents a thematic analysis of the multifaceted ethical challenges encountered in COVID-19-related research in South Korea, as revealed through a systematic scoping review of the relevant literature. Utilizing the scoping review methodology of Tricco et al. (2018), the analysis identified ten recurring ethical themes reflecting the complex interplay of factors impacting research during the pandemic, focusing on how Korean COVID-19 responses are relevant to the 2024 revision of the DoH [50]. The themes, detailed below, highlight the need for adaptive ethical frameworks, robust governance structures, and effective community engagement within the South Korean research landscape.

1. Prioritizing Participant Safety: The pandemic heightened concerns about participant safety, particularly for those with preexisting conditions. While the existing regulations emphasized participant safety, the pandemic necessitated additional measures to minimize risks. This included guidelines for minimizing hospital visits, promoting remote monitoring, and facilitating alternative methods of drug delivery (relevant paragraphs of the 2024 DoH (same hereafter): para. 7, 8).
2. Ensuring Effective National Research Governance: The pandemic underscored the importance of a robust national research governance system capable of responding swiftly to emerging public health threats. The government's response involved establishing a centralized IRB and promoting interagency collaboration to streamline ethical reviews, but challenges remain in ensuring timely and consistent review processes across various institutions (para. 23).
3. Fostering Meaningful Community Engagement: The need for community engagement in research was emphasized to ensure culturally sensitive approaches. However, establishing effective community networks and integrating community perspectives into research design and implementation faced significant challenges, particularly during the initial phase of the pandemic (para. 6).
4. Promoting Global Solidarity and Cooperation: The pandemic highlighted the crucial role of global collaboration in research. While South Korea actively participated in international research initiatives and information sharing, the need for effective mechanisms for harmonizing regulatory frameworks and facilitating cross-border collaborations required further development.
5. Striving for Reliable Research and Avoiding Waste: The rapid-response nature of the pandemic heightened concerns regarding research waste and the risk of producing unreliable results. The government's emphasis on rigorous scientific

design and efficient resource allocation helped mitigate these risks, but concerns regarding the potential for duplicated research efforts persisted (para. 21).

6. Ensuring Rapid and Thorough IRB Review: The urgency of the situation necessitated expediting IRB review processes. While the government promoted this through various mechanisms (such as joint IRB operations and expedited review committees), challenges remained in maintaining the independence and integrity of IRB review across different institutions and ensuring timely and efficient review processes (para. 8).
7. Adapting Consent Procedures for Pandemic Conditions: The pandemic posed significant challenges to obtaining informed consent, necessitating adaptations to accommodate physical distancing and quarantine measures. The government's allowance of remote consent methods and alternative consent arrangements was an important step, but the long-term ethical implications and appropriate safeguards required further consideration (para. 26, newly including electronic obtaining informed consent).
8. Fair and Responsible Inclusion of Vulnerable Groups: The pandemic highlighted the need for a fairer and more responsible inclusion of vulnerable groups in research. While the Bioethics and Safety Act already addressed vulnerable populations, the pandemic required adjustments to ensure that the existing protections were adequately enforced and that these groups were not inadvertently excluded from research opportunities due to pandemic-related challenges (para. 19. 20).
9. Addressing the Ethical Issues of Human Challenge Trials: The pandemic sparked discussions on the ethical considerations of conducting human challenge trials to evaluate vaccines and therapeutics. While no such trials were conducted in Korea, the preparedness for possible future trials, in light of ethical considerations and safety standards, became a focal point in discussions (para. 8).
10. Addressing Ethical Concerns Regarding the Secondary Use of Human Material Collected During the Pandemic: The pandemic presented ethical complexities surrounding the use of data or samples obtained during the pandemic response without explicit consent for secondary research purposes. The government's response addressed this through mechanisms such as IRB review and data anonymization to allow secondary use, although this approach remains a subject of ongoing ethical debate (para. 32).

This analysis reveals that while the South Korean government implemented several measures to address the ethical challenges arising from pandemic-related research, significant challenges remain regarding research governance, community engagement, and the long-term ethical considerations of pandemic research.

Table 6.1 shows the thematic discussion of ethical challenges to integrate the government's responses for a structured overview of the analysis.

Table 6.1 Alignment of government responses with ethical challenges in pandemic research

Ethical challenge (relevant paragraphs of the 2024 DoH)	Description	Government response	Effectiveness assessment
1. Prioritizing participant safety (Para. 7, 8)	Protecting participant Well-being, especially those with preexisting conditions, during the pandemic	Guidelines for minimizing hospital visits, remote monitoring, and alternative drug delivery methods	Partially effective; remote methods improved safety but access to care for some remained difficult. Further research is needed on the long-term impact on participation and data integrity
2. Ensuring effective National Research Governance (Para. 23)	Establishing a robust national research governance system for rapid and ethical responses to public health threats	Joint IRB operations, expedited review committees, designation of a supervising agency: KAMS for central IRB and clinical trial safety support institution	Improved efficiency and consistency but not full elimination of delays centralization may require more effective communication and coordination between institutions
3. Fostering meaningful community engagement (Para. 6)	Ensuring culturally sensitive engagement with communities throughout the research process	Limited direct government response	Limited effectiveness; highlights the need for future policies focusing on building robust community networks and engaging diverse community stakeholders in research planning and execution
4. Promoting global solidarity and cooperation	Facilitating international collaboration in research through information sharing and harmonization of regulatory frameworks	Active participation in international research initiatives and information sharing	Effective in advancing scientific knowledge, but challenges remain in harmonizing regulatory frameworks globally further development of mechanisms needed for seamless cross-border collaboration
5. Striving for reliable research and avoiding waste (Para. 21)	Ensuring rigorous research design, efficient resource allocation, and preventing duplicated research efforts	Research prioritization, emphasis on rigorous methodology, and efforts to coordinate research efforts	Partially effective; duplication still occurred in some instances, underscoring a need for improved coordination and project management

(continued)

Table 6.1 (continued)

Ethical challenge (relevant paragraphs of the 2024 DoH)	Description	Government response	Effectiveness assessment
6. Expedited IRB review (Para. 8)	Accelerating IRB review processes to handle time-sensitive research during a public health crisis	Prioritization of COVID-19-related research by IRBs, the establishment of expedited review committees, and guidelines for non-face-to-face meetings	Significant improvement in review times, but challenges remained in ensuring consistency and maintaining ethical standards across all IRBs
7. Adapting informed consent (Para. 26)	Adapting consent procedures to account for pandemic-related restrictions on in-person interactions	Permission for non-face-to-face methods (telephone, digital transmission of consent forms)	Effective in facilitating research during the pandemic, but the long-term ethical implications and optimal safeguards need further consideration and evaluation
8. Fair inclusion of vulnerable groups (Para. 19. 20)	Ensuring equitable research participation for vulnerable populations	Government response was less direct during this initial phase. Primarily addressed through existing legislation and guidance from IRBs.	Partially effective; existing policies needed reinforcement to ensure that vulnerable groups were not inadvertently excluded due to pandemic-related restrictions
9. Ethics of human challenge trials (Para. 8)	Addressing the ethical implications of human challenge trials for vaccine/therapeutic development	No specific government action during this phase, but preparedness measures would be essential for ethical oversight of future trials	Minimal effectiveness during this phase; highlights the need for long-term preparedness
10. Secondary use of data collected without consent (Para. 32)	Addressing ethical challenges related to the use of data or samples collected during the pandemic response without informed consent for secondary research	Guidelines for IRB review and data anonymization to allow secondary use (under specific conditions)	Partially effective; the issue remains subject to ongoing ethical debate, necessitating clearer guidelines to ensure full compliance with ethical principles

Source: This table is consolidated from some parts of the findings from the previous papers [25, 50]

5 Building a Resilient Research Ethics Ecosystem: Long-Term Preparedness in Korea

The second phase of the Korean government's response to research ethics challenges during the COVID-19 pandemic shifted from immediate crisis management to the establishment of robust, long-term frameworks and infrastructure to enhance preparedness for future public health emergencies. This period (January 2021–February 2023) saw a focus on strengthening national research governance, promoting ethical research practices, and bolstering institutional capacity. Several key initiatives directly addressed challenges encountered during the pandemic's initial phase [25]:

1. Strengthening National Research Governance: Recognizing the limitations of existing research governance structures exposed during the pandemic's early stages, the government implemented several key reforms. This included revising the Pharmaceutical Affairs Act to establish:

   - A Central IRB: This centralized ethics review body significantly reduced the burden on individual institutional IRBs, streamlining review processes and promoting efficiency in pandemic-related research.
   - A Clinical Trial Safety Support Institution: This institution focused on providing support and guidance for conducting clinical trials, facilitating safe and effective trial operations. These structures were designed to not only address the immediate pandemic situation but also to improve the long-term efficiency and effectiveness of research governance.

2. Enhancing Institutional Capacity for Research Ethics: Recognizing the crucial role of Institutional Bioethics Committees (IBCs) in upholding ethical standards, the government introduced an ongoing evaluation and accreditation system for IBCs. This formalized assessment process, coupled with continuous support for staff training and educational initiatives, aimed to enhance the capabilities of IBCs nationwide and establish consistent minimum standards across institutions. This directly addressed some of the inconsistencies in ethical oversight revealed during the pandemic's initial response.
3. Emergency Medical Product Development and Supply: Acknowledging the vulnerability of the healthcare system exposed by the COVID-19 pandemic, the government established a comprehensive legal framework for the rapid development and emergency supply of medical products [51]. This encompassed not only pharmaceuticals but also non-pharmaceutical items and medical devices. This proactive strategy sought to create a more resilient system capable of handling future health emergencies. The creation of the Emergency Medical Products Safety and Supply Committee within the Ministry of Food and Drug Administration played a key role in this initiative.

4. Strengthening Research Integrity and Preventing Misconduct: A National Framework.

   The COVID-19 pandemic underscored the critical need for a robust national system to promote research integrity and prevent misconduct within South Korea's research ecosystem. While preexisting legislation addressed research ethics, the pandemic revealed the need for a more comprehensive and unified approach. This need was met through the enactment of the National Research and Development Innovation Act (2020), which came into effect on January 1, 2021 [52]. This landmark legislation represented a significant step in establishing a cohesive national framework for research integrity. Prior to the Act's enactment, the oversight of research ethics and integrity was fragmented, varying across different government agencies and research institutions [53]. This decentralized approach created inconsistencies and potential vulnerabilities within the research system. The Act addressed this by consolidating existing guidelines and regulations under a single, unified legal framework. Key features include:

- Centralized oversight: The act moved away from a fragmented system to one with centralized oversight, thereby reducing inconsistencies and enhancing the efficacy of efforts to promote ethical conduct and prevent misconduct. The emphasis on establishing a comprehensive, national-level system for research integrity responded directly to the inefficiencies and inconsistencies exposed during the pandemic's early stages.
- Proactive measures: The Act instituted proactive measures to promote responsible research practices, preventing potential misconduct before it occurred. This emphasis on proactive measures was a direct response to the increased scrutiny of research ethics and integrity in the wake of the pandemic and represents a paradigm shift toward fostering a culture of transparency and accountability within the South Korean research landscape.
- Clear definitions and procedures: The Act provided clear definitions of research misconduct and detailed procedures for investigation and sanctioning of such conduct. The inclusion of clear definitions and procedures helped to clarify expectations and ensure consistency in addressing research misconduct, addressing a previous lack of clarity that had proven problematic in handling research integrity issues during the pandemic.
- Enhanced accountability: The Act strengthened accountability by linking funding decisions to compliance with research integrity standards. This ensured that institutions and researchers held greater responsibility for upholding ethical conduct. This enhanced accountability aimed to foster a culture of responsible research by making compliance with research integrity standards a prerequisite for continued national research funding.

To sum up, the National Research and Development Innovation Act played a pivotal role in establishing a comprehensive and unified national system for research integrity in South Korea. This strengthened approach directly responded to weaknesses exposed during the COVID-19 pandemic, fostering a more robust and resilient research environment. This systemic reform reflects a proactive commitment to maintaining the highest standards of ethical conduct and ensuring transparency and accountability within the nation's research ecosystem.

These long-term initiatives reflect a broader commitment to enhance the ethical conduct of research within South Korea. They were implemented to ensure the country's preparedness for future pandemic situations, incorporating lessons learned from the challenges and successes of the initial pandemic response. The improvements in research governance, institutional capacity, and research ethics education directly addressed many of the vulnerabilities exposed during the early stages of the COVID-19 pandemic, creating a more resilient and ethically sound research ecosystem.

6 Harmonizing Korean Bioethics with Global Standards: Implications of the 2024 DoH Revision

The 2024 revision of the World Medical Association's Declaration of Helsinki (DoH) offers a valuable lens through which to examine the current state and future direction of South Korea's bioethics framework. While the 2013 Bioethics and Safety Act represents a substantial advancement in aligning Korean research ethics with international guidelines, the 2024 DoH necessitates further adaptations to address emerging challenges and ensure continued alignment with global standards. The DoH necessitates a comprehensive reassessment of South Korea's bioethics framework, particularly its alignment with evolving international ethical standards. While the 2013 Bioethics and Safety Act represented a substantial step forward, significant gaps remain, necessitating further legislative adaptations to fully incorporate the updated DoH principles and address the unique challenges of contemporary biomedical research.

A critical area needing immediate attention is the terminology employed in the Bioethics and Safety Act. The frequent use of "research subjects," appearing over 40 times in the Act, portrays research participants as passive entities, undermining their autonomy and agency [31]. This terminology reflects a paradigm that contrasts sharply with the participant-centric ethos underpinning the 2024 DoH, where individuals are viewed as active collaborators rather than mere recipients of research interventions. This is not just a semantic issue; it reflects a fundamental difference in legal and ethical perspectives on the nature of research participation. The 2024 DoH strongly advocates for a shift to "research participants," explicitly emphasizing their rights, agency, and active roles in all stages of the research process. A parallel shift in Korean terminology is therefore vital to ensure consistency with international ethical norms.

Beyond terminology, the Bioethics and Safety Act's approach to vulnerable populations requires careful reconsideration. While acknowledging the need to protect vulnerable individuals, the Act's overly cautious approach mandates proxy consent even for minimal-risk research involving children or individuals with diminished decision-making capacity. This creates unnecessary barriers to much-needed research that directly impacts the well-being of these groups, thus exacerbating existing health disparities. This contrasts with the 2024 DoH's emphasis on the fair and responsible inclusion of vulnerable groups in research, as long as their specific needs and priorities are addressed, and the research design incorporates appropriate safeguards and protections. The DoH clearly advocates for minimizing the harms of exclusion while balancing these against the potential harms of inclusion.

The revised DoH underscores the need for a more participant-centric approach and for addressing structural inequalities within research. This requires a closer examination of the existing Korean legal framework. The key question is how the 2024 DoH revision necessitates changes to the Korean Bioethics and Safety Act to ensure alignment with evolving international ethical standards. Key areas of consideration include informed consent mechanisms, particularly concerning the use of human-derived materials and data; the equitable inclusion of vulnerable populations; and the facilitation of meaningful community engagement in research. Amendments to the Act strengthened oversight of IRBs and capacity-building initiatives are necessary to address these issues effectively, ensuring the ethical conduct and global competitiveness of South Korean research. The 2024 DoH's emphasis on data governance and the ethical use of biological materials necessitates a careful review of the Bioethics and Safety Act's provisions to ensure compliance with evolving international standards, especially in the areas of big data, AI-driven research, and data sharing.

The limitations of the Bioethics and Safety Act are particularly evident in its treatment of data governance in the era of big data, genomics, and AI-driven research. While the Act incorporates elements of broad consent, its provisions—predating the current data-intensive research landscape—are inadequate for addressing the complexities of global research collaboration and data sharing. This reliance on "specific consent," requiring narrowly defined research purposes, limits research scope, weakens data governance, and compromises data protection. This restricted approach contrasts with the globally accepted principles of data protection, as exemplified in frameworks such as the General Data Protection Regulation (GDPR) [54] and the US Common Rule, both of which explicitly permit broad consent mechanisms while maintaining robust data protection safeguards. The 2024 DoH, recognizing these advancements, connects ethical principles concerning the use of human-derived biological materials to the Declaration of Taipei (DoT) [55]. Thus, the DoH is to be revised to strongly advocate for broader consent models and advanced data protection measures, such as pseudonymization, to ensure ethical data governance in the face of rapidly advancing data-intensive research techniques.

To achieve a comprehensive alignment with the 2024 DoH, South Korea must embark on a substantial revision of the Bioethics and Safety Act. This necessitates not only changes in terminology and legislation but also a profound shift in perspective and a dedicated commitment to fostering a truly participant-centric research culture. This includes implementing effective strategies for community engagement, strengthening IRB capacity and oversight, and providing comprehensive education programs for researchers and IRB members alike. By enacting such measures,

South Korea can ensure that its research practices uphold the highest ethical standards, contributing to both the advancement of biomedical research and the protection of individual rights and well-being. The Korean experience in the context of the COVID-19 pandemic, with its specific challenges, should inform these revisions. Future developments in Korean bioethics must prioritize amending the Bioethics and Safety Act to reflect these updated guidelines, improving and standardizing ethical review processes (including IBC functionality), and fostering meaningful community participation in research. This ongoing process will not only strengthen the ethical integrity of Korean research but also enhance its global competitiveness and facilitate broader international collaborations.

7 Conclusion

In conclusion, this chapter has demonstrated the significant impact of the COVID-19 pandemic on research ethics in South Korea, highlighting the urgent need for adaptation and alignment with evolving international standards. While the 2013 Bioethics and Safety Act provided a robust foundation, the pandemic exposed areas requiring further development, particularly concerning expedited review processes, informed consent in a digital age, and the fair inclusion of vulnerable populations. The Korean government's swift and multifaceted response, from immediate emergency measures to the establishment of long-term preparedness frameworks, played a crucial role in navigating these challenges. However, ensuring continued alignment with the revised 2024 Declaration of Helsinki and other global ethical guidelines requires ongoing efforts to amend the Bioethics and Safety Act, strengthen the functionality of Institutional Bioethics Committees (IBCs), enhance data governance, and foster meaningful community engagement. To enhance research governance, promote community engagement, and strengthen ethical research conduct in Korea, a harmonization process is vital for promoting research integrity and enhancing the nation's standing within the global research landscape.

Acknowledgments As a delegate of the Korean Medical Association (KMA), the author participated in the WMA Regional Meeting in Washington, DC, in August 2024, and the WMA General Assembly in Helsinki in October 2024.

Conflict of Interest The author declares no competing interests relevant to the content of this chapter.

References

1. World Medical Association. WMA Declaration of Helsinki: ethical principles for medical research involving human participants. Accessed 25 Oct 2024. https://www.wma.net/policies-post/wma-declaration-of-helsinki/.

2. Ndebele P. The declaration of Helsinki, 50 years later. JAMA. 2013;310(20):2145–6. https://doi.org/10.1001/jama.2013.281316.
3. Kim OJ. Significance and implications of the 2024 revision of the declaration of Helsinki. J Korean Med Assoc. 2024;67(11):668–73.
4. Nature Medicine. Declaration of Helsinki embraces health equity. Nat Med. 2024;30(12):3383. https://doi.org/10.1038/s41591-024-03433-5.
5. O'Grady C. Key global bioethics guidelines get 'dramatic' update. Science Published October 29, 2024. Accessed 20 Nov 2024. https://www.science.org/content/article/key-global-bioethics-guidelines-get-dramatic-update.
6. Resneck JS Jr. Revisions to the declaration of Helsinki on its 60th anniversary: a modernized set of ethical principles to promote and ensure respect for participants in a rapidly innovating medical research ecosystem. JAMA. 2025;333(1):15–7. https://doi.org/10.1001/jama.2024.21902.
7. Reis AA, Upshur R, Moodley K. Future-proofing research ethics—key revisions of the declaration of Helsinki 2024. JAMA. 2025;333(1):20–1. https://doi.org/10.1001/jama.2024.22254.
8. Wise J. Helsinki declaration: rights of study participants are put at the heart of latest update. BMJ. 2024;387:q2357. https://doi.org/10.1136/bmj.q2357.
9. Bloom L. Revisiting the declaration of Helsinki—a patient-centered perspective. JAMA. 2025;333(1):22–3. https://doi.org/10.1001/jama.2024.22077.
10. Bierer BE. Declaration of Helsinki-revisions for the 21st century. JAMA. 2025;333(1):18–9. https://doi.org/10.1001/jama.2024.22281.
11. Solbakk JH, Bentzen HB, Holm S, Heggestad AK, Hofmann B, Robertsen A, et al. Back to WHAT? The role of research ethics in pandemic times. Med Health Care Philos. 2021;24(1):3–20. https://doi.org/10.1007/s11019-020-09984-x.
12. Park S, Lim HJ, Park J, Choe YH. Impact of COVID-19 pandemic on biomedical publications and their citation frequency. J Korean Med Sci. 2022;37(40):e296. https://doi.org/10.3346/jkms.2022.37.e296.
13. Faust A, Sierawska A, Krüger K, Wisgalla A, Hasford J, Strech D. Challenges and proposed solutions in making clinical research on COVID-19 ethical: a status quo analysis across German research ethics committees. BMC Med Ethics. 2021;22(1):96. https://doi.org/10.1186/s12910-021-00666-8.
14. Singh S, Cadigan RJ, Moodley K. Challenges to biobanking in LMICs during COVID-19: time to reconceptualise research ethics guidance for pandemics and public health emergencies? J Med Ethics. 2022;48(7):466–71. https://doi.org/10.1136/medethics-2020-106858.
15. De Sutter E, Lalova-Spinks T, Borry P, Valcke P, Kindt E, Negrouk A, et al. Rethinking informed consent in the time of COVID-19: an exploratory survey. Front Med (Lausanne). 2022;9:995688. https://doi.org/10.3389/fmed.2022.995688.
16. Barroga E, Matanguihan GJ. Fundamental shifts in research, ethics and peer review in the era of the COVID-19 pandemic. J Korean Med Sci. 2020;35(45):e395. https://doi.org/10.3346/jkms.2020.35.e395.
17. Sisk BA, DuBois J. Research ethics during a pandemic: a call for normative and empirical analysis. Am J Bioeth. 2020;20(7):82–4. https://doi.org/10.1080/15265161.2020.1779868.
18. Ten Have H. The COVID-19 pandemic and global bioethics. Cham: Springer; 2022.
19. Zhang H, Shao F, Gu J, Li L, Wang Y. Ethics committee reviews of applications for research studies at 1 hospital in China during the 2019 novel coronavirus epidemic. JAMA. 2020;323(18):1844–6. https://doi.org/10.1001/jama.2020.4362.
20. Rothwell E, Brassil D, Barton-Baxter M, Brownley KA, Dickert NW, Ford DE, et al. Informed consent: old and new challenges in the context of the COVID-19 pandemic. J Clin Transl Sci. 2021;5(1):e105. https://doi.org/10.1017/cts.2021.401.
21. Jones XM, Zimba O, Gupta L. Informed consent for scholarly articles during the COVID-19 pandemic. J Korean Med Sci. 2021;36(3):e31. https://doi.org/10.3346/jkms.2021.36.e31.
22. Tosoni S, Voruganti I, Lajkosz K, Mustafa S, Phillips A, Kim SJ, et al. Patient consent preferences on sharing personal health information during the COVID-19 pandemic: "the more

informed we are, the more likely we are to help". BMC Med Ethics. 2022;23(1):53. https://doi.org/10.1186/s12910-022-00790-z.
23. Voit K, Skuban-Eiseler T, Orzechowski M, Steger F. Informed consent in COVID-19-research: an ethical analysis of clinical studies performed during the pandemic. Healthcare (Basel). 2023;11(12):1793. https://doi.org/10.3390/healthcare11121793.
24. Ma X, Wang Y, Gao T, He Q, He Y, Yue R, et al. Challenges and strategies to research ethics in conducting COVID-19 research. J Evid Based Med. 2020;13(2):173–7. https://doi.org/10.1111/jebm.12388.
25. Park YS, Kim OJ. Government initiatives for research ethics during COVID-19 pandemic in Korea. J Korean Med Sci. 2024;39(12):e116. https://doi.org/10.3346/jkms.2024.39.e116.
26. Korea Institute of Science & Technology Evaluation and Planning (KISTEP). Status of scientific and technological paper publications and citations in Korea, 2022. Eumseong: Korea Institute of S&T Evaluation and Planning. p. 2024.
27. Ministry of Food and Drug Safety. Approval for clinical trials of medicinal products in 2023 increased by 10.1% compared to 2022. Cheongju: Ministry of Food and Drug Safety; 2024.
28. Kim OJ. The 2024 revision of the declaration of Helsinki and the future directions of Korea's bioethics and safety act. J Korean Med Sci. 2025;40(9):e99. https://doi.org/10.3346/jkms.2025.40.e99.
29. Ryu YJ. Between truth and profit: scientific misconduct case of human cloned embryonic stem cell and revisiting cases during COVID-19 pandemic. In: Kurihara C, Greco D, Dhai A, editors. Ethical innovation for global health: pandemic, democracy and ethics in research. Springer; 2023. p. 69–104.
30. Kim J, Park K. Ethical modernization: research misconduct and research ethics reforms in Korea following the Hwang affair. Sci Eng Ethics. 2013;19(2):355–80. https://doi.org/10.1007/s11948-011-9341-8.
31. Kim OJ. Bioethics and research ethics in Korea after the Hwang Woo-Suk scandal. In: Ten years ago from the present: the Hwang Woo-Suk scandal. Seoul: National Academy of Medicine of Korea; 2018. p. 37–49.
32. Kim OJ, Park BJ, Sohn DR, Lee SM, Shin SG. Current status of the institutional review boards in Korea: constitution, operation, and policy for protection of human research participants. J Korean Med Sci. 2003;18(1):3–10. https://doi.org/10.3346/jkms.2003.18.1.3.
33. United Nations Educational, Scientific and Cultural Organization. Universal Declaration on Bioethics and Human Rights. 19 October 2005. Available at: https://www.unesco.org/en/legal-affairs/universal-declaration-bioethics-and-human-rights?hub=66535.
34. Dighe A, Cattarino L, Cuomo-Dannenburg G, Skarp J, Imai N, Bhatia S, et al. Response to COVID-19 in South Korea and implications for lifting stringent interventions. BMC Med. 2020;18(1):321. https://doi.org/10.1186/s12916-020-01791-8.
35. Issac A, Stephen S, Jacob J, Vr V, Radhakrishnan RV, Krishnan N, et al. The pandemic league of COVID-19: Korea versus the United States, with lessons for the entire world. J Prev Med Public Health. 2020;53(4):228–32. https://doi.org/10.3961/jpmph.20.166.
36. Jeong E, Hagose M, Jung H, Ki M, Flahault A. Understanding South Korea's response to the COVID-19 outbreak: a real-time analysis. Int J Environ Res Public Health. 2020;17(24):9571. https://doi.org/10.3390/ijerph17249571.
37. Kim SW. COVID-19 outbreak in Daegu City, Korea and response to COVID-19: how have we dealt and what are the lessons? J Korean Med Sci. 2022;37(50):e356. https://doi.org/10.3346/jkms.2022.37.e356.
38. Park K, Baek HJ. Contextual response to the COVID-19 pandemic from the experience of South Korea. Public Health. 2023;222:e7–8. https://doi.org/10.1016/j.puhe.2022.07.015.
39. Lee D, Heo K, Seo Y, Ahn H, Jung K, Lee S, et al. Flattening the curve on COVID-19: South Korea's measures in tackling initial outbreak of coronavirus. Am J Epidemiol. 2021;190(4):496–505. https://doi.org/10.1093/aje/kwaa217.
40. Yang TU, Noh JY, Song JY, Cheong HJ, Kim WJ. How lessons learned from the 2015 Middle East respiratory syndrome outbreak affected the response to coronavirus disease 2019 in

the Republic of Korea. Korean J Intern Med. 2021;36(2):271–85. https://doi.org/10.3904/kjim.2020.371.
41. Park J, Chung E. Learning from past pandemic governance: early response and public-private partnerships in testing of COVID-19 in South Korea. World Dev. 2021;137:105198. https://doi.org/10.1016/j.worlddev.2020.105198.
42. Wu F, Hu Q, Zhu C, Wang H, Yu Q, Sun H. New structural economic analysis of anti-COVID-19 pandemic model of BEST region. Int J Environ Res Public Health. 2021;18(15):7822. https://doi.org/10.3390/ijerph18157822.
43. Lee S, Wong R. COVID-19 responses of South Korea as hybrids of governance modes. Front Public Health. 2021;9:654945. https://doi.org/10.3389/fpubh.2021.654945.
44. Jeon J, Kim H, Yu KS. The impact of COVID-19 on conduct of clinical trials for medical products in Korea. J Korean Med Sci. 2020;35(36):e329. https://doi.org/10.3346/jkms.2020.35.e329.
45. Yoo SH, Sim JA, Shin J, Keam B, Park JB, Shin A. The impact of COVID-19 on cancer care in a tertiary hospital in Korea: possible collateral damage to emergency care. Epidemiol Health. 2022;44:e2022044. https://doi.org/10.4178/epih.e2022044.
46. Moon MJ. Fighting COVID-19 with agility, transparency, and participation: wicked policy problems and new governance challenges. Public Adm Rev. 2020;80(4):651–6. https://doi.org/10.1111/puar.13214.
47. Oh J, Lee JK, Schwarz D, Ratcliffe HL, Markuns JF, Hirschhorn LR. National response to COVID-19 in the Republic of Korea and lessons learned for other countries. Health Syst Reform. 2020;6(1):e1753464. https://doi.org/10.1080/23288604.2020.1753464.
48. Choi JY, Lee SH, Jamal T. Smart Korea: governance for smart justice during a global pandemic. J Sustain Tour. 2021;29(2–3):541–50.
49. Lee S. Steering the private sector in COVID-19 diagnostic test kit development in South Korea. Front Public Health. 2020;8:563525. https://doi.org/10.3389/fpubh.2020.563525.
50. Park YS, Doris S, Kim OJ. Research ethics challenges in pandemic Korea and their implications for the revised 2024 declaration of Helsinki. J Korean Med Sci. 2025;40(47): Forthcoming
51. You K, Kim D, Kim OJ. Application and expansion of the harm principle to the restrictions of liberty in the COVID-19 public health crisis: focusing on the revised bill of the March 2020 「infectious disease control and prevention act「. Korean Soc Law Med. 2020;21(2):105–62.
52. National Research and Development Innovation Act. Accessed 13 Mar 2025. https://www.law.go.kr/LSW/eng/engLsSc.do?menuId=2&query=National%20Research%20and%20Development%20Innovation%20Act%20.
53. Lee MJ, Kim OJ, Cho EH, Jeon JH, Lee DH, Lee S, et al. Introduction of the National R&D Innovation Act and research ethics in Korea. Asia Pac J Health Law Ethics. 2021;14(2):155–92. https://doi.org/10.38046/apjhle.2021.14.2.007.
54. Regulation (EU) 2016/679 of the European Parliament and of the Council of 27 April 2016 on the protection of natural persons with regard to the processing of personal data and on the free movement of such data, and repealing Directive 95/46/EC (General Data Protection Regulation). Official Journal of the European Union. 2016.4.5.2016: L 119/1-88.
55. The World Medical Association. WMA Declaration of Taipei on ethical considerations regarding health databases and biobanks. Adopted in October 2002, revised in October 2016. https://www.wma.net/policies-post/wma-declaration-of-taipei-on-ethical-considerations-regarding-health-databases-and-biobanks/.

Chapter 7
Scientific Misconduct and the Waste of Research Funding: Can Existing Ethical Principles and Legal Frameworks Be Sufficient to Prevent Intentional Violations?

Young-Joon Ryu

Abstract There is scarcely any need to invoke the inhumane war crimes committed by Japan's Unit 731 or the Auschwitz concentration camp operated by Nazi Germany during World War II. Why does research misconduct persist and continue to recur, rather than being eradicated? What impact does the prevention of research misconduct, first addressed in the 2024 revision of the Declaration of Helsinki, have on the eradication of such misconduct? Previous studies have primarily attributed the causes of research misconduct to external factors such as research environments, the competitive structure of academia, and evaluation methods for research outcomes. These studies tend to conclude that misconduct results from the vulnerability of individual researchers to various temptations. However, this chapter raises a critical issue with such existing analyses, arguing that they fail to adequately account for instances of deliberate and premeditated research misconduct. The author has examined several major scandals of research misconduct that have gained international notoriety over the past 30 years. This analysis reveals that the immense financial waste, direct loss of human lives, and social disruption caused by these cases are by no means trivial, even when compared to other criminal offenses. It is imperative to move beyond the perception of researchers as noble truth-seekers who momentarily succumb to temptation. Research misconduct must be explicitly defined as a criminal act, and stringent punitive measures should be established to deter such offenses.

Keywords Research integrity · Scientific misconduct · Hwang Woo-Suk scandal · Embryonic stem cell research · Somatic cell nuclear transfer · Criminal punishment

Y.-J. Ryu (✉)
Department of Medical Humanities and Medical Ethics, Department of Pathology, College of Medicine, Kangwon National University, Chun-Cheon, South Korea
e-mail: mindmade@kangwon.ac.kr

C. Kurihara et al. (eds.), *The 2024 Declaration of Helsinki*,
https://doi.org/10.1007/978-981-96-9294-1_7

1 Unresolved Questions in Major Cases of Research Misconduct

Scientific misconduct is by no means a novel issue. From the war-crime experiments conducted by Nazi Germany at Auschwitz during World War II and by Japan's Unit 731 to the numerous high-profile cases of research fraud that have emerged in contemporary times, history has repeatedly demonstrated that unethical practices in scientific research persist across generations [1, 2]. Despite the accumulation of profound personal and societal harm over the course of history, and despite numerous efforts to establish and reinforce research ethics, instances of scientific misconduct continue to recur as if no lessons have ever been learned.

Examining major research misconduct cases that have sparked international controversy over the past three decades provides a clearer understanding of the gravity of this issue. In 1998, British physician Andrew Wakefield published a paper in *The Lancet* claiming a causal link between the MMR vaccine and autism [3]. As the study gained media attention, it provided seemingly credible support for those already skeptical of vaccines. This triggered a global wave of vaccine hesitancy, particularly among parents deeply concerned about their children's health. As a result, essential childhood vaccinations were increasingly refused, leading to the resurgence of diseases once believed to have been eradicated. Shockingly, it was later revealed that Wakefield's study had been entirely fabricated. In 2002, Jan Hendrik Schön, a German physicist at Bell Labs, published groundbreaking research on nanoscale organic single-molecule transistors in *Nature* [4], followed by several other prestigious journals. However, it was later revealed that 16 out of his 21 published studies had been deliberately fabricated, causing significant damage to the credibility of the physics community. In 2005, Hwang Woo-Suk of Seoul National University (SNU) in South Korea published a paper in *Science* claiming to have successfully derived the first human-cloned embryonic stem cell line. However, it was later revealed that the research data had been deliberately fabricated [5]. Shortly after its publication, it was discovered that no cloned stem cells had actually been produced and that the data had been entirely falsified. This revelation sent shock waves not only through South Korea but also across the global biotechnology community. Diederik Stapel, a social psychologist at Tilburg University in the Netherlands, published over 55 research papers on human behavior. His studies made bold claims, such as "unclean environments promote discrimination," "meat advertisements increase selfish behavior," and "white people have more honest-looking faces than other races." However, in 2011, it was revealed that all of his data had been entirely fabricated. Haruko Obokata of Japan's RIKEN Institute published a groundbreaking study on stimulus-triggered acquisition of pluripotency (STAP) cells, claiming that fully differentiated cells could be easily reprogrammed into pluripotent stem cells simply by immersing them in an acidic solution. However, as scientists worldwide failed to replicate her results, suspicions grew. In 2014, it was ultimately revealed that the data had been intentionally fabricated from the outset [6]. In 2016, Paolo Macchiarini, an Italian thoracic surgeon at Sweden's Karolinska

Institute, attempted the world's first transplant of an artificial trachea. His approach involved using a three-dimensional (3D)-printed plastic scaffold coated with cultured stem cells, which was then implanted into living patients. However, a whistleblower later revealed that his research and surgical outcomes had been falsified, with evidence dating back to 2008. The scandal ultimately led to the deaths of at least seven patients, making it one of the most devastating cases of medical misconduct in history [7].

As evidenced by the aforementioned cases of research misconduct, the recurring nature of such large-scale fraudulent incidents inevitably raises fundamental questions. Are these ethical breaches simply the result of the absence of well-established research ethics principles? Do they stem from a lack of ethical training in the researcher development process? Or, more critically, are they perpetuated by the insufficiency of legal frameworks and punitive measures to deter such misconduct? These pressing concerns demand deeper examination. Furthermore, as previous studies on research ethics suggest, one cannot help but question whether scientific misconduct is merely the result of momentary lapses in judgment driven by fleeting temptation. The recurring and systematically orchestrated nature of many high-profile cases calls into doubt the notion that such actions are merely accidental or isolated ethical failures. If the consequences of scientific misconduct are not merely minor infractions but instead lead to profound social, ethical, and academic harm—comparable to other serious crimes—then it is imperative that we engage in a comprehensive, in-depth, and far-reaching discussion on the current state of research integrity and its broader implications.

These persistent and unresolved questions suggest that research misconduct may stem from more fundamental factors beyond research environments, institutional structures, academic competition, and performance evaluation systems. The author submitted only one opinion responding to the World Medical Association's public consultation for the 2024 revision of the Declaration of Helsinki (DoH) to include a text to prevent scientific misconduct. This was accepted and this concept was for the first time included in the DoH (para. 12) [8]. The author believes it is a prerequisite in internationally recognized research ethics principle; however, description as "principle" is never enough. Therefore, this study aims to analyze specific cases of scientific fraud to uncover its underlying causes and, ultimately, to explore effective and practical measures for its prevention.

2 Why Does Research Misconduct Persist Despite Ample Experience and Lessons from the Past?

2.1 *Research Misconduct Does Not Occur Due to the Absence of Ethical Guidelines*

2.1.1 Robert K. Merton's Four CUDOS Principles

In 1942, Robert K. Merton proposed the "Normative Structure of Science," which defines the normative framework that scientific activities should adhere to and the ethical and social norms that scientists must follow [9]. He established four fundamental norms that must be upheld to ensure the development and credibility of science, emphasizing its inherently social nature: Communism, Universalism, Disinterestedness, and Organized Skepticism. Later, they became collectively known by the acronym "CUDOS," derived from the initials of each principle.

Merton first argued that scientific knowledge should not be privately owned but rather considered a public good, a principle he termed "Communism." He asserted that scientists must not monopolize their research findings but should actively share them to foster scientific progress. His core argument emphasized that scientific discoveries should not be used as tools for the benefit of specific individuals or institutions but should be openly accessible for the advancement of society as a whole. Second, Merton argued that the value of scientific knowledge should not be influenced by a researcher's personal characteristics—such as race, gender, class, or nationality—but should instead be assessed solely based on objective criteria. He termed this principle "Universalism," emphasizing that scientific inquiry must remain free from social and cultural biases and undergo a universal and objective verification process. Third, Merton asserted that scientists must maintain independence from personal interests and external political or economic pressures, dedicating themselves solely to the pursuit of truth. He defined this principle as "Disinterestedness" and emphasized its crucial role in enhancing the credibility of scientific activity and preventing research misconduct. To safeguard science from being distorted by commercial interests or political agendas, researchers must uphold a stance free from personal biases or conflicts of interest. Fourth, Merton asserted that all scientific claims and research findings must undergo rigorous critical scrutiny to validate their reliability, a principle he termed "Organized Skepticism." This principle emphasizes that research results should be verified through reproducible methodologies and that the legitimacy of scientific knowledge is established through meticulous peer review and critique. Furthermore, research misconduct is often uncovered through this process of systematic skepticism, highlighting the inherent risk of compromised scientific integrity when conflicts of interest interfere with objective evaluation.

The four normative principles proposed by Merton have become essential elements in maintaining the institutional and social integrity of scientific practice. They have long served as a fundamental theoretical framework for discussing research ethics and preventing misconduct. His theory functions as a crucial standard for emphasizing ethical responsibility within the scientific community and continues to exert significant influence as a foundational principle of research ethics today.

2.1.2 An Overabundance of Research Ethics Guidelines

After World War II, the Nuremberg Trials exposed the inhumane human experimentations conducted by Nazi Germany, bringing the ethical issues of scientific research to the forefront of international discourse [10]. This historical experience led to the Declaration of Helsinki, issued by the World Medical Association (WMA) in 1964, which expanded upon the human rights protection principles established in the Nuremberg Code. The declaration provided more detailed ethical guidelines specifically for medical research involving human subjects [11]. However, despite this broad consensus formed through humanity's historical experiences, cases of scientific misconduct—causing significant harm and loss—have continued to occur persistently. Major cases of research misconduct have been continuously reported, particularly in leading scientific nations such as the United States, Europe, China, Australia, Japan, and South Korea. In response, governments, academic institutions, and research organizations in these countries have successively established research ethics guidelines as part of their efforts to manage these incidents and prevent future violations. Representative international research ethics guidelines include the Belmont Report [12], which establishes ethical principles for research involving human subjects; the International Committee of Medical Journal Editors (ICMJE) Research Ethics Guidelines [13], which set ethical standards for medical journal publications; and the Committee on Publication Ethics (COPE) Guidelines [14], which provide directives on ethical concerns in academic publishing. Additionally, the National Institutes of Health (NIH) Research Integrity Principles in the United States [15], the Organisation for Economic Co-operation and Development (OECD) Research Ethics Guidelines [16], and the United Nations Educational, Scientific, and Cultural Organization (UNESCO) Recommendation on Science and Scientific Researchers further delineate global standards for responsible research practices [17]. In parallel, numerous nations and institutions have established "Research Ethics Education and Training Programs" to cultivate a culture of ethical awareness and integrity in scientific inquiry.

The ICMJE Guidelines define research ethics in the fields of medicine and life sciences, establishing fundamental principles such as integrity, transparency, objectivity, and accountability. These guidelines emphasize the roles and responsibilities of researchers, ensuring that ethical standards are upheld throughout the research process. The COPE is an international organization dedicated to academic publishing ethics. It establishes ethical standards that researchers must adhere to when presenting and publishing their findings, providing concrete principles and guidelines to uphold research integrity and transparency throughout the publication process. The NIH Research Integrity Guidelines focus on preventing research misconduct and ensuring fairness in scientific inquiry. By emphasizing ethical responsibility, transparency, and reliability in research, these guidelines have significantly influenced researchers receiving NIH funding, reinforcing the principles of integrity and accountability in scientific endeavors. The OECD Research Ethics Guidelines establish ethical principles for scientific research while emphasizing transparency, fairness, and objectivity. By doing so, they provide a foundational framework for fostering international research collaboration based on ethical integrity. The UNESCO Code of Ethics underscores human dignity, fairness, and equality in research, placing particular emphasis on ethical principles for studies involving human subjects. It has played a pivotal role in establishing international standards for research ethics, ensuring the protection of human rights and integrity in scientific inquiry. Additionally, Research Ethics Education programs offered by various research institutions and universities play a crucial role in preventing misconduct by instilling fundamental ethical principles such as integrity, objectivity, transparency, and accountability. These educational initiatives ensure that researchers affiliated with these institutions adhere to rigorous ethical standards throughout their scientific endeavors.

In South Korea, various research ethics guidelines have been established since 2005 to reinforce ethical standards in scientific research. Notable research ethics guidelines in South Korea include the 2014 Guidelines for Ensuring Research Ethics established by the Korea Research Foundation (KRF) [18]; the 2022 Research Ethics Guidelines formulated by the National Research Council of Science & Technology (NST) in accordance with Article 31 of the National Research and Development Innovation Act and Article 58 of its enforcement decree [19]; and the 2007 Code of Ethics for Scientists and Engineers, enacted by the Korean Federation of Science and Technology Societies (KOFST) [20]. Additionally, the Korea Citation Index (KCI) Research Ethics Guidelines, the Korea Researcher Conduct Association (KRCA), the Korea Research Ethics Association (KREA), as well as institutional research ethics policies at major universities, Institutional Review Boards (IRB), and research ethics review guidelines have been established. The sheer number of these ethical frameworks is so vast that listing them all is nearly impossible, reflecting the extensive efforts to institutionalize research integrity in South Korea.

In particular, KOFST, a government-affiliated organization, established its first Research Ethics Code in response to the 2005 Hwang Woo-Suk scientific fraud scandal involving fabricated research published in *Science*. This initiative was driven by the growing recognition of the need to define the social responsibility and ethical consciousness of scientists and engineers in South Korea. This code primarily emphasized the need for researchers to uphold professional integrity and receive ethics education, but it lacked concrete provisions for sanctions against researchers involved in misconduct, rendering it largely declarative rather than enforceable. As a result, it faced criticism for being insufficient in addressing research fraud. However, at the time, South Korea had no established guidelines for research ethics, and the introduction of this first-ever Research Ethics Code marked a significant milestone in laying the foundation for ethical standards in scientific research.

Meanwhile, the establishment of IRBs at universities and research institutions gained momentum. In particular, IRBs were introduced and began operating under the supervision of the Ministry of Health and Welfare, primarily within university-affiliated hospitals where clinical research is actively conducted. Since 2010, most universities in South Korea have established their own IRBs. In May 2015, a total of 77 universities joined forces to launch the Korea University Council for Research Ethics (KUCRE), marking the beginning of a coordinated institutional effort to strengthen research ethics at the university level. This development was largely driven by the comprehensive revision of the "Bioethics and Safety Act" in 2022. Additionally, to evaluate compliance with research ethics during the publication process and to ensure ethical integrity in academic publishing, the Korea Citation Index (KCI) also established its own research ethics regulations. These guidelines reinforce ethical standards for academic journals and authorize sanctions such as delisting journals from the index if cases of research misconduct are identified. Furthermore, KRCA and various academic societies have been continuously issuing research ethics guidelines, providing ethics education, and consulting services to support researchers in adhering to ethical standards in their work.

Despite initial resistance from the Korean scientific community to the newfound emphasis on research ethics, over time, a broader consensus emerged within Korean society regarding the necessity of establishing and upholding ethical standards in scientific research. Thanks to the dedicated efforts of many scholars in the field of ethics, the 2021 National Research and Development Innovation Act explicitly defined the scope of research ethics and, for the first time, made the establishment of research ethics regulations a legal obligation. Subsequently, a comprehensive research ethics guideline was established in February 2022, followed by the Ministry of Education's directive on ensuring research ethics, issued on July 17, 2023. These ongoing institutional reforms reflect a significant and encouraging advancement in the systematic reinforcement of research integrity. Despite various government measures, the persistence of both large scale and smaller, deliberate cases of research misconduct prompted a further escalation in regulatory action. In 2023, the Ministry of Education, through an official directive, incorporated for the first time specific penalty provisions for researchers found guilty of misconduct. These provisions were based on the "Guidelines for Securing Research Ethics" issued by the KRF, which oversees the distribution of national research funding. Article 32 of this directive outlines the procedures and subsequent actions in cases of research ethics violations. It explicitly mandates sanctions such as the revocation of state research funding and restrictions on future funding eligibility for researchers found guilty of research misconduct.

However, this is where progress stands. The strong resistance within the scientific community against the perception of researchers as potential criminals and the imposition of criminal penalties on scientific misconduct remains a significant barrier. Moreover, 20 years is still too short a time to fully uncover and establish a comprehensive framework for justice in research ethics. Although still insufficient, these changes mark the culmination of 18 years of continuous efforts by both the

South Korean scientific community and the government to uphold research integrity since the 2005 Hwang Woo-Suk scandal.

In this way, South Korea has gradually established ethical and legal frameworks aligned with international research ethics guidelines to strengthen research integrity. This evolution signifies a clear shift beyond merely emphasizing individual virtues such as honesty, objectivity, reliability, transparency, and accountability. Instead, it reflects an increasing focus on punitive measures for research misconduct, reinforcing the legal consequences of fraudulent practices within the scientific community.

2.2 *Major Cases of Research Misconduct Are Not Merely Momentary Lapses in Judgment by Researchers Succumbing to Temptation*

In 2023, the SNU Research Integrity Committee released its findings from an investigation into cases of research misconduct that occurred within the university between 2020 and 2022. The report identified 28 new cases of research misconduct, including falsification and fabrication of authorship and data, manipulation and alteration of research data, plagiarism, improper citations, and duplicate publications. Additionally, according to an investigation by the KRF, a total of 41 cases were officially identified as research misconduct between 2019 and September 2023. The research projects involved in these cases received a staggering 61.4 billion KRW (approximately 46 million USD) in funding [21]. Despite having experienced the 2005 Hwang Woo-Suk scandal, a globally renowned case of research misconduct, the Seoul National University continues to face recurring instances of fraudulent research practices. Even more striking is the fact that no significant punitive measures have been imposed on those involved. This reality raises critical questions about the effectiveness of existing research integrity measures and highlights the persistent vulnerabilities in addressing scientific misconduct.

In South Korea, newly adapted forms of research misconduct have recently been emerging and adding to the existing patterns of unethical research practices. Beyond simple manipulation of research papers, research misconduct in South Korea has increasingly become entangled with the structural issue of intense competition in university admissions, reflecting a more complex and systemic problem. A notable example is the recent occurrence of two high-profile cases in South Korea involving fabricated credentials and manipulated research papers to gain unfair advantages in university admissions for the children of high-ranking officials. A former Minister of Justice was found guilty of forging certificates, awards, academic records, and research papers to create favorable credentials for their child's university admission. As a result, in December 2024, the Supreme Court upheld a guilty verdict, sentencing them to prison. Additionally, another former Minister of Justice, a political rival of the aforementioned figure, was also found to have violated research ethics by artificially fabricating credentials "SPEC(qualification)" to benefit their high school-aged child's university admission. However, they avoided criminal punishment on the grounds that the fraudulent research materials were never actually used in the admission process [22]. These cases illustrate not only the persistent prevalence of research misconduct in South Korea but also a troubling shift in societal norms, where unethical practices for personal gain are becoming increasingly normalized and are now extending beyond universities to high schools.

Research misconduct in science and medicine is by no means a phenomenon unique to the modern era. Since the Scientific Revolution, as science has come to exert immense social and economic influence, research achievements have evolved beyond mere pursuits of truth to become tools that determine individual social success, generate economic profit, and even serve as instruments of national competition. With this elevated status, the temptation for researchers to engage in misconduct in pursuit of social prestige and power has been inherently present.

However, the general public has long maintained a romanticized perception of researchers as highly trained individuals devoted to societal progress, often neglecting to scrutinize their work with critical oversight. This tendency, coupled with Korea's entrenched elite corruption, has created an environment where as long as fellow scientists turn a blind eye, researchers can exploit the system for personal gain without external accountability.

Scholars who study cases of research misconduct in the scientific field commonly identify the following key factors as underlying causes of fraudulent practices [23–25]. First, a performance-driven research environment exerts excessive pressure on researchers, potentially compromising research integrity. Second, the traditional apprenticeship model in academia leads to a lack of structured ethics education, making it a significant factor contributing to research misconduct. Third, the overuse of deductive reasoning in hypothesis-driven research can encourage researchers to manipulate or selectively report data to achieve desired outcomes, increasing the likelihood of fraudulent practices. Fourth, moral blindness occurs when researchers either fail to recognize the unethical nature of their actions or, even when aware, rationalize their misconduct, perpetuating unethical behavior. Fifth, questionable research practices (QRPs)—such as data dredging, selective reporting, and hypothesis formulation based on results—reduce the replicability of scientific findings and may ultimately lead to serious research misconduct.

However, such analyses suggest that research misconduct occurs when an otherwise ethically conscious researcher is driven beyond moral boundaries by external environmental factors. While this explanation is not entirely incorrect, it is primarily applicable to isolated, relatively minor cases of misconduct. This perspective, however, fails to account for the large-scale research fraud cases outlined earlier in this chapter. These cases do not arise from a single instance of dishonesty but rather emerge through multiple layers of premeditated deception, carefully constructed over time.

The 2005 Seoul National University human embryonic stem cell research fraud case serves as a stark example of how research misconduct can escalate beyond mere ethical lapses or momentary transgressions into a deliberately orchestrated and systematic act of deception. Even after his violations of research ethics were exposed, Hwang Woo-Suk refused to acknowledge his wrongdoing. Instead, he perpetuated false claims and engaged in manipulative rhetoric, plunging both the academic community and the broader society into turmoil. His actions extended beyond scientific fraud, as he leveraged his influence within the scientific community and even exhibited ambitions to wield political power, demonstrating how research misconduct can evolve into a far-reaching crisis of academic integrity and institutional trust.

Rather than conducting genuine scientific research, Hwang Woo-Suk strategically leveraged his public recognition to strengthen his political standing and secure financial support. He pursued projects designed to evoke nationalistic and political sentiments, such as the cloning of the Baekdu Mountain tiger, yet utterly disregarded scientific principles in the actual experimental process. For instance, he attempted to clone a Baekdu Mountain tiger using pig oocytes, a scientifically dubious approach, while simultaneously manipulating research data and concealing key aspects of the experimental process. His actions went far beyond mere violations of research ethics—they constituted a deliberate propaganda effort, using science as a tool for deception and manipulation.

The research misconduct committed by Hwang Woo-Suk was not a mere isolated ethical lapse but rather a highly sophisticated and systematically orchestrated act of scientific fraud, requiring meticulous planning and sustained effort. Through a series of fabricated research projects, he sought to bolster his scientific reputation while simultaneously expanding his political influence and securing financial gain. Ultimately, his deception was exposed through internal whistleblowing, serving as a watershed moment in South Korea's awareness of research ethics and prompting a nationwide reassessment of academic integrity and ethical oversight in scientific research.

These cases, along with the major research misconduct scandals outlined at the beginning of this paper, suggest that such fraudulent acts are not merely the result of momentary lapses in judgment or personal temptation. Rather, they represent deliberate, long-term schemes, driven by a convergence of structural factors and individual motivations. Research misconduct, therefore, extends beyond individual ethical lapses; it is often intentionally committed in the pursuit of academic power, social prestige, political influence, and financial gain. Given this reality, it is insufficient to view research misconduct merely as an issue of moral failure. Instead, it must be recognized as a systemic problem that demands reinforced ethical oversight in academia and society at large, along with early detection and preventive mechanisms. A shift in perspective is necessary—one that regards serious research misconduct not just as an ethical violation, but as a criminal act requiring rigorous intervention and accountability.

2.3 Scientific Research Misconduct: A Crime with Consequences Comparable to Criminal Offenses

Scientific research misconduct thrives in proportion to society's complacency, which is rooted in the inherent trust placed in scientists. However, its consequences are far-reaching, causing severe damage to academic, social, and national foundations. Research fraud is not merely a matter of individual ethics; rather, its repercussions extend deeply into the scholarly community and society at large, often resulting in long-term and sometimes irreversible harm [26, 27].

Research misconduct erodes trust among scientists, wastes valuable time and resources, disrupts the academic careers of young researchers, and diminishes the overall research capacity of institutions. It undermines the very foundation of the scientific community's credibility, creating structural challenges that hinder the development of future generations of scholars. On a societal level, it weakens public trust in the scientific and expert community, discouraging government and private sector support for research, and ultimately destabilizing the research ecosystem. On a national scale, it damages international confidence in a country's scientific and technological advancements, tarnishes its global reputation, and squanders public funds and national resources, leading to a decline in national competitiveness. Furthermore, the process of investigating and rectifying research fraud often exacerbates social divisions, and in extreme cases, it can even result in direct harm to human life. The profound consequences of research misconduct become especially evident when examining notorious cases of scientific fraud, which clearly illustrates its devastating impact.

In 1998, British physician Andrew Wakefield published a study in *The Lancet* claiming that the measles-mumps-rubella (MMR) vaccine caused autism. The paper sparked global controversy, leading to widespread public distrust of vaccines and a sharp decline in essential immunization rates. As a direct consequence, preventable infectious diseases such as measles and mumps resurged, resulting in outbreaks that had previously been under control. Tragically, cases were reported where children who had not been vaccinated contracted these diseases and died. Even more alarming was the revelation that Wakefield's study was entirely fraudulent, based on manipulated data. This case remains one of the most damaging instances of scientific misconduct, demonstrating how fabricated research can have catastrophic real-world consequences, including the loss of human life.

In 2002, Jan Hendrik Schön, a German physicist at Bell Labs, published over 30 papers, including in *Nature* and *Science*, based on fabricated data without conducting actual experiments. His research was initially hailed as a breakthrough in polymer electronics and nanotechnology, earning him widespread recognition. However, it was later revealed that his data had been entirely falsified, leading to the retraction of all his papers. This scandal not only discredited Schön himself but also severely damaged the reputation of the entire physics community, highlighting the far-reaching consequences of large-scale scientific fraud.

In 2005, Hwang Woo-Suk of SNU published a paper in *Science* claiming to have successfully created 11 human embryonic stem cell lines. His research garnered global attention and secured hundreds of billions of Korean won in funding from both the South Korean government and private institutions. However, subsequent verification efforts revealed that his findings were entirely fabricated, constituting one of the most egregious cases of scientific fraud in modern history. As a result, South Korea's scientific community suffered a severe loss of international credibility. The confirmed financial damage from government grants alone—including funding from the Ministry of Science and Technology, the Ministry of Information and Communication, and the Ministry of Education—amounted to 28.75 billion KRW (approximately $25 million USD). Media reports suggested that if planned research facilities at SNU were included, the total losses could exceed 100 billion KRW (around $90 million USD) [28, 29].

In 2014, Haruko Obokata of Japan's RIKEN Institute published a paper in *Nature*, claiming that stimulus-triggered acquisition of pluripotency (STAP) cells could be generated simply by immersing somatic cells in a mildly acidic solution, enabling them to differentiate into any cell type. This discovery was expected to revolutionize stem cell research and received substantial funding from the Japanese government. However, subsequent studies failed to reproduce the results, leading to allegations of data manipulation. Investigations ultimately confirmed research misconduct, and the paper was retracted. The scandal took an even more tragic turn when Yoshiki Sasai, the deputy director of RIKEN's Center for Developmental Biology, who had supervised Obokata, fell into severe distress over the controversy. On August 5, 2014, overwhelmed by guilt and shame, he took his own life, marking one of the most devastating consequences of research fraud in modern science.

In 2016, Italian thoracic surgeon Paolo Macchiarini performed artificial trachea transplants using stem cells on human patients at Sweden's Karolinska University Hospital. His work initially earned him international acclaim as a pioneer in regenerative medicine. However, subsequent investigations revealed that the implanted artificial tracheas were inadequately prepared and that he had used unethical and scientifically unproven treatment methods. The consequences were catastrophic—his experimental surgeries resulted in the deaths of at least three patients, including Hannah Warren, Yulia Tuulik, and Christopher Lyles. This case stands as a grim example of medical research misconduct, demonstrating how fraudulent scientific claims, when applied in clinical practice, can lead to fatal consequences.

As demonstrated by these cases, research misconduct does not simply end with the retraction of papers; rather, it leads to widespread societal turmoil, national financial losses, setbacks in scientific progress, and even loss of human life. These incidents underscore the reality that research fraud is not merely a breach of scientific ethics but a serious criminal act with irreversible social, economic, and national consequences—ones that extend far beyond the scientific community. Merely emphasizing personal ethical responsibility, flaws in peer review systems, or a lax research culture is insufficient to prevent deliberate scientific fraud. It is no longer viable to regard research misconduct as an individual failing; instead, it demands systematic oversight and reinforced ethical frameworks to fundamentally reform the research ecosystem. Just like other criminal offenses, scientific fraud can lead to deaths, reputational destruction, and financial ruin. Allowing scientists alone to evade accountability for such actions is not only an injustice but also a violation of the principles of fairness and equity in legal responsibility. The time has come to recognize major research misconduct as a criminal act and implement rigorous enforcement mechanisms to ensure the integrity of science.

2.4 *Lack of Legal Sanctions to Prevent Deliberate Research Misconduct: South Korea's Regulatory Framework and Its Limitations*

In South Korea, laws that serve to caution against and regulate research misconduct include the Higher Education Promotion Act (administered by the Ministry of Education), the Guidelines for Securing Research Ethics (issued by the Ministry of Education), the Industrial Technology Innovation Promotion Act (amended in 2006 under the Ministry of Trade, Industry, and Energy), the Bioethics and Safety Act (enacted in 2005 under the Ministry of Health and Welfare), and the National Research and Development Innovation Act (implemented in 2021). These legal frameworks were commonly established in response to the 2005 Hwang Woo-Suk scandal, with a primary focus on strengthening research ethics and developing mechanisms to address research misconduct.

The "Guidelines for Securing Research Ethics" is an enforcement regulation delegated under Article 15 of the Higher Education Promotion Act, issued by the Ministry of Education. Its primary objective is to establish research ethics among researchers and universities and prevent research misconduct by providing fundamental principles and directions. These guidelines apply to universities and research institutions under the jurisdiction of the Ministry of Education and mandate compliance with research ethics throughout the entire research and development process.

A particularly noteworthy aspect of these guidelines is Chap. 3, which, from Article 12 to Article 26, outlines key provisions regarding the definition of research misconduct, verification procedures, whistleblower protection, and rights of the accused, composition of investigative committees, judgment procedures, appeals, and measures against research misconduct. Additionally, Article 19, Clause 3 of the Higher Education Promotion Act stipulates that individuals found guilty of research misconduct may face suspension of research funding and exclusion from academic support programs. Furthermore, Article 20 includes provisions for the imposition and collection of penalty surcharges, reinforcing accountability in cases of research ethics violations. However, this legislation does not specify concrete punitive measures for research misconduct. Instead, it contains ambiguous provisions stating that the head of each institution is responsible for taking appropriate actions against those found guilty of misconduct. Moreover, there are no binding clauses or penalties to enforce compliance should an institution fail to implement these measures.

This lack of enforceability has been widely criticized as a key limitation, undermining the law's effectiveness in ensuring accountability and deterring research misconduct.

The Industrial Technology Innovation Promotion Act, amended in 2006, aims to promote industrial technology innovation, enhance industrial competitiveness, and strengthen national innovation capacity. This legislation addresses ethical issues in research and development, and Article 11–2, Clause 1, Subclause 6 explicitly states that individuals found guilty of research misconduct—such as falsification, fabrication, or plagiarism of research data and results—may be restricted from participating in national R&D projects. However, this law also lacks direct punitive provisions for research misconduct, meaning that no specific legal sanctions are imposed on violators. In contrast, Article 11–3 stipulates that misuse of government funding for purposes unrelated to research and development will result in penalty surcharges and recovery of funds. While financial mismanagement is subject to legal consequences, no similarly stringent penalties exist for research misconduct itself, highlighting a critical gap in the legal framework.

The Bioethics and Safety Act, enacted in 2005, was originally designed to grant Hwang Woo-Suk exclusive rights to conduct human-cloned embryonic stem cell research. However, in an ironic turn, following the Hwang Woo-Suk scandal, the law became a key regulatory framework for research ethics in South Korea. This legislation is particularly significant as it is the only law that explicitly regulates embryonic research procedures and includes punitive measures for violations. Articles 65 to 68 outline penalties, Article 69 establishes dual liability provisions, and Article 70 includes clauses on administrative fines. However, despite its regulatory role, this law does not contain provisions that directly penalize research misconduct. It fails to criminally prosecute or impose fines on researchers engaged in fraudulent research practices. Consequently, while the law serves as an important ethical safeguard for biomedical research, it remains insufficient in preventing and punishing broader research misconduct.

The National Research and Development Innovation Act, enacted in 2021, is the first law in South Korea to include concrete punitive measures for research misconduct. This legislation aims to reform the national R&D management system, reinforce researchers' autonomy and accountability, and establish a strong ethical foundation for research. Article 31, Clause 1 explicitly defines types of research misconduct, including falsification, fabrication, plagiarism, and improper authorship attribution related to research materials and outcomes. Article 32 (Sanctions for Misconduct, Clause 1, Subclause 3) stipulates that researchers found guilty of research misconduct may face up to a 10-year ban from participating in national R&D projects and may be required to pay penalty surcharges up to five times the amount of government research funding received. Furthermore, Article 36 allows for damage claims against researchers who have engaged in intentional or grossly negligent misconduct, thereby establishing clear legal accountability for research fraud. This law holds significant importance as the first legal framework introduced in 16 years since the 2005 Hwang Woo-Suk scandal to formally penalize research misconduct. It represents a crucial step toward strengthening research ethics and enabling substantive legal sanctions against fraudulent practices in South Korea. However, despite this progress in defining and penalizing research misconduct, criticism remains regarding the lack of strong enforcement and the relatively lenient penalties, raising concerns about the government's commitment to effectively implementing the law.

3 Preventing Deliberate and Systematic Research Misconduct: Necessary Actions for the Present

In 2005, the Hwang Woo-Suk human cloned embryonic stem cell research fraud case at SNU, which involved the fabrication of a paper published in *Science*, became the first instance in South Korea where research misconduct led to a criminal investigation. This case was considered exceptional for two reasons: (1) at the time, South Korean law had no provisions for the criminal prosecution of researchers for any research misconduct and (2) until then, no instance of research fraud within a university had ever resulted in criminal prosecution. This situation reflected the distinctive perception that the public held toward researchers in South Korean society. Furthermore, within academia, there was strong resistance to the notion that research misconduct emerging in the realm of academic inquiry should be subject to legal sanctions. Upon the revelation of his research fabrication, Hwang Woo-Suk officially acknowledged his wrongdoing. Only after the exposure of such flagrant misconduct and the unveiling of his true character did voices among the public begin to demand punitive measures against him. However, since there was no legal basis at the time for directly penalizing research misconduct, he was indicted on other charges under criminal law, such as embezzlement and fraud. Lawmakers, whose duty is to enact laws, not only took photographs with Hwang Woo-Suk and pledged their support when he attracted significant media attention but also actively defended him even after the controversy erupted. Consequently, these very lawmakers, who should have intervened, were, in a broad sense, complicit in condoning or even endorsing the manipulation of research. In addition, this political reluctance to establish strict legal penalties for research misconduct stemmed from self-preservation—many within the political sphere had themselves engaged in academic fraud, including plagiarism and falsification of research records. Despite numerous recurring cases of research misconduct, legislative efforts to strengthen research ethics laws remained stagnant, revealing a deep-seated conflict of interest within South Korea's political establishment.

Twenty years later, the National Research and Development Innovation Act has finally redefined research misconduct not just as an ethical violation but as a legal offense—a significant and welcome advancement. However, the absence of stringent sentencing guidelines for research misconduct, coupled with the lack of prosecutorial resolve within the judicial authorities, constitutes a significant shortcoming. Furthermore, even if such legislation were enacted, it would prove ineffective without robust law enforcement. Indeed, in 2024, reports emerged from Korea of instances in which research reports—required to be submitted by prosecutors following overseas training—were manipulated even within judicial institutions, including the prosecution. These circumstances indicate that, under the current conditions, the prospect of rigorous law enforcement to eradicate research misconduct is bleak, suggesting that much progress remains to be made.

Research misconduct, as examined earlier, is not merely an individual mistake or an academic mishap; rather, it can be a deliberate and systematic act of deception that inflicts significant harm on society. While researchers deserve respect for their academic dedication, society must also remain vigilant against the potential for scientific fraud. This necessitates shifting away from unconditional trust in researchers and enhancing oversight mechanisms to ensure research integrity. Instead of neglecting research ethics under the assumption that researchers will always act responsibly, it is essential to establish a systematic verification system that guarantees the credibility of scientific research. Furthermore, clear legal provisions for the criminal prosecution of research fraud must be implemented to hold violators accountable under the law. To prevent further errors in handling research misconduct cases, legal frameworks must be revised to treat research fraud with the same severity as other criminal offenses.

Such legal measures will be particularly crucial in cases where research misconduct extends beyond academic fraud to organized crimes at national and international levels. A historical parallel can be drawn from Unit 731, the biological warfare research unit established by the Imperial Japanese Army during World War II in Manchuria. Despite conducting human experiments that led to the deaths of thousands, or even tens of thousands, the unit's commanders and leading scientists escaped prosecution by negotiating with the US government. Many of them later assumed prominent roles in Japan's scientific community. This historical case is not fundamentally different from contemporary failures in addressing research misconduct—where fraudulent researchers are shielded from accountability due to their scientific influence or societal status. Failing to prosecute research misconduct and granting immunity to researchers based on their social standing will ultimately erode trust in scientific research. To break this cycle of impunity, strong legal and institutional reforms must be established to uphold justice in scientific research. Research fraud must no longer be dismissed as an internal academic issue but rather recognized as a criminal act that undermines public trust and causes national harm. Accordingly, legal sanctions must be enforced with the same rigor applied to other crimes of comparable impact. While researchers should rightfully be respected members of society, it is also undeniable that some may engage in research misconduct. Protecting the social status of researchers and strictly penalizing research misconduct are not mutually exclusive; rather, they are complementary efforts in safeguarding the fundamental integrity of scientific inquiry. To preserve the core values of science, it is imperative to eliminate undue privilege, establish a transparent research environment, and implement a stringent punishment system for research misconduct. Only through the enforcement of strict penalties can research misconduct be prevented and academic trust restored. The continued occurrence of scientific fraud serves as a resounding call for urgent legislative reform and active enforcement—a call that can no longer be ignored.

Conflict of Interest The author has no competing interests to declare that are relevant to the content of this chapter.

References

1. Segal NL. Twins living apart: behavioral insights/twin study reviews: managing monochorionic-diamniotic twin pregnancies; paternity testing in multiple pregnancies; twin research on resilience; trisomies in twin pregnancies/human interest: reunited Brazilian twins; website for twins with disabled co-twins; twins separated in secret of the Nile series; Mengele: unmasking the angel of death; twins helping others. Twin Res Hum Genet. 2020;23:300–5. https://www.cambridge.org/core/journals/twin-research-and-human-genetics/article/twins-living-apart-behavioral-insightstwin-study-reviews-managing-monochorionicdiamniotic-twin-pregnancies-paternity-testing-in-multiple-pregnancies-twin-research-on-resilience-trisomies-in-twin-pregnancieshuman-interest-reunited-brazilian-twins-website-for-twins-with-disabled-cotwins-twins-separated-in-secret-of-the-nile-series-mengele-unmasking-the-angel-of-death-twins-helping-others/87914896D019F172175044E4F674307B.
2. Johnson KJ. A scientific method to the madness of unit 731's human experimentation and biological warfare program. Hist Med Allied Sci. 2022;77:24–47. https://pubmed.ncbi.nlm.nih.gov/34897467/.
3. [RETRACTED], Wakefield AJ, Murch SH, Anthony A, Linnell J, Casson DM, Malik M, Berelowitz M, Dhillon AP, Thomson MA, Harvey P, Valentine A, Davies SE, Walker-Smith JA. Ileal-lymphoid-nodular hyperplasia, non-specific colitis, and pervasive developmental disorder in children. Lancet. 1998;351:637–41. https://pubmed.ncbi.nlm.nih.gov/9500320/.
4. [RETRACTED] Schön JH, Meng H, Bao Z. Self-assembled monolayer organic field-effect transistors. Nature. 2001;413(6857):713–6. doi: https://doi.org/10.1038/35099520. Retraction in: Nature. 2003 Mar 6;422(6927):92. doi: 10.1038/nature01464. Erratum in: Nature 2001 Nov 22;414(6862):470. https://www.nature.com/articles/nature01464.
5. [RETRACTED], Hwang WS, et al. Patient-specific embryonic stem cells derived from human SCNT blastocysts. Science. 2005;308:1777–83. https://www.science.org/doi/10.1126/science.1112286.
6. [RETRACTED], Obokata H, et al. Stimulus-triggered fate conversion of somatic cells into pluripotency. Nature. 2014;505:641–7. https://www.nature.com/articles/nature12968.
7. Silva JA. Ethical perspectives and ramifications of the Paolo Macchiarini case. Indian J Med Ethics. 2017;2:270–5. https://pubmed.ncbi.nlm.nih.gov/28343147/.
8. World Medical Association. WMA declaration of Helsinki—ethical principles for medical research involving human participants. First adopted in 1964, last amended in 2024. https://www.wma.net/policies-post/wma-declaration-of-helsinki/.
9. Robert K. Merton. The sociology of science. Theoretical and empirical investigations. The University of Chicago Press Chicago and London. 1973.
10. Nuremberg Trials. The nuremberg code. Law Med Health Care. 1991;19(3–4):266.
11. Ashcroft RE. The declaration of Helsinki. The Oxford textbook of clinical research ethics. Oxford University Press; 2008. p. 141–8. https://academic.oup.com/book/54519/chapter-abstract/422591213?redirectedFrom=fulltext.
12. Sims J. A brief review of the Belmont report. Dimens Crit Care Nurs. 2010;29:173–4. https://pubmed.ncbi.nlm.nih.gov/20543620/.

13. International Committee of Medical Journal Editors. Recommendations for the conduct, Reporting, Editing, and Publication of Scholarly Work in Medical Journals Updated January 2025. https://www.icmje.org/icmje-recommendations.pdf.
14. Wager E. The Committee on Publication Ethics (COPE): objectives and achievements 1997–2012. Presse Med. 41:861–6. https://www.sciencedirect.com/science/article/abs/pii/S0755498212002874.
15. Kretser A, Murphy D, Dwyer J. Scientific integrity resource guide: efforts by federal agencies, foundations, nonprofit organizations, professional societies, and academia in the United States. Taylor & Francis Press; 2017. https://pubmed.ncbi.nlm.nih.gov/27748637/.
16. OECD Policy Paper. Research ethics and new forms of data for social and economic research 2016. https://www.oecd.org/en/publications/research-ethics-and-new-forms-of-data-for-social-and-economic-research_5jln7vnpxs32-en.html.
17. UNESCO. Recommendations on Science and Scientific Researchers. 2017. https://www.unesco.org/en/recommendation-science.
18. Korea Research Foundation Guidelines for Securing Research Ethics. 2014. https://www.nrf.re.kr/biz/notice/view?menu_no=44&nts_no=232580&biz_no=248.
19. National Research Council of Science & Technology. NST Guideline 2022. https://www.nst.re.kr/www/selectBbsNttList.do?bbsNo=7&key=41&searchCtgry=%EC%82%AC%EC%97%85%EA%B4%80%EB%A6%AC.
20. The Korean Federation of Science and Technology Societies. KOFST Code of Ethics for Scientists and Engineers 2007. https://www.kofst.or.kr/main.bit?sys_type=0000.
21. "Seoul National University Identified 28 Cases of Research Misconduct Over Three Years, Yet Issued No Severe Disciplinary Actions" Hankyoreh, October 3, 2022. https://www.hani.co.kr/arti/society/society_general/1061081.html?utm_source=chatgpt.com.
22. Ryu YJ. Between truth and profit: scientific misconduct case of human cloned embryonic stem cell and revisiting cases during COVID-19 pandemic. In: Kurihara C, Greco D, Dhai A, editors. Ethical innovation for global health: pandemic, democracy and ethics in research. Springer; 2023. p. 69–104. https://link.springer.com/chapter/10.1007/978-981-99-6163-4_5.
23. Fanelli D. How many scientists fabricate and falsify research? A systematic review and meta-analysis of survey data. PLoS One. 2009;4(5):e5738. https://doi.org/10.1371/journal.pone.0005738.
24. Martinson BC, Anderson MS, de Vries R. Scientists behaving badly. Nature. 2005;435(7043):737–8. https://doi.org/10.1038/435737a.
25. Stroebe W, Postmes T, Spears R. Scientific misconduct and the myth of self-correction in science. Perspect Psychol Sci. 2012;7:670–88. https://pubmed.ncbi.nlm.nih.gov/26168129/.
26. Steen RG. Retractions in the scientific literature: do authors deliberately commit research fraud? J Med Ethics. 2011;37(2):113–7. https://doi.org/10.1136/jme.2010.038125. Epub 2010 Nov 15.
27. Fang FC, Steen RG, Casadevall A. Misconduct accounts for the majority of retracted scientific publications. Proc Natl Acad Sci U S A. 2012;109(42):17028–33. https://doi.org/10.1073/pnas.1212247109. Epub 2012 Oct 1. Erratum in: Proc Natl Acad Sci U S A. 2013 Jan 15;110(3):1137.
28. Hwang Woo-seok criminal procedure "false purchase of pigs, duplicate receipt of research funds, and use 63 borrowed-name accounts." Law Newspaper Article 12 May, 2006. Korean. https://www.lawtimes.co.kr/news/20292?utm_source=chatgpt.com.
29. "King of Clone" Netflix documentary. 23 June 2023 on air.

Chapter 8
From Professional Freedom to Professional Autonomy: Ethical Principles for the Use of Unproven Interventions

Takeo Saio and Chieko Kurihara

Abstract Paragraph 37 of the Declaration of Helsinki (DoH) issued by the World Medical Association (WMA) covers the clinical usage of unproven interventions. The 2024 amendment of the DoH added some provisions for ineligible patients to reconcile with ongoing clinical trials on the same intervention and warning against physicians' continuous usage of unproven interventions to elude the burden to follow ethical principles.

While the newly added provisions are right enough, our proposal to request continuous monitoring of safety and efficacy with accumulated data in clinical applications was not embraced by the WMA. This proposal was based on the policy of the World Health Organization (WHO), referred to as "monitored emergency use of unregistered and experimental intervention (MEURI)," which reflects the experiences of the Ebola outbreak and the COVID-19 pandemic.

This chapter lays out this stringent monitoring policy as one of the core elements of physicians' professional autonomy based on virtue ethics. Furthermore, in concert with the professional autonomy of physicians, it would contribute to establishing healthcare democratization.

Keywords Unproven intervention · Professional autonomy · Virtue ethics · Monitored emergency use of unregistered and experimental intervention (MEURI) · Evidence-based medicine (EBM) · Patient and public involvement (PPI)

T. Saio
Department of Internal Medicine and Psychiatry, Fuji Toranomon Orthopedic Hospital, Shizuoka, Japan

C. Kurihara (✉)
Kanagawa Dental University, Yokosuka, Kanagawa, Japan
e-mail: chieko.kurihara@nifty.ne.jp

C. Kurihara et al. (eds.), *The 2024 Declaration of Helsinki*,
https://doi.org/10.1007/978-981-96-9294-1_8

1 Discussions at the WMA Regional Meetings and the Revisions in Paragraph 37 of the 2024 DoH

1.1 Reflections on the Opinions and the Remaining Chasm

A series of discussions at the World Medical Association (WMA)'s regional meetings led to the 2024 amendment of the Declaration of Helsinki (DoH) [1]. These discussions were not generous in Paragraph 37, which allows physicians' conditional discretion to apply unproven interventions. Eventually, some provisions were added to require the reconcilement of ongoing clinical trials involving ineligible patients, as well as to warn against misemployment of physicians' discretion to continue the clinical application of unproven interventions to elude conducting research to evaluate the safety and efficacy of these interventions.

It is obvious that physicians must strictly adhere to these revisions, reflecting the voices of interested parties in the WMA regional meetings. In addition, the chasm between the most stringent opinions expressed at the regional meetings and the softened revision in the new DoH should be further examined.

This chapter presents the implications of the authors' proposal to bridge this chasm at the boundary between practice and research in order to pursue the best interests of the patient.

1.2 Opinions to Restrict the Physicians' Discretion

At the WMA's regional meeting in Tokyo in December 2023, Dr. Yasuhiro Fujiwara, Chief Executive of the Japan's drug review authority Pharmaceuticals and Medical Devices Agency (PMDA), stated that unproven interventions should be conducted as clinical trials adhering to the Good Clinical Practice (GCP) Ordinance ("GCP trial" hereafter) under the Pharmaceutical and Medical Devices Act of Japan. Not limited to the argument in his lecture, he also responded to a question from a patient in the audience seeking a framework for patient-initiated research. In his answer, he stated that such research should be conducted as GCP trials. More than half a year later, at the meeting in Washington, DC, in August 2024, the world-renowned bioethicist Prof. Arthur Caplan clearly stated his view that unproven interventions should be conducted as research.

Contrarily in Japan during the COVID-19 pandemic, the Prime Minister at that time promoted the off-label usage of a drug for COVID-19. The drug, favipiravir, was approved for other indications and sold by a Japanese company [2]. Subsequently, the Japanese government provided funding to observational studies to collect data on this off-label usage of the drug. This was strongly criticized by the

Japan Medical Association and a citizen's group demanding randomized clinical trials to prove the safety and efficacy of this drug for COVID-19. As a result, the efficacy of this drug was not demonstrated. These studies would be classified as "research waste" which was newly clarified as something to be avoided (paragraph 21) in the 2024 DoH.

1.3 Our Opinion to Bridge over the Chasm: Safety and Efficacy Monitoring in Clinical Practice

The two main amendments in the adopted paragraph 37 of the DoH and the authors' proposal in response to the public consultation can be summarized as follows and are presented in Table 8.1.

- The already existing provision in paragraph 37 "*This intervention should subsequently be made an object of research designed to evaluate safety and efficacy*" was reinforced by the new provision ***"These interventions must never be undertaken to circumvent the protections for research participants set forth in this Declaration."***
- The new provisions for ineligible patients to reconcile with ongoing clinical trials on the same intervention were added as conditions: ***"Enrolment in a clinical trial is not possible"*** **and** ***"avoid compromising clinical trials."***
- Authors' proposal not adopted by the WMA: the need for **monitoring** and **accumulation of data** on **safety and efficacy** of unproven interventions.

This proposal for paragraph 37 of the DoH was first described in a chapter written by a patient and public group [3] which Kurihara co-authored, clarifying what lacks in the DoH 2013 version. They made a comparison between the DoH 2013 version and the policy proposed by the WHO referred to as "monitored emergency use of unregistered and experimental intervention (MEURI) [4]," which reflects the experiences of the Ebola outbreak and the COVID-19 pandemic. The patient and public representatives submitted their opinion to the WMA in response to the WMA's public consultation. In accordance with these opinions, one of the authors of this chapter, Saio, included this proposal in his comments to the WMA, in consensus with the coauthors of the comment, Prof. Dirceu Greco, an expert clinician in infectious diseases and bioethics who is also one of the editors of this book, and the Brazilian Society of Bioethics. Furthermore, this proposal was also incorporated in the comments submitted to the WMA by members of the International Federation of Associations of Pharmaceutical Physicians and Pharmaceutical Medicine (IFAPP), composed of physicians and nonphysician experts of pharmaceutical medicine (IFAPP's opinion was led by Kurihara).

Table 8.1 Changes of the principles for the clinical use of unproven intervention in the DoH since 1964 to 2013; comparison between the principles between 2013 and 2024, and author's proposal based on WHO-MEURI

Years of revisions	Changes of the principles for the clinical use of unproven intervention
1964	・*In the treatment of the sick person, **the doctor must be free to use a new therapeutic measure**, if in his judgment it offers **hope of saving life, reestablishing health, or alleviating suffering**.* ・Freely given ***consent*** from a patient or a representative.
1975	・"therapeutic"→"***diagnostic and therapeutic***" ・"his"→"***his or her***" (Second sentence on consent in the above in the 1964 DoH was deleted. The first text was included in the set of principles for "clinical research".)
1983	"doctor" → "***physician***"
1989; 96	No change.
2000	・"the sick person"→"a patient" ・The principle since the 1964 version was replaced to: ・"a new"→"unproven or new"; add "prophylactic" to "diagnostic.." Newly added conditions and principles: ・"***where proven prophylactic, diagnostic and therapeutic methods do not exist or have been ineffective***" ・"Where possible, ***these measures should be made the object of research, designed to evaluate their safety and efficacy***." ・"***In all cases, new information should be recorded and, where appropriate, published.***" (The added principle that the other principles in the DoH should be followed was deleted in 2008 and out of the scope of this analysis.)
2008	・"*proven prophylactic, diagnostic and therapeutic methods*" → "*interventions*" ・Add "***after seeking expert advice***" ・"must be free to use" → "***may use***" ・"where appropriate, published" → "***where appropriate, made publicly available***"
2013	・"patient" → "an individual patient" ・"Where possible" was deleted in the text to recommend that the unproven intervention should be made object of research.

Conditions in 2013 DoH	Conditions in 2024 DoH
"*In the treatment of an individual patient, where proven interventions do not exist or other known interventions have been ineffective*";	"*When an unproven intervention is utilized in an attempt to restore health or alleviate suffering for an individual patient because approved options are inadequate or ineffective*"

(continued)

Table 8.1 (continued)

N/A	"***enrollment in a clinical trial is not possible***"
"*This intervention should subsequently be made the object of research, designed to evaluate its safety and efficacy*";	"*it should subsequently be made the object of research designed to evaluate safety and efficacy*"
After seeking expert advice	"*first seek expert advice*"
"*In the physician's judgment it offers hope of saving life, re-establishing health or alleviating suffering*";	"***weigh possible risks, burdens, and benefits***"
"*With informed consent from the patient or a legally authorized representative*"	"*obtain informed consent*"
"*In all cases, new information must be recorded and, where appropriate, made publicly available.*"	"*They must also record and **share data** when appropriate*"
N/A	"***avoid compromising clinical trials***"
N/A	"***These interventions must never be undertaken to circumvent the protections for research participants set forth in this Declaration.***"
a proposal of additional conditions by authors, based on the WHO-MEURI	
Data accumulation and safety and efficacy monitoring.	

(Texts in italic are quoted from the DoH. Bold and underlines are made by authors.)

2 Ethos of Physicians to Monitor Safety and Efficacy of Individual Patients

2.1 Professional Autonomy Versus Company-Oriented Clinical Trials

This proposal was based on the premise that such behavior is an essential factor of professional autonomy. In clinical practices, it is a prerequisite to monitor the efficacy and safety of individual patients while assuring the patient's right to access unproven intervention.

Indeed, it is important that the utilization of unproven interventions be limited to cases where the patient cannot participate in clinical trials of the intervention. As suggested in the revised DoH, it is unethical to use unproven interventions for eluding research regulations and it might compromise ongoing clinical trials. Moreover, the added sentence "*These interventions must never be undertaken to circumvent the protections for research participants set forth in this Declaration*" gives a warning to physicians lacking professional autonomy who intend to bypass the tasks required by research regulations and seek their own financial benefit under the pretext that the unproven interventions are performed in the best interests of the patients.

The new provisions related to clinical trials are a reminder of the reality that clinical trials conducted by pharmaceutical companies set narrow eligibility criteria aiming to prove successfully the efficacy of the investigational drugs; thus, ineligible patients may seek access to those interventions outside the clinical trials. However, the new principle in the DoH related to clinical trials is much too procedural, resembling regulatory guidance, and is not appropriate for a high-level code of conduct of physicians. It might also reflect the subordinance of physicians to pharmaceutical companies in the usage of unproven interventions. Such an assumption that clinical trials initiated by companies are more credible than treatment by physicians upholding professional autonomy for the best interests of patients would downgrade the physicians' professionalism. If data are accumulated by even individual physicians treating individual patients with continuing rigorous monitoring of safety and efficacy, then such actions may be prioritized to clinical trials by pharmaceutical companies.

Pharmaceutical companies often worry that information of adverse events in clinical application, e.g., in expanded access program or individual physicians' usage, may have a detrimental influence on the regulatory approval of this drug. This must rather be recognized as the duty of physicians to share the information of safety and efficacy of unproven interventions that can contribute to regulatory decision-making and the society at large, through rigorously peer-reviewed publications.

2.2 Evidence-Based Autonomy

The authors' proposal sincerely calls for the professional autonomy of physicians both as individuals and as members of professional societies independent from governments and pharmaceutical companies. They should unyieldingly monitor the efficacy and safety of the treatments that they provide for the best interest of patients. Professional autonomy was first proposed by the WMA in the Madrid Declaration on Professional Autonomy and Self-Regulation in 1987 [5], following the Declaration on Professional Freedom in1986 rescinded in 2005 [6], and then replaced separately by the Declaration of Seoul on Professional Autonomy and Clinical Independence in 2008 [7] and the Declaration of Madrid on Professionally led Regulation in 2009 [8]. These Declarations require physicians to perform evidence-based practices in the best interests of the individual patient, which is the fundamental premise set forth by the Geneva Declaration in 1948 [9] and the International Code of Medical Ethics in 1949 [10], both referred in the DoH, since its first version.

Amplifying the theory of the French physiologist Claude Bernard, which was proposed in his masterpiece entitled "An Introduction to Experimental Medicine" (*Introduction à l'étude de la médecine expérimentale*) [11], physicians are continuously conducting experiments of new methods while intending to provide the best possible treatment for the individual patient. This can be interpreted in the context of evidence-based medicine (EBM) that all physicians are inevitably accumulating the efficacy and safety data within their own medical decision-making mechanisms in each of their daily practices. Alternatively, medical professional bodies may collect and build databases composed of data derived from unapproved interventions performed as a practice of the individual physicians.

Some jurisdictions require the application of unproven interventions to get approvals from regulatory authorities or the ethics committee. Even if those kinds of approvals were not required by laws, each institutional policy may require the approval of a clinical ethics committee. In any case, it is recommended to accumulate and monitor the data on the efficacy and safety from the physician's experience. Our argument to add the requirement of safety and efficacy monitoring to the DoH could be replaced with the insufficient requirements in the 2013 DoH: "*new information must be recorded and, where appropriate, made publicly available*" ("made publicly available" was revised in 2024 to "share data").

In addition, such accumulation of the safety and efficacy data in clinical practice is also required for the planning of research as the next step, according to the principle of the DoH.

2.3 Autonomy and Self-Regulation Led by Virtue Ethics

The above-mentioned proposal also adds the essence of virtue ethics to the WMA's framework of deontology: the WMA has issued a series of deontological declarations and statements as principles of self-regulation of medical societies. Such framework is based on the virtue ethics that seeks the morality which guides human beings to righteous behaviors; the WMA's system has the premise of the existence of virtuous physicians based on the supposition of the inherent goodness of physicians.

In other words, the professional autonomy of the WMA, which presupposes physicians as virtue bearers, inevitably entails the assumption that any physician who performs unproven interventions shall continue to accumulate and monitor data constantly and tirelessly to confirm their efficacy and safety to an individual patient. Adding this assumption to paragraph 37 of the DoH envisions the WMA's professional autonomy to the patients and the public.

3 Next Challenge: Patient and Community Engagement in Professional Autonomy

3.1 Historical Development of "Autonomy" in the DoH

As shown in Table 8.1, the principle of the physicians' discretion was already included in the first version of the DoH in 1964 described as "*the doctor* ***<u>must be free to use</u>*** *a new therapeutic measure*." Subsequent amendments of the DoH in

1975, 1983, 1989, and 1996 did not change or add any principles, while the 2000 version added two new important requirements: "*these measures should be made the object of research, designed to evaluate their safety and efficacy*" and "*new information should be recorded and, where appropriate, published*." The latter was changed in 2008 to "*made publicly available*." This additional requirement for "clinical practice" aligned with the requirements for "research" added in 2000 of positive/negative result publication and in 2008 the registration to publicly available databases. However, it was softened in 2024 to just stating "share data." Along with these changes, the phrase "*must be free to use*" indicates that professional freedom had been maintained until 2000 but was toned down in 2008 to "***may use***."

Behind this were the debates that led to the WMA Declarations on professional autonomy mentioned above. This is partly due to criticisms against the professional autonomy of medical societies from the general public.

3.2 From Criticism to Co-creation

Needless to say, there have been unfriendly relationships between physicians and society attributable to unremitting injuries of patients with medical errors or unproven interventions. But the 2024 DoH enables such concerns to be resolved through the envisioned collaboration between physicians and patients which can be called "co-creation" [12]. The revised DoH includes a new principle to engage participants and their community in all stages of the research as well as an amendment of the term "research subject" to "research participant." Although paragraph 37 of the DoH does not reflect this principle of co-creation yet, the weighty influence of patient public involvement (PPI) on medical professionalism was apparent in the process of the revision of the DoH.

The opinion from a patient public group expressed in paragraph 37 is one of the examples [13]. Their interpretation is that the professional autonomy of physicians might be materialized by the monitoring and accumulation of the safety and efficacy data of unproven interventions and that such treatments are also the effusion of altruism of conscientious patients. Patients duly share decisions with trustable physicians on their acceptance of unproven interventions. The altruism of the patients would contribute to the advancement of treatments, even if their therapeutic experience will not result in their own cure. The expectations of the physician trying what Bernhardt called an experiment in practice and the hopes of the patients when they are making treatment choices as "participants" rather than mere objects must be perfectly aligned.

The term "research participant" and community engagement principle in the 2024 DoH must be also extended to the practice of care, including the application of unproven interventions. In other words, the challenge for the future is to shape a "co-creation" in healthcare that embraces the dynamic creative activities motivated from the professional autonomy of all scientists and experts including clinicians, together with patients and the public sometimes taking the role of research participants. This would lead to the democratization of healthcare, not limited to research. It will be one of the key issues that must be considered in future revisions of the WMA's core documents, not limited to the DoH but also other Declarations for medical ethics.

4 Conclusion

The amendments to paragraph 37 in the 2024 DoH reflect the strong opinions expressed at the regional meetings that unproven interventions should be conducted as research, with additional provisions for consideration of ongoing clinical trials and a warning that conducting unproven interventions should not be performed as a means of evading research regulations. Two remaining issues must be examined for future revisions of the DoH as well as in the clinical practice of physicians with professional autonomy.

1. It is imperative that the requirement for individual physicians to accumulate data for monitoring the safety and efficacy of unproven interventions with individual patient. Virtue ethics shall be bestowed upon the WMA's professional autonomy.
2. In future revisions of the DoH, the principle on clinical usage of unproven interventions should be consistent with (1) the alteration of the term "research subject" to "research participant"; (2) calling for community engagement at all stages of research; and (3) "co-creation" with the patient and the public, which leads to democratization of healthcare, not limited to the use of unproven intervention.

Acknowledgments CK appreciates the International Federation of Associations of Pharmaceutical Physicians and Pharmaceutical Medicine (IFAPP) for granting her travel expenditure for her in-person participation in the WMA Regional Meeting in Tel Aviv (December 2022) and the WMA for granting her travel expenditure for her in-person participation and presentation as a panelist in WMA Regional Meeting in Washington, DC (August 2024), and waiver of registration fee from WMA as a journalist to attend the General Assembly in Helsinki (2024 October). She participated in all the other WMA Regional Meetings at her own expense or online.

Conflict of Interest The authors declare no competing interests relevant to the content of this chapter.

References

1. The World Medical Association. WMA Declaration of Helsinki—ethical principles for medical research involving human participants. First adopted in 1964, last amended in 2024. https://www.wma.net/policies-post/wma-declaration-of-helsinki/
2. Kurihara C, Saio T. Therapeutic misconception as the basis for vaccine nationalism of Japan: a historical reflection and perspectives for global public health. In: Kurihara C, Greco D, Dhai

A, editors. Ethical innovation for global health: pandemic, democracy and ethics in research. Singapore: Springer; 2023. p. 105–30.
3. Kurihara C, Inoue K, Kai H, Suzuki K, Saeki H, Funabashi Y, Kishi N, Kuge A, Murakami T, Saito Y, Uchida E, Tsutsumi N, Imamura K. Our "WMA Declaration of Helsinki": opinions and proposals from patient and public for research ethics. In: Kurihara C, Greco D, Dhai A, editors. Ethical innovation for global health: pandemic, democracy and ethics in research. Springer; 2023. p. 243–69.
4. World Health Organization. Emergency use of unproven clinical interventions outside clinical trials: ethical considerations. 2022 Apr 12. https://www.who.int/publications/i/item/9789240041745
5. World Medical Association. Archived: WMA Declaration of Madrid on Professional Autonomy and Self-Regulation. Adopted October 1987, rescinded October 2009. https://www.wma.net/policies-post/wma-declaration-of-madrid-on-professional-autonomy-and-self-regulation/
6. World Medical Association. Archived: WMA Declaration on Physician Independence and Professional Freedom. Adopted October 1986 and rescinded October 2005. https://www.wma.net/policies-post/wma-declaration-on-physician-independence-and-professional-freedom/
7. World Medical Association. WMA Declaration of Seoul on Professional Autonomy and Clinical Independence. Adopted October 2008 and amended October 2018. https://www.wma.net/policies-post/wma-declaration-of-seoul-on-professional-autonomy-and-clinical-independence/
8. World Medical Association. WMA Declaration of Madrid on Professionally-Led Regulation. Adopted October 2009 and revised October 2019. https://www.wma.net/policies-post/wma-declaration-of-madrid-on-professionally-led-regulation/
9. The World Medical Association. The Declaration of Geneva. 1948. Last amended 2017. https://www.wma.net/policies-post/wma-declaration-of-geneva/
10. The World Medical Association. The International Code of Medical Ethics. 1949. Last revised in 2022. https://www.wma.net/policies-post/wma-international-code-of-medical-ethics/
11. Bernard C. An introduction to the study of experimental medicine, transl. by Greene HC. New York: Schuman; 1949. [Introduction à l'étude de la méde- cine expérimentale. 1865]
12. Kurihara C, Matsuyama K, Baroutsou V. World Medical Association's Declaration of Helsinki, 2024 revision: celebrating the 60th anniversary, at Helsinki. Clin Eval. 2025;52(3):W1–27. http://cont.o.oo7.jp/52pop/52pop_contents_e.html
13. Kurihara C, Saito Y, Kai H, Funabashi Y, Inoue K, Kishi N, Kuge A, Murakami T, Suzuki K, Takahashi H, Uchida E, Imamura K. Patient Public Declaration of Research Ethics (1st edition): research ethics of the people, by the people, for the people-expanding the impact of the 2024 revision of the Declaration of Helsinki. Clin Eval. 2024;52(3):W28–39. http://cont.o.oo7.jp/52pop/52pop_contents_e.html

Part II
Forward-Looking Perspectives for International Ethical Principles

Chapter 9
Bridging Between the CIOMS and the WMA: Common Ground, Similarities, Differences, and Next Steps

Dominique Sprumont

Abstract This chapter describes important points of the common ground, similarities, and differences between the two authoritative documents of research ethics principles: the 2024 Declaration of Helsinki of the World Medical Association and the 2016 International Ethical Guidelines for Health-related Research involving Humans of the Council for International Organizations of Medical Sciences, considering the latest revision of the Declaration of Helsinki, adopted in October 2024.

The author has been involved in the previous revisions of both documents since the 1990s. While there have been plentiful discussions about them, some points deserve in-depth consideration. This chapter highlights new items and controversial points in the 2024 revision of the Declaration of Helsinki, namely, good governance practice of research teams or institutions, community engagement and vulnerability, research waste and social value, and placebo-controlled trials.

Keywords Institutional governance · Community engagement · Vulnerability · Research waste · Social value · Placebo-controlled trial

1 Introduction: Bridging the WMA and the CIOMS

1.1 The DoH 2024 Revision and the 2016 CIOMS Guidelines

The Declaration of Helsinki (DoH) [1] of the World Medical Association (WMA) is the best-known guiding document at the international level on ethical principles applied to medical research involving human participants. There is also another well-recognized document of almost the same scope, issued by the Council for International Organizations of Medical Sciences (CIOMS), namely the International Ethical Guidelines for Health-Related Research Involving Humans [2] (CIOMS

D. Sprumont (✉)
Institute of Health Law, University of Neuchâtel, Neuchâtel, Switzerland
e-mail: Dominique.Sprumont@unine.ch

C. Kurihara et al. (eds.), *The 2024 Declaration of Helsinki*,
https://doi.org/10.1007/978-981-96-9294-1_9

Guidelines). The DoH was first issued in 1964 and the latest revision was adopted in October 2024. Meanwhile, the CIOMS Guidelines were first issued in 1982 and last updated in 2016. The CIOMS, in collaboration with the World Health Organization (WHO), prepared its guidelines to indicate how the ethical principles of the DoH could be applied, particularly in low-resource settings (see 1982 CIOMS Proposed International Guidelines for Biomedical Research Involving Human Subjects).

This chapter focuses on how the differences from the CIOMS Guidelines were discussed during the 2024 revision process of the DoH while also referring to the common ground and similarities of both documents. There have been several controversies, some of which were shared with interested external parties, while others were discussed only inside of the Working Group of WMA, which included representatives of National Medical Associations, most of them being physicians. However, designated nonphysician advisory experts were engaged in the process without voting privilege. Meanwhile, the CIOMS Guidelines Working Group was composed of multidisciplinary members, including a patient representative. This may be reflected in some of the differences between the two documents.

Those differences also reflect the diversity of opinions, cultural backgrounds, and philosophies of those involved in research or research review in several regions of the world. Respecting, considering, and discussing differences and common grounds would contribute to building consensus among the various parties and to the future improvement of both documents. From the perspective of a "bridge maker" between these organizations, this chapter describes the common ground, similarities, and differences between the 2024 revision of the DoH and the 2016 CIOMS Guidelines and provides a brief vision for the next steps.

1.2 As a Bridge Maker

I have tried to play the role of "bridge maker" in both organizations as a representative of the WMA in the CIOMS Executive Committee. I have been involved in the revision of the CIOMS Guidelines since its 1993 version. I have also participated in revisions of the DoH since 2000, more actively in the versions adopted in 2008, 2013, and 2024, As a legal expert I have been designated academic partner of WMA. Additionally, I was involved in the drafting of the 2016 version of the WMA's Declaration of Taipei (DoT) on health databases and biobanks [3], complementing the DoH. If the DoT has been formally included in the DoH in the latest revision, the secondary use of data and biological material for research is already covered in the 2016 CIOMS Guidelines.

I received my Doctorate in Law (Dr iur) from the University of Fribourg (Switzerland) in 1993 on the protection of research participants in biomedical research. In my research, I have had the privilege of being invited by Prof. Jay Katz and Prof. Robert Levine at Yale Law School and Yale Medical School and benefited from their support in drafting my doctorate. Both professors are immensely renowned physician-bioethicists. Prof. Levine was one of the main contributors to the internationally well-known US government policy to present the theory of boundary between research (to generalizable knowledge) and routine practice (to provide patient care) and the principles of research ethics (respect for autonomy, beneficence, non-maleficence, and justice) introduced in the Belmont Report issued in 1979 [4]. He had continuously kept dominant power in developing research ethics policy documents including CIOMS Guidelines until the 2002 version, as well as in the 2000 revision of the DoH. Prof. Jay Katz has also been instrumental in the development of research ethics and is particularly recognized for his contribution to the rule of informed consent both in medical practice and in research.

In 1993, after returning to Switzerland, I started my career as a legal advisor to the Swiss Drug Agency and drafted the 1993 regulation on clinical trials with medicinal products. Back then, I was contributing to the drafting of many pieces of legislation on research involving human participants, patients' rights, and public health in Switzerland, Europe, and internationally. My latest full commitment to the CIOMS achievement was to act as the Chairperson of the Working Group on Good Governance Practice for Research Institutions (GGPRI), published in November 2023 [5].

In the revision process of the DoH starting in April of 2022, the WMA held eight regional meetings collaborating with the AMA, which were led by the Working Group as well as the National Medical Association of each host country, in Tel Aviv (December 2022), Sao Paulo (February 2022), Copenhagen (September 2023), Tokyo (December 2023), Vatican City State (January 2024), Johannesburg (February 2024), Munich (May 2024), Washington, DC (August 2024), with each focusing on a specific topic. It was a great honor for me to be able to actively participate in five of these regional meetings (Sao Paulo, Copenhagen, Vatican City State, Johannesburg, and Munich) as a speaker in most and remotely in the remaining meetings. Lastly, I attended the General Assembly in Helsinki, in October 2024, where the new DoH was adopted.

2 From Physicians' Code to Ethical Principles for Organizational Governance

2.1 CIOMS 2023 Good Governance Practice for Research Institutions (GGPRI) and Paragraph 2 of the DoH

The GGPRI reflects the historical transformation of the clinical research landscape from a private, individual activity induced by physicians' interests in patient care to large projects with enormous financial investment requiring multidisciplinary teams, collaborating internationally and seeking clear governance practices. The GGPRI was published as a CIOMS report and influenced the 2024 revision of the DoH. The DoH was created by physicians and intended for physicians, thereby reflecting the original nature of the WMA as a physician's association. Nevertheless, research ethics is not solely intended for the medical profession.

On this premise, the 2024 revision included a significant modification in its paragraph 2 which states that "*While the Declaration is adopted by physicians, the WMA holds that these principles should be upheld by* **all individuals, teams, and organisations** *involved in medical research, as these principles are fundamental to respect of and protection of all research participants, including both patients and healthy volunteers*" (italic, bold by author, hereafter).

Such an amendment does not imply the compulsory application of the WMA views to professionals other than medical practitioners. Instead, it underlines that the principles embedded in the DoH reflect fundamental ethical principles of necessary application regardless of who applies them. The CIOMS GGPRI document shares a similar rationale as it applies to any research institution conducting health-related research involving human participants. The focus here is set on sharing common values and principles rather than distinguishing between professions and types of institutions.

2.2 History of the Birth of the Physician's Authoritative Code

This amendment is also important from a historical standpoint, especially considering that the DoH celebrated its 60-year anniversary in 2024. The WMA was founded in 1947, following a heavy deliberation on the atrocities committed by Nazi German physicians abusing patients and fellow human beings, including merciless human experimentations during World War 2. Their acts were judged as war crimes at the Nuremberg Doctors trial. A part of the judgment, known today as the "Nuremberg Code," was proclaimed in 1947 and recognized as the basic moral, ethical, and legal conditions of permissible human experimentation. The WMA continuously discussed such ethically questionable medical practices even after the war. They issued two fundamental codes of conduct for physicians: the Declaration of Geneva in 1948 [6] and the International Code of Medical Ethics in 1949 [7]. Both are quoted in the introductory paragraphs of the DoH from its very first version and constitute a core principle on the primacy of the individual which always requires prioritizing patient well-being, rights, and dignity against the sole interest of research, which is to generate new knowledge, and society.

In such a context, prior to the first draft of the DoH, human rights lawyers attempted to develop an international treaty on human experimentation expected to be legally binding [8]. Physicians expressed reluctance to the idea of having a legally binding tool, and the initiative was rejected by the WMA. Partially in reaction to this, the WMA produced a Draft Code of Ethics on Human Experimentation in 1962. Its title was changed to Ethical Principles to Recommendations Guiding Doctors in Clinical Research during the General Assembly in Helsinki in 1964, resulting in less restrictive guidelines and suppressing rules on the prohibition of research involving institutionalized children, asylum inmates, and prisoners [8].

Thus, the birth of the physician's authoritative code of self-regulation was partially brought about by rejecting legal experts' proposals with a stronger human rights approach. As shown in the 2024 version of the DoH, the WMA clearly sets its priorities in favor of respect and protection of research participants.

3 Meaningful Community Engagement and Vulnerability

3.1 Community Engagement and Ubuntology

The new paragraph promoting community engagement follows the second sentence of paragraph 6 on the recognition of "structural inequities" in research. The third sentence recommends: "*Meaningful* **engagement with potential and enrolled participants and their communities** *should occur* **before, during and following** *medical research. Researchers should enable potential and enrolled participants and their communities to share their priorities and values; to participate in research* **design, implementation, and other relevant activities**, *and to engage in* **understanding and disseminating results.**"

Historically, the Belmont Report may have influenced the DoH's core foundation regarding "respect for autonomy" to develop the principles with an individualistic approach and strengthen informed consent. Yet we have seen a major shift from individualism to a collective approach. This concept of community engagement was intensively discussed at the regional meeting in Johannesburg, where the audience learned much about the philosophy of "Ubuntology/Umuntology"—an African communitarian worldview to define a human being in the context of others and their community. John Mbti defines a human being as "*We are therefore I am, I am therefore we are*" [9]. Such a view is in opposition with Descartes's principle of autonomy "I think, therefore I am." Ubuntology is also in line with the Western concepts of "solidarity" and "reciprocity," and with the philosophy of Confucianism in Asia, as well as with the Japanese culture of cherishing harmony among people despite their differences and particularities.

3.2 Community Engagement in CIOMS Guidelines

In the CIOMS Guideline, the concept of community engagement has been included from its first version in 1982 as a recommendation for external sponsors to collaborate with local institutions and the scientific community regarding service provision and training. This concept has been further deepened in subsequent revisions through the recommendation of capacity development within the study host community. During the 2016 revision, this was established as guideline 7 "Community engagement." Most importantly, the principle of community engagement was developed with a clear objective to avoid the exploitation of resource-limited communities by external sponsors. It aimed to impede them from conducting research without respecting communities' values or without ensuring affordable access to the results and benefits of their research.

3.3 Reformed Concept of Vulnerability

Another prominent modification of the DoH's 2024 revision is the concept of vulnerability reframing. Traditionally, vulnerable groups are categorized as needing strengthened protection, which leads to the exclusion of these groups from research. As a result, the WMA reached a consensus on considering vulnerability as "*fixed or contextual and dynamic*" and on the implication of excluding people in a vulnerable situation—it "*can potentially perpetuate or exacerbate their disparities*"—in the new DoH. Therefore, the DoH states that: "*the harms of exclusion must be considered and weighed against the harms of inclusion*." (Paragraph 19) The promotion of responsible inclusion must be associated with strengthened safeguards depending on the nature of vulnerability while ensuring benefits resulting from research for all.

This concept of reframing was already introduced in the CIOMS Guidelines in 2016. However, extensive and repeated discussions during regional meetings in Vatican City State (January), Johannesburg (February), and Munich (May) in 2024, which ultimately led to a consensus, seem to have had a great impact on the global community engaged in research ethics.

4 Research Waste and Social Value

4.1 Avoiding Research Waste

In the 2024 revision of the DoH, a new statement about avoiding "research waste" (paragraph 21) was introduced. Several lessons were drawn from COVID-19, particularly on the limited or missing social value of most research conducted during the pandemic. It is critical to evaluate the value of research, and not solely for the protection of its participants. Such a paternalistic idea that overemphasizes protection ("protectionism") is outdated. Beyond the necessity of guaranteeing the protection of human participants in medical research, research involving human beings must also respond to the common interest.

Greater attention must be paid to the specific needs and expectations of research participants and their communities when developing a research project. In other words, nothing for the patients without the patients or nothing for the participants without the participants. The priority should always be to respond to their needs and interests, to produce common goods not only in terms of scientific knowledge but also specific value for the communities involved.

4.2 Implications of Social Value

The term "social value" was clearly stated in the CIOMS guidelines as a requirement that does not supersede respect for human rights, and whose scientific validity is not sufficient per se (Guideline 1). There has been a critical discussion among the Working Group of WMA on whether to include the term "social value" in the 2024 revision. As many of the interested parties know, this term was included in the draft shared in the first official public consultation organized by WMA from January 13 to February 7, 2024. However, in the next draft for the second public consultation, from June 3 to 24, 2024, it was deleted. The original proposal was to include "social value" as the goal of research, in addition to generating knowledge. While the expression has not been introduced in the adopted version, it did strengthen the purpose of medical research as being "*ultimately to advance individual and public health*" (paragraph 7). This idea is already included in the CIOMS Guidelines and serves as an explanation for the concept of "social value."

The official rationale described in the document for the second public consultation was that the concept was considered too vague, according to the first public consultation feedback. Besides these external opinions, a few national medical associations also expressed objections within the Working Group. The term "social value" reminded them of the idea of communism or socialism. In international discussions in research ethics, social value means avoiding the exploitation of resource-limited communities while promoting research aimed at respecting community values and sharing the benefits of the research. During the 2024 DoH revision process, we learned that experiences with the concept of social value vary according to historical, political, and personal contexts and sensitivities. In the end, if the expression "social value" is not used in the DoH, it does not influence the fact that the concept behind it is anchored in the DoH, which is well aligned with the CIOMS Guidelines on this issue.

5 Placebo Debate

5.1 Wordings of Acceptable Risk of Placebo Arm Where There Is a Proven Intervention

I have been involved in debates around placebo for a long time. In the late 1990s, the HIV/AIDS pandemic raised international controversies on the justification of placebo-controlled trials in resource-limited regions when few interventions were proven to be safe, effective, and only available in wealthier regions. From such controversies, we learned that finding an agreement can be difficult and that it is important to prioritize some topics over others depending on their concrete impact in practice.

During the 2024 revision process of the DoH, we explained, in the name of the CIOMS, to the WMA Working Group that there is no major difference between the wordings for acceptable risk of participants in the placebo arm—when there is a proven intervention—in the CIOMS Guidelines (minor increase above minimal risk) and the DoH (no increase of risk of serious or irreversible harm and extreme care is made to prevent abuse of the use of placebo). The regional meeting in Sao Paulo addressed this issue head-on, but no one presented explicit evidence that the DoH's language has caused any ethically problematic placebo-controlled trials.

5.2 Motion from the Uruguay Medical Association

The Uruguay Medical Association submitted a motion to delete the words "serious or irreversible" from paragraph 33. Their argument was to go back to the 2000 version of the DoH, when a placebo-controlled trial was permissible even without a proven intervention. Deleting "serious or irreversible" means making no difference between the risk of harm between comparative harms, including placebo harm. The motion's rationale was that "*Deleting "serious and irreversible" would help the paragraph to be more consistent with what is established in the CIOMS guide, contributing to the concept of risk minimization.*"

In theory, the Uruguay Medical Association's argument is supported. The DoH has taken a pragmatic stance in its choice of words, bearing in mind the political challenge of maintaining a highly influential document coexisting with all legislation and research ethics principles worldwide. Topics like these, raising conflicting views globally, should be addressed in a practical decision by research ethics committees (RECs) and regulatory authorities in each community, on a case-by-case basis. Paragraph 33's last sentence should be considered with serious attention: "*Extreme care must be taken to avoid abuse of this option.*" When in doubt, the use of a placebo should be avoided and is prohibited.

6 Conclusion: Next Steps

This chapter briefly compared the 2024 DoH and the 2016 CIOMS Guidelines exploring common ground, similarities, and differences. Controversy can sometimes spark, provoking emotional conflicts; however, recognizing that all are working toward a common goal despite differing views is crucial for understanding each other. In practice, our passionate debates do contribute to improving the quality of health research involving human participants and research review.

We have learned many lessons from this revision process of the DoH. So what should be the next step? The WMA revises key documents every 10 years; the 2024 revision was a due process from the previous 2013 version. Should we start updating the 2016 CIOMS Guidelines in response to the discussions and results of the DoH's 2024 revision? Or should the DoH's modifications not be limited, particularly in this recent ever-changing global landscape of research ethics? Whether such a revision appears necessary or not, it should be discussed among the relevant parties. The WMA will start a revision process of the Declaration of Taipei in April 2025. The CIOMS would then be able to update the Guidelines, add more guidelines to the current version, or develop other guidelines (and its corresponding working groups). We wish that all interested parties exchange views and opinions regarding the next steps, with transparency, accountability, a collaborative partnership mindset, and respect as a priority.

Acknowledgments The author wishes to thank Anahita de Lafond, research assistant and PhD student at the Institute of Health Law, University of Neuchâtel, who deserves all the gratitude for editing this chapter.

Conflict of Interest WMA and CIOMS have covered my participation in all regional meetings that I attended, as well as the WMA General Assembly in Helsinki.

References

1. The World Medical Association. WMA Declaration of Helsinki—ethical principles for medical research involving human participants. First adopted in 1964, last amended in 2024. https://www.wma.net/policies-post/wma-declaration-of-helsinki/
2. Council for International Organizations of Medical Sciences. International ethical guidelines for health-related research involving humans. 2016. https://cioms.ch/publications/product/international-ethical-guidelines-for-health-related-research-involving-humans/
3. The World Medical Association. The Declaration of Taipei on Ethical Considerations Regarding Health Databases and Biobanks. 2002. Last revised 2016. https://www.wma.net/policies-post/wma-declaration-of-taipei-on-ethical-considerations-regarding-health-databases-and-biobanks/
4. The National Commission for the Protection of Human Subjects of Biomedical and Behavioral Research. The Belmont report: ethical principles and guidelines for the protection of human subjects of research. April 18, 1979. https://www.hhs.gov/ohrp/regulations-and-policy/belmont-report/read-the-belmont-report/index.html
5. Council for International Organizations of Medical Sciences. International guidelines on good governance practice for research institutions. 2023. https://cioms.ch/publications/product/international-guidelines-on-good-governance-practice-for-research-institutions/
6. World Medical Association. WMA Declaration of Geneva. Adopted in September 1948, last amended in October 2017. https://www.wma.net/policies-post/wma-declaration-of-geneva/
7. World Medical Association. WMA International Code of Medical Ethics. Adopted in October 1949, last amended in October 2022. https://www.wma.net/policies-post/wma-international-code-of-medical-ethics/

8. Schmidt U, Frewer A, Sprumount D, editors. Ethical research: the Declaration of Helsinki, and the past, present and future of human experimentation. Oxford: Oxford University Press; 2020.
9. Nsengiyumva JC, Muhenda J, Njuguna S, Nyabul PO. UBUNTU philosophy: a contemporary version of the traditional African communalism. IJSAR. 2019;6(9):16–21. https://www.ijsar.in/Admin/pdf/ubuntu-philosophy-a-contemporary-version-of-the-traditional-african-communalism-.pdf

Chapter 10
Justice and Equity of Placebo-Controlled Trials: What Has Changed?

Chieko Kurihara, Dirceu Greco, Ames Dhai, and Takeo Saio

Abstract In the 2024 revision of the Declaration of Helsinki (DoH) by the World Medical Association (WMA), paragraph 33 on placebo-controlled trials was one of the focuses, but no meaningful amendments were made. Particularly, the discrepancy between the Council of International Organizations of Medical Sciences (CIOMS) guidelines and the DoH in terms of the acceptable risk for research participants in a placebo arm when there is a proven intervention was a key issue: the CIOMS defines "minor increase above minimal risk" while the DoH defines "no increase of serious or irreversible harm." Although the Uruguayan Medical Union made a motion to amend it for consistency with the CIOMS, it was voted down.

The concern is that vulnerable study participants may be disproportionately exposed to the risk defined to be "acceptable" by the 2024 DoH. This is against the principle of justice. Furthermore, the WMA states the fundamental (core) principle that the rights and interests of research participants take precedence over the goal of the research. The contradiction between this fundamental principle and the placebo clause must be resolved. The continuing debates over the placebo clause in the DoH are a testament to the commitment and aspiration of the global research ethics community seeking the highest ethical standards that simultaneously meet the demands of ethics and science.

Keywords Placebo-controlled trial · Justice · Double-standard · Ethics dumping · Resource-limited settings · Vulnerability

---

C. Kurihara (✉)
Kanagawa Dental University, Yokosuka, Kanagawa, Japan
e-mail: chieko.kurihara@nifty.ne.jp

D. Greco
School of Medicine, Federal University of Minas Gerais, Belo Horizonte, Brazil

A. Dhai
School of Clinical Medicine, University of the Witwatersrand, Johannesburg, Gauteng, South Africa

T. Saio
Department of Internal Medicine and Psychiatry, Fuji Toranomon Orthopedic Hospital, Shizuoka, Japan

C. Kurihara et al. (eds.), *The 2024 Declaration of Helsinki*,
https://doi.org/10.1007/978-981-96-9294-1_10

1 Introduction: Justice and Inequity in Placebo Studies

1.1 Reflection on Inequity Without Change in Placebo Clause

The ethical justifiability of placebo-controlled trials is one of the most important propositions in research ethics for implementing the principle of justice [1]. The 2024 revision of the World Medical Association (WMA)'s Declaration of Helsinki (DoH) [2], adopted on October 19, 2024, introduced a remarkable new text ("para." means "paragraph"):

> *Since medical research takes place in the context of various structural inequities, researchers should carefully consider how the benefits, risks, and burdens are distributed.* (para. 6)

However, there was no meaningful change in the placebo clause (para. 33), in line with the above important reflection. High-risk threshold for placebo[1]-controlled trials when there is a proven intervention, i.e., "No *increase of risk of serious or irreversible harm*" as defined by the DoH could lead to a disproportionate distribution of risk to vulnerable people. This is because vulnerable people may participate in placebo studies due to insufficient understanding of the nature of the study or for economic reasons, i.e., that participation in the study is the only way to access care. This is counter to the principle of justice [3, 4].

The Council of International Organizations of Medical Sciences (CIOMS)'s International Ethical Guidelines for Health-related Research Involving Humans set another risk threshold in the 2016 revision [5]: "*minor increase above minimal risk*." The discrepancy between these two well-recognized international policy documents has not yet been resolved.

This chapter examines the WMA's review process on the placebo item of the 2024 DoH, which would help improve research and research review practices, in preparation for the next revision of the DoH.

1.2 The Motion of the Uruguayan Medical Union

At the final stage of the revision process at the WMA Medical Ethics Committee in Helsinki (October 16, 2024), a motion by the Uruguayan Medical Union [6] attempted to reinstate the principle once established in the 2000 version of the DoH [7], allowing placebo-controlled trials only when there is no proven intervention (Table 10.1). This principle in the 2000 DoH was regarded as "a perfect, prima facie norm" by Dr. Eitaka Tsuboi [8], who was President of the WMA at the time of the 2000 revision, representing the Japan Medical Association. Latin American

[1] Accurately, meaning of "placebo" here is described in the DoH in the term "other than the best proven one(s), the use of placebo, or no intervention." For simplification, this chapter uses the term "placebo" in this meaning.

Table 10.1 Changes in the placebo clause in the 2024 revision of the Declaration of Helsinki and the motion from the Uruguayan Medical Union

Changes from the 2013 version of the DoH in the finally adopted version in the 2024 version
33. The benefits, risks, burdens, and effectiveness of a new intervention must be tested against those of the best proven intervention(s), except in the following circumstances: ~~Where~~ **If** no proven intervention exists, the use of placebo, or no intervention, is acceptable; or ~~Where~~ **If** for compelling and scientifically sound methodological reasons the use of any intervention other than the best proven one**(s)**, the use of placebo, or no intervention is necessary to determine the efficacy or safety of an intervention; and the ~~patients~~ **participants** who receive any intervention ~~less effective~~ **other** than the best proven one**(s)**, placebo, or no intervention will not be subject to additional risks of serious or irreversible harm as a result of not receiving the best proven intervention. Extreme care must be taken to avoid abuse of this option.
The proposed change in the motion of the Uruguayan Medical Union
"and the participants who receive any intervention other than the best proven one(s), placebo, or no intervention will not be subject to additional risks of ~~serious or irreversible~~ harm as a result of not receiving the best proven intervention."
Comment and reasoning in the motion of the Uruguayan Medical Union
COMMENTS: Without prejudice to the comments already made in the public consultation and in the working group to improve the wording of the 2013 version of paragraph 33, and maintaining the position that the use of placebo or no treatment is acceptable only when there is no proven intervention, the Sindicato Médico del Uruguay[a] proposes to amend the working group proposal by removing the adjectives **serious and irreversible** from the phrase "…, they will not be subject to additional risks of harm.". **REASONING:** Deleting "serious and irreversible" would help the paragraph to be more consistent with what is established in the CIOMS guide, contributing to the concept of risk minimization

Authors' note: The text in the motion is included under the permission of the Uruguayan MedicalUnion. The document Number: MEC 227/DoH/Oct2024

[a]Sindicato Médico Del Uruguay = Uruguayan Medical Union

countries have so far maintained a consensus on this more restrictive policy of the 2000 DoH, in their legal frameworks [9, 10] and in a series of declarations [11–16]. This placebo clause in 2000 was consistent with the following fundamental (core) principles of the DoH. The two quotations from other WMA core documents have been maintained with some updates from the first DoH in 1964 to the present day:

- "*The health and well-being of my patient will be my first consideration*" (para. 3, quoted from the Declaration of Geneva, 1948, last updated 2017 [17].)

- "*The physician must commit to the primacy of patient health and well-being and must offer care in the patient's best interest.*" (para. 3, quoted from the International Code of Medical Ethics, 1949, last updated 2022 [18].)
- Rights and interests of individual research participants take precedence over the goal of the research. (para. 7, the negative sentence in the DoH was reworded here into a positive sentence.)

However, the motion of the Uruguayan Medical Union was regrettably voted down, at the WMA Medical Ethics Committee. Members of two northern European Medical Associations expressed explicit objections to the motion, arguing that such a principle would impede needed research. In the version finally adopted, only small editorial adjustments were made to the wording (Table 10.1).

2 Debate Toward the Final Decision

2.1 Discrepancy Between the DoH and CIOMS Guidelines

The CIOMS guidelines in the 2016 revision provided another risk threshold stating that the use of a placebo could allow only a "*minor increase above minimal risk.*" Whether or not the DoH follows the CIOMS principle was one of the important focuses of the 2024 revision [19]. This is because the CIOMS guidelines are prepared to indicate how the ethical principles in the DoH "*could be effectively applied, particularly in low-resource settings*" [5]. Nevertheless, the DoH did not follow the CIOMS policy on the use of a placebo. The risks acceptable by the DoH but not acceptable in CIOMS are shown in Fig. 10.1: continuing, significant pain, burden,

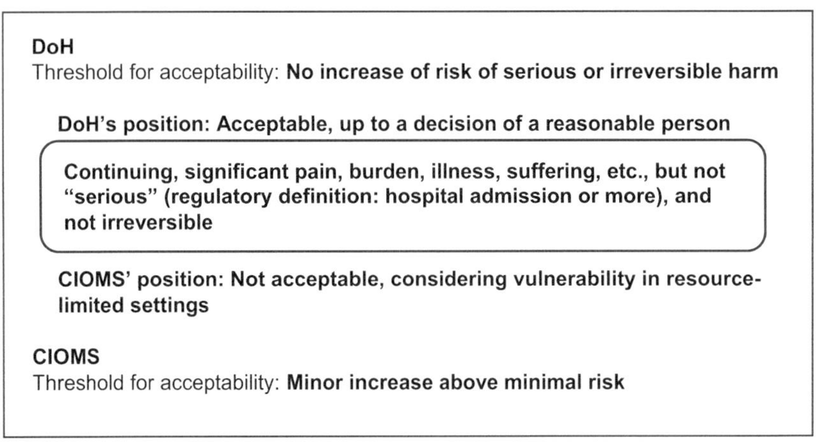

Fig. 10.1 Risk level of placebo-controlled study when there is a proven intervention, accepted by the DoH but not by CIOMS Guidelines. (Revised from the presentation by CK in the webinar, Ref. [19])

illness, suffering, etc., but not "serious" (regulatory definition of "serious" is such event to need hospital admission or more), and not irreversible (although severity and irreversibility are very difficult to determine a priori).

How this important issue was discussed in the WMA's Workgroup for the 2024 revision has not been officially disclosed by the WMA. According to some opinions made available which do not necessarily reflect the positions of all national associations, it is purported that the difference between the two documents is not significant, and the acceptability of the level of risk can be decided by the Research Ethics Committee (REC) and by each research participant [20]. This idea is also based on the reasoning that if all the other paragraphs of the DoH are complied with, research participants will be adequately protected [20–23]. The problem with this reasoning is that it is sometimes difficult for research participants to understand the risk of participating in a randomized placebo-controlled trial [24, 25]. In addition, some participants may participate in the trial because of limited access to care. This occurs not only in resource-limited regions but also in high-income countries.

The DoH has maintained the principle to permit only minimal risk and burden in research without any direct benefit to particularly vulnerable participants who are incapable to give consent:

> *In addition to receiving the protections for the particularly vulnerable, those incapable of giving consent must only be included if the research is likely to either personally benefit them or if it entails only minimal risk and minimal burden.* (para. 28)

However this principle must apply not only to the "particularly vulnerable," with an apparent inability to give consent, but also to other cases of vulnerability, due to dynamic and changing contextual situations, as the new paragraph suggests:

> *Some individuals, groups, and communities are in a situation of more vulnerability as research participants due to factors that may be fixed or contextual and dynamic, and thus are at greater risk of being wronged or incurring harm.* (para. 19)

Furthermore, any clinical trial should not tolerate the anticipated (or even intended) risks that are permitted by the DoH but must seek to minimize the research-induced risks at least to the CIOMS level, which would be the maximum tolerable compromise in view of human dignity. This is in line with the fundamental principle of medical ethics "do no harm" which must also apply to studies with nonvulnerable participants.

2.2 Consultations with Latin American Communities

The DoH revision process began in April 2022 and eight regional meetings were held in various locations around the world. The second regional meeting was held in Sao Paulo, Brazil, in February 2023, and focused on "placebo" to be discussed with Latin American communities. Most of the discussions were not directly about the DoH texts, but rather about methodologies of controlled trials and several examples of different studies. There were criticisms by some of the presenters on the

alternative screening died whereas 254 of 140,000 women of the control group receiving no screening died, both due to cervical cancer during the long-term follow-up. Eric Suba [37] in response to Sankaranarayanan et al. [38], authors of this study, argued that "*these India screening studies may become regarded, together with the Tuskegee Syphilis Study, as among the most scientifically and ethically problematic clinical studies ever supported by US taxpayers*" [37]. He submitted a complaint about these studies to the US Government's Office of Human Research Protections (OHRP). The OHRP response concluded that the study can be ethically justified although there were some issues in the ethical review and the informed consent process. CIOMS position on this case [36] is explained in this report citing its 2016 Guidelines [5] (Guideline 5) "*The CIOMS ethical guidelines take the stance that any potential new intervention should be tested against an established effective intervention, and that researchers may only deviate from this rule when withholding or delaying such interventions is methodologically necessary and exposes participants to no more than a minor increase above minimal risk.*" This example also presents the difference of ideas for protecting participants between the US government and CIOMS.

3.2 Downgrading in Placebo Clause

The WMA's deontological policy to ensure "best proven" intervention in any clinical study, including controlled trials was established in its 1975 version. During the debates of the 1990s, the American Medical Association (AMA) submitted its proposal to the WMA to eliminate this deontological view [1]. This proposal was not adopted and the 2000 DoH did not allow the use of placebo when there is a proven intervention. However, especially based on economics, which means it is less expensive to conduct a placebo-controlled study, the US Food and Drug Administration (FDA), global pharma industries, and some academic researchers engaged in international studies were in favor of the use of placebo even when there is a proven intervention [39].

In 2008, the US FDA dropped the requirement to adhere to the DoH for studies conducted outside the United States to be submitted for New Drug Application in the United States [40]. Instead, the DoH was substituted by mandatory adherence to ICH-GCP [41]. This was criticized by the scientific community [42, 43] and public media [44, 45]. This is because although some aspects of ethics are addressed in the GCP guidelines (e.g., ethics committees, informed consent), it mainly deals with procedural issues.

In 2002, 2 years after the approval of the 2000 DoH, the WMA added a Note of Clarification, to allow a placebo-controlled study if there is a compelling scientific reason, linking this condition via the conjunction "or" with another condition "no increase of risk of serious or irreversible harm." In 2008 "or" was changed to "and," and this was incorporated into the main text of the DoH. This condition was retained until 2024. This seemed to be influenced by the 2000 ICH-E10 guidelines on the choice of the control group [46] and the 2002 CIOMS Ethical Guidelines, which once had almost the same position as the current DoH. However, the CIOMS position of 2002 was modified in the 2016 revision.

Shortly before the adoption of the 2008 DoH, the Brazilian Code of Medical Ethics prohibited placebo studies when there is a proven intervention [9]. At the WMA General Assembly in 2008, the President of the Brazilian Federal Medical Council (Conselho Federal de Medicina, CFM), Dr. Edson Andrade, expressed concern, asking what the scientific justification was for modifying the 2000 placebo restrictions without eroding the duties of physicians [47]. This was completely contrary to the objections to the Uruguayan motions by two Medical Associations at the 2024 Medical Ethics Committee session. In 2008, Brazilian delegates received no response with scientific evidence, and the WMA has yet to provide any.

3.3 Deontology Versus Utilitarianism: Any Debate for the 2024 DoH?

The differences in the ongoing debate (but not scrutinized for the 2024 DoH revision process) can be summarized as differences between deontological and utilitarian views on these three discussion points [1]: (1) standard of care (whether "proven intervention" to be compared with tested intervention means universal standard or locally available standard); (2) validity of "clinical equipoise" (whether the condition of the comparative study is the situation that there is no consensus on which comparative arm is better); and (3) acceptable risk threshold in placebo arm when there is a proven intervention (Table 10.2) [1].

Table 10.2 Deontological and utilitarian views and main points of contentious issues in the placebo debates

	Deontological view		Utilitarian (pragmatic) view
(1) Standard of care (choice of control in comparative study)	Universal standard: Intervention has proven to be the best in the world: "universalism" (DoH, not limited to specific version)		Local standard: Intervention available in the host community of the study: "pluralism"
(2) Clinical equipoise	Clinical equipoise is a necessary condition for comparative study		It is deceptive to regard clinical equipoise to be a condition for comparative study (US bioethicists)
(3) Acceptable risk threshold where there is a proven intervention	Setting a risk threshold is not acceptable (DoH 2000; Uruguayan motion; Brazilian system; CONFEMEL)	"Minor increase above minimal risk" (CIOMS 2016)	"No increase of risk of serious or irreversible harm" (CIOMS 2002; DoH 2008–2024, ICH-E10)

Updated from the Table 3 in the Ref. [1]

Regarding (1) standard of care, the WMA has not changed its position from the beginning in support of universal standard (universalism, against the "double standard" [48, 49]), despite the pros and cons before and after the 2000 revision of the DoH: "best-proven intervention" means the best proven in the world, not "best locally available" (local standard, plurarism). This was confirmed at several academic meetings [20–23] including the Sao Paulo regional meeting in 2023 [20]. The authors of this chapter have argued that this position should be clarified in the DoH text [50, 51], but WMA did not present it as a central topic of discussion for the 2024 revision.

Regarding (2) clinical equipoise, there have been substantial debates in the literature, but it has not been sufficiently discussed in the decades-long DoH revision processes. Some have argued that the justification for controlled clinical trials is "clinical equipoise" [52, 53] or "uncertainty" [54] (when it is not known which arm of the trial is better in comparative arms). Others have argued that it is "deceptive" because a comparative study is designed with the intention to prove the efficacy of some intervention knowing the risks in the controlled arm may be greater than the investigational intervention [55–57]. The point of their argument is that the high risk related to research can be acceptable as long as research participants give informed consent with a full understanding of that risk.

Regarding (3) acceptable risk threshold, this topic was also not set as an official agenda for the 2024 DoH; however, it was obvious that there is a clear discrepancy between the 2013 DoH and the 2016 CIOMS guidelines. The fundamental idea in the motion of the Uruguayan Medical Union, the Brazilian legal system, and COFEMEL is the position of the 2000 version of the DoH. Furthermore, the Uruguayan motion states that their position is consistent with the 2016 CIOMS (Table 10.1, "REASONING"). In essence, this type of debate on risk management theories (e.g., making theoretical definitions of "negligible," "acceptable," or "tolerable" death ratio for the general public or workers, etc.) [58] is not appropriate for high-level ethical principles such as DoH and CIOMS. Discussions should focus on the issue of clinical equipoise from the perspective of the physician's duty of care.

Utilitarian views have been repeatedly expressed by the US ethicists. In June 2024, just before the final regional meeting in Washington, DC, United States, which was in August 2024, the immediate past director of the US OHRP published his view in the *Journal of American Medical Association (JAMA)* [59]. The argument was to abolish the core principle of the DoH, which is to give priority to the patient's interest over the research goal of generating new knowledge (para. 8 in 2013, 7 in 2024 DoH), suggesting that it causes "therapeutic misconception" (misunderstanding of research as treatment). He recommended that this principle derived from the physician's obligation should be abolished, and instead should be changed to the researcher's duty to adequately explain the risks of research to prospective participants. He also points out what he sees as the inconsistency between this core principle and the placebo clause. His argument was clearly countered by one of the authors (TS), from the perspective of a Japanese physician, to defend this fundamental principle of DoH [60].

The same argument was expressed during the WMA Washington, DC meeting by a representative of the US Centers for Disease Control and Prevention. This was also countered by a Korean bioethicist who is the author of another chapter in this book (Chap. 6). She is a medical professor and stated that she always emphasizes this core principle in her educational lectures [20].

4 Health Crisis and Real-World Data

4.1 Opportunities for Reconsideration (1): Ebola Outbreak

The global community has experienced two prominent health crises between the 2013 and 2024 revisions of the DoH: the Ebola outbreak (2014) and the COVID-19 pandemic (2020). These were key opportunities to acknowledge health inequality and revisit the placebo clause in the DoH.

During the Ebola outbreak, several views emerged against placebo-controlled trials, even in the absence of a proven intervention. Some argued that equipoise could not be established when a large number of people wanted to receive an experimental intervention for a deadly disease, even if it was unproven [61]. These authors also argued that placebo-controlled trials lack feasibility when there is insufficient technical infrastructure for implementation and of an adequate social, cultural background to understand the scientific methodology. There have been other opinions defending that placebo-controlled trials are essential and justified in the absence of intervention with proven efficacy and safety. The US Presidential Commission for the Study of Bioethical Issues published a report to gather both opinions [62], while the World Health Organization (WHO) has published an expert consensus report to allow the clinical usage of unproven intervention in emergency situations [63]. The WHO has expanded its policy called "Monitored emergency use of unregistered and experimental interventions (MEURI)" including the use of unproven intervention, based on the experiences of Ebola and COVID-19 [64]. This framework expands the possibility of establishing a robust framework for compassionate use or expanded access program.

This was the opportunity to reconsider the placebo clause in the DoH which leaves no doubt about placebo use in the absence of a proven intervention (the first condition in para 33 of the 2024 DoH).

4.2 Opportunities for Reconsideration (2): COVID-19 Pandemic

During the COVID-19 pandemic, the two typical arguments supporting "local standard" were expressed by some French doctors and a WHO Adhoc Expert Group: At the beginning of the pandemic, French doctors stated that vaccine trials should be conducted in Africa where there are "no masks, no treatments, no resuscitation." This was keenly criticized especially by the WHO [65], and the African community [66]. However, when some experimental vaccines were shown to be safe and effective, a WHO ad hoc Group stated that it would be ethically permissible to conduct placebo-controlled trials in the countries where access to the established vaccine is limited and that this position was in accordance with the principles in the DoH and CIOMS guidelines [67]. Both positions of the French doctors and WHO Adhoc Group are clear examples of double standards [49]. To avoid such misconceptions, the WMA should have clarified that "best proven" does not mean the best available in a local community, but rather the best proven in the world [50, 51]. Nontheless, this was not clarified in the text of the DoH.

As a global collaboration for fair allocation of vaccines, the COVAX initiative was established to facilitate programs of joint purchasing and donations to the most-disadvantaged regions. However, bilateral contracts between companies and governments in wealthier countries led to the terms of this initiative not being complied with. As a result, many communities, especially in low- and middle-income countries (LMICs), including the communities that participated in placebo-controlled trials, did not have timely access to approved COVID-19 vaccines. While Japan has not participated in global placebo-controlled trials of vaccines, it purchased vaccines for seven times the total population, leading to vaccine "hoarding" and a substantial amount being discarded [68]. In the later phase of the pandemic, a Japanese company conducted placebo-controlled studies in Vietnam and began selling the developed vaccine in Japan in 2024.

4.3 Scientific Challenges Including Real-World Evidence

On the other hand, United States [69] and European [70] regulatory authorities as well as CIOMS [71] have put forward the idea of facilitating the use of real-world data, which lead to an agreement in 2025 on Annex 2 of ICH-GCP(R3) [72].

Several examples of the use of external controls for regulatory approval of a new intervention have now been reported [73]. This methodology has been used primarily in cases of life-threatening diseases or rare diseases, but accumulated experiences may increase the credibility of using real-world data as external controls. The policy regarding placebo-controlled randomized trials as the gold standard may become obsolete. Risk-minimization strategies have already been recommended in the ICH-E10 guidelines [46], which could become more sophisticated, with the development of science and technology that prioritize ethics (Table 10.3). Placebo-controlled trials may be a gold standard in terms of "scientific value"; however, replacing the "scientific rationale" with an "ethically compelling rationale" will generate an outcome of greater "social value" without compromising the scientific value. Scientific value is a necessary condition for social value, but not a sufficient condition, according to the 2016 CIOMS ethical guidelines [5].

On the other hand, technological innovations for the use of real-world data could be achieved by setting lower thresholds of risk in acceptable placebo controls. Furthermore, promoting the use of real-world data will facilitate standardization and improvement of the reliability of electronic data generated in routine care. This

Table 10.3 Required considerations for the ethical conduct of the placebo-controlled trial

<table>
<tr><td colspan="2">■ Trade-off considerations in clinical trial designs</td></tr>
<tr><td colspan="2">● Comparison between traditional methodologies</td></tr>
<tr><td>Equivalence/non-inferiority or superiority test with active control
Advantage: ethical “gold standard”; higher social value if superior to a truly established intervention, not under discussion, is proven.
Disadvantage: higher cost.</td><td rowspan="2">Superiority test against placebo
Advantage: scientific “gold standard” to find genuine effect of the tested intervention; lower cost.
Disadvantage: less ethical justifiability; less social value if it is superior only to placebo, even if there is a proven intervention; difficulty in recruitment.</td></tr>
<tr><td>Equivalence/non-inferiority or superiority test with “add-on” vs placebo, both adding to standard of care
Advantage: ethical, social value is to find the efficacy/safety of combination treatment.
Disadvantage: higher cost.</td></tr>
<tr><td colspan="2">● Comparison with emerging methodologies</td></tr>
<tr><td>Using external control
Advantage: acceptability in ethical view and feasibility especially in case of serious conditions or rare diseases.
Disadvantage: lower scientific validity.</td><td rowspan="3">Traditional placebo-controlled trial
Advantage: scientific “gold standard” in terms of experimental methodology; higher cost.
Disadvantage: less ethical justifiability; less social value due to lower external validity because of limited eligibility criteria, in order to eliminate bias as much as possible; difficulty in recruitment.</td></tr>
<tr><td>Case report using Bayesian statistics
Advantage: acceptability in ethical view and feasibility especially in case of serious conditions or rare diseases.
Disadvantage: lower (or challenging) scientific validity.</td></tr>
<tr><td>Pragmatic trial
Advantage: mainly greater external validity, feasibility in some scenarios including public health emergency (e.g., WHO Solidarity trial [74, 75]); more relevant in later phases (e.g., phase 4 of approved drug).
Disadvantage: limited scientific/social value; sometimes not relevant in pre- approval phases.</td></tr>
<tr><td colspan="2">■ Consideration of universal standard/clinical equipoise issues</td></tr>
<tr><td colspan="2">Following cases can be considered as “there is NO proven intervention in the world”, so the justifiability of a placebo study in these settings cannot reject the ethical gold standard that “placebo study can be permitted only when there is no proven intervention in the world”.
• The established intervention is being questioned to its effectiveness (question of “assay sensitivity” as a control).
• Specific sub-group is not responding to proven intervention.
• The intervention whose effectiveness is proven but safety issues arise.
• The proven intervention is not relevant to some specific reason due to intrinsic and/or extrinsic factors, e.g., ethnicity (including cultural perception of ethnic difference), variety of environmental factors, including differences in viruses, background of medical practice, but not including economic reason, e.g., insufficient monetary support for the study.</td></tr>
<tr><td colspan="2">■ Risk minimization strategy (described in ICH-GCP E10 and other literatures). These should be considered in order not to allow the level of risk accepted by the DoH.</td></tr>
<tr><td colspan="2">• Establishment of a data safety monitoring committee.
• Early termination with rescue therapy by intensive observation.
• Post-trial care after termination or completion of the study.</td></tr>
</table>

improvement in research utilization will also contribute to improving the use of health data in daily clinical practice.

This is also related to the 2024 revision of the DoH linked to the WMA Declaration of Taipei (DoT) for health databases and biobanks [76]. If the personal data and biological materials obtained in research may be subject to secondary use in future research or other utilization, we must adhere to the DoT. Recently, a drug for a rare disease was approved using external controls from earlier phase clinical trial data [77]. This reinforces the importance of linking to the DoT in the 2024 DoH revision, which defines that secondary use of accumulated research data should be compliant with the DoT.

5 Participation of Patients, Public, and Vulnerable Communities

5.1 Patients, Public, and Community Participation

For the patients, research participants, and the public, it is difficult to find justification for a placebo-controlled study if there is a proven intervention [78]. Even if there is no proven intervention, it is still difficult to understand it adequately. In case patients have access to a proven intervention, they would not participate in the study, opting instead to receive the proven intervention. If they do not have access to an existing proven intervention, they may agree to participate in a placebo study with some probabilistic expectation to access to therapy of unknown efficacy and safety. The real implication of the DoH placebo clause may facilitate such a double standard in research. We need to be aware of double standards in two different dimensions:

- A double standard is applying a high ethical standard to a resource-sufficient community while a lower ethical standard is applied to a resource-limited community [49].
- A double standard is to use an exception to the ethical principle, assuming that a reasonable decision will be made by ethics committees and study participants, even though it is known that vulnerable individuals are likely to make the decision to accept risks without sufficient understanding of the risk; or, even with sufficient understanding of the risk, but with limited access to proven interventions. In addition, there may be an intrinsic vulnerability of the research ethics committee to external pressure.

From the authors' experiences of ethics committee meetings, comparative studies are approved confirming that we do not know which arm is better, and the study does not deprive the opportunity of patients to access to proven interventions. This means "equipoise" or "uncertainty" between study arms as well as between the options of participating and not participating in the study. For the public members of the ethics committee, it becomes critical that they find the acceptability of a placebo study from the perspective of patients. Lay members of the ethics committee will become the final gatekeeper not to permit exploitative placebo study for their community. A Japanese group of patients and the public expressed their willingness to participate in global discussions on the acceptability of a placebo study after the 2024 revision of the DoH [79].

5.2 Helsinki Statement: Global South, Mediterranean, and Asian Perspectives

Although the DoH and a group of physicians did not consider the counterargument from CIOMS, in which its working group included nonphysicians, and given the resource-limited settings, it will be possible to find an alternative way to form an international consensus on research ethics. One example is the "Helsinki Statement," agreed upon by interested parties, mainly from the Global South and Asia [80]. This statement was issued on October 18, 1 day before the final adoption of the 2024 DoH, acknowledging 10 points of improvements, expressing five points of disagreement, including the fact there was no change in the placebo clause. Three of the authors of this chapter (CK, DG, TS) and two other contributors to this book (Baroutsou, Matsuyama) are corresponding authors who were involved in the planning, management, and leadership of the writing of the Statement, having twice broadcast virtual web meetings from Helsinki to the world. These were accompanied by e-mail discussions and calls for signatories until the end of November 2024. Of the 125 individual signatories, 74 were from the global south, 44 from Asia, 6 from Europe, and 1 from Australia [20]. Despite the bias of dissemination of information calling for signatories, it is significant that this consensus, formed quickly suggesting disagreements with the current WMA policy including the one on placebo-controlled trials.

It was also worth noting that the 32 members of the Egyptian Network of Research Ethics Committees (ENREC) agreed with the Helsinki Statement and provided an Arabic translation [81], a week after the Statement was first announced (see Chap. 3). Furthermore, it was reported that the change in the DoH from the term "research subject" to "research participant" is not new in the Arabic translation of the DoH. This is because they already used the Arabic term meaning "participant" instead of "subject" which is used for "inanimate objects" [20]. At the time of writing this chapter, a Greek translation of the Helsinki Statement is also under preparation for publication. The selection of the term "research participant" in Egypt is the same as in Greece. Our experience has led to the hypothesis that cultures in the Global South, Mediterranean, and Asian regions, while differing from Anglo-Saxon empiricism by emphasizing experimental medicine, would be a basis for strict limits on placebo-controlled trials, since this is a typical study design for treating humans as subjects of experimentation.

Recognizing the differences in values, historical backgrounds and cultures in different communities is important as a process of seeking common values while respecting the diversity of each community. The placebo clause in the current DoH does not reflect a truth in ethical theory or a consensus reached among diverse interested parties. It is simply a compromise between the competing power politics of science, ethics, and economical imperatives, determined by a group of physicians. Therefore, it is necessary to continually seek principles that reflect diverse values, as well as the highest ethical standard.

6 Conclusion: Never Give Up

Although the long-running placebo controversy has caused distress, it is a crucial point of discussion in research ethics to avoid exploitation and inequality. In-depth discussions on the ethics of placebo-controlled trials stimulate deliberations on other important topics, such as justice and injustice, effective protection of vulnerable groups, valid informed consent processes, use of real-world data, strengthened ethical review, the scientific rigor of emerging clinical trial designs, risk management during the conduct of research and the necessary care to be provided after the completion of trials, and so on. All of these could be considered for future research ethics standards, including the next DoH revision, as well as improving the practice of protocol design, ethical review, conduct of research, and appropriate publication and implementation of results.

The ongoing debates on the placebo clause, post-trial access rights, and other points of disagreement in the DoH are a testament to the commitment and aspirations of the global research ethics communities to the highest ethical standards that simultaneously meet the demands of science and ethics.

Acknowledgments CK appreciates the International Federation of Associations of Pharmaceutical Physicians and Pharmaceutical Medicine (IFAPP) for granting her travel expenditure for her in-person participation in the WMA Regional Meeting in Tel Aviv (2022 December) and the WMA for granting her travel expenditure for her in-person participation and presentation as a panelist in WMA Regional Meeting in Washington, DC (2024 August), and waiver of registration fee from WMA as a journalist to attend the General Assembly in Helsinki (2024 October). She participated in all the other WMA Regional Meetings at her own expense or online.

DG was officially engaged in discussions for the revisions of the Declaration of Helsinki from 2000 to 2008 as a member of the Brazilian Medical Association and Brazilian Medical Council teams; he also participated in the CIOMS 2016 guidelines as a member of the working group; he was invited to the regional meeting in Sao Paulo by the WMA to discuss the ethics of placebo-controlled trials; and he participated, with partial funding from CIOMS, in the 2024 WMA General Assembly that adopted the 2024 version of the DoH.

Conflict of Interest AD was a Board member of the South African Medical Association, which was a member of the Workgroup of the World Medical Association for the 2024 revision of the Declaration of Helsinki. The authors declare no other competing interests relevant to the content of this chapter.

References

1. Kurihara C, Greco D, Dhai A, Saio T, Tsubaki H. Ethics of placebo-controlled trials: historical analysis including experiences during the COVID-19 pandemic. In: Kurihara C, Greco D, Dhai A, editors. Ethical innovation for global health: pandemic, democracy and ethics in research. Singapore: Springer; 2023. p. 195–224.
2. The World Medical Association. WMA Declaration of Helsinki—ethical principles for medical research involving human participants. First adopted in 1964, last amended in 2024. https://www.wma.net/policies-post/wma-declaration-of-helsinki/
3. The National Commission for the Protection of Human Subjects of Biomedical and Behavioral Research. The Belmont report: ethical principles and guidelines for the protection of human subjects of research. 1979. https://www.hhs.gov/ohrp/regulations-and-policy/belmont-report/read-the-belmont-report/index.html
4. Rawls J. A theory of justice. Cambridge: Belknap Press; 1971.
5. Council for International Organizations of Medical Sciences. International ethical guidelines for health-related research involving humans. 2016. https://cioms.ch/publications/product/international-ethical-guidelines-for-health-related-research-involving-humans/
6. Sindicato Médico del Uruguay. World Medical Association Submission of a Motion to Amend (on Paragraph 33, submitted to) 227th Council Session and General Assembly, Helsinki 2024, 8 Oct 2024.
7. World Medical Association. Declaration of Helsinki: ethical principles for medical research involving human subjects. 2000. Rescinded. https://www.wma.net/wp-content/uploads/2018/07/DoH-Oct2000.pdf
8. Tsuboi E. Kurihara C, Interview. The 2000 revision of the Declaration of Helsinki and its implications in medical ethics: interview with the WMA immediate past president, Eitaka Tsuboi. Clin Eval. 2002;30(1):99–107. Japanese. http://cont.o.oo7.jp/30_1/p99-107.pdf
9. Brazilian National Health Council (CNS), Brazilian National Research Ethics Committee (CONEP/CNS/MS). Guidance manual: frequent pending issues in clinical research protocols. Version 1.0. 2015. https://abracro.org.br/images/legislacao/guidance_manual_frequent_pending_issues_in_clinical_research_protocols_CONEP_V1_2015.pdf
10. Brazilian Research Ethics Commission Resolution 466/2012, which succeeded Resolution 404/2008.
11. Homedes N, Ugalde A. Clinical trials in Latin America: where ethics and business clash. Cham: Springer; 2014.
12. The Buenos Aires Declaration on Ethics and Clinical Trials. Buenos Aires, May 13, 2008. In: Homedes N, Ugalde A, editors. Clinical trials in Latin America: where ethics and business clash. Cham: Springer; 2014.
13. Red Latinoamericana y Del Caribe de Bioetica—Redbioetica/UNESCO. Carta de Córdoba sobre ética en investigaciones con seres humanos. Córdoba, Argentina, 14 de noviembre de 2008. https://www.psi.uba.ar/academica/carrerasdegrado/psicologia/sitios_catedras/obligatorias/723_etica2/material/casuistica/carta_de_cordoba_08.pdf
14. Solbakk JH. National ethics advisory bodies and committees in Norway: history, lessons learnt, and common challenges ahead. National Bioethics Committees in Action. United Nations Educational, Scientific and Cultural Organization; 2010. https://unesdoc.unesco.org/ark:/48223/pf0000189548
15. Confederation Medica latinoamericana Y del Caribe (Latin American and Caribbean Medical Confederation). Declaration of Pachuca 2013.
16. Greco DB. Ethical dilemmas, research, and access to treatment. In: International encyclopedia of the social & behavioral science, vol. 8. 2nd ed. Amsterdam: Elsevier; 2015. p. 31–6.
17. World Medical Association. WMA Declaration of Geneva. Adopted in Sept 1948, last amended in Oct 2017. https://www.wma.net/policies-post/wma-declaration-of-geneva/

18. World Medical Association. WMA International Code of Medical Ethics. Adopted in Oct 1949, last revised in Oct 2022. https://www.wma.net/policies-post/wma-international-code-of-medical-ethics/
19. The Brazilian Society of Bioethics (SBB); International Federation of Associations of Pharmaceutical Physicians and Pharmaceutical Medicine (IFAPP); Clinical Evaluation Inc.; The 129th Pharmaceutical Study Group, Co-organizers. The 2024 Declaration of Helsinki: taking forward bioethics and human rights. Clin Eval. 2025;52(3):375–452. http://cont.o.oo7.jp/52_3/p375-452.pdf
20. Kurihara C, Matsuyama K, Baroutsou V. World Medical Association's Declaration of Helsinki, 2024 revision: celebrating the 60th anniversary, at Helsinki. Clin Eval. 2025;52(3):W1–27. http://cont.o.oo7.jp/52pop/52pop_contents_e.html
21. Human D, Kurihara C, Mitsuishi T, Interview. Interview with Delon Human, Secretary General of the World Medical Association—discussion on the 2000 revision to the Declaration of Helsinki and onwards. Clin Eval. 2002;29(2):307–13. Japanese. http://cont.o.oo7.jp/29_23/p307-13.pdf
22. Kloiber O, Kurihara C, Saio T, Interview. Discussion toward the 50th anniversary of the Declaration of Helsinki: Interview with Dr. Otmar Kloiber, Secretary General, World Medical Association. Clin Eval. 2013;41(2):351–72. http://cont.o.oo7.jp/41_2/p351-72eng.pdf
23. Kloiber O, Greco D, Watanabe H, Imamura K, Yamamoto Y, Matsuyama K, Saio T, Kurihara C. International collaborative research and new trends of research ethics: follow-up session. Clin Eval. 2020;48(1):233–65. http://cont.o.oo7.jp/48_1/w55-w87.pdf
24. Lidz CW, Appelbaum PS, Grisso T, Renaud M. Therapeutic misconception and the appreciation of risks in clinical trials. Soc Sci Med. 2004;58(9):1689–97. https://doi.org/10.1016/S0277-9536(03)00338-1.
25. Shiono YN, Zheng YF, Kikuya M, Kawai M, Ishida T, Kuriyama S, Ohuchi N. Participants' understanding of a randomized controlled trial (RCT) through informed consent procedures in the RCT for breast cancer screening, J-START. Trials. 2014;(15):375. https://doi.org/10.1186/1745-6215-15-375.
26. CONFEMEL. Intervención de la Dra. Zaida Arteta en la Asamblea General de la AMM. 16 octubre, 2024. https://www.confemel.com/intervencion-de-la-dra-zaida-arteta-en-la-asamblea-general-de-la-amm/
27. CONFEMEL. CONFEMEL considera adecuado y oportuno el proceso de Revisión de la DoH llevado adelante por la AMM. 24 April 2023. https://www.confemel.com/confemel-considera-adecuado-y-oportuno-el-proceso-de-revision-de-la-doh-llevado-adelante-por-la-amm/
28. Sindicato Medico del Uruguay. SMU propone cambios en la Declaración de Helsinki. 16 October 2024. https://www.smu.org.uy/el-smu-en-la-asamblea-de-la-asociacion-medica-mundial-propone-cambios-en-la-declaracion-de-helsinki/
29. Connor EM, Sperling RS, Gelber R, et al. Reduction of maternal-infant transmission of human immunodeficiency virus type 1 with zidovudine treatment. Pediatric AIDS Clinical Trials Group protocol 076 study group. N Engl J Med. 1994;331(18):1173–80.
30. Lurie P, Wolfe SM. Unethical trials of interventions to reduce perinatal transmission of the human immunodeficiency virus in developing countries. N Engl J Med. 1997;337(12):853–6.
31. Angell M. The ethics of clinical research in the Third World. N Engl J Med. 1997;337(12):847–9. https://doi.org/10.1056/NEJM199709183371209.
32. Reverby SM. 'Normal exposure' and inoculation syphilis: PHS 'Tuskegee' doctors in Guatemala, 1946–1948 and at Sing Sing prison, Ossining, New York, 1953–1954. 2010.
33. Varmus H, Satcher D. Ethical complexities of conducting research in developing countries. N Engl J Med. 1997;337(14):1003–5. https://doi.org/10.1056/NEJM199710023371411.
34. [Editorial]. Declaration of Helsinki embraces health equity. Nat Med. 2024;30(12):3383. https://doi.org/10.1038/s41591-024-03433-5.
35. Wendland CL. Research, therapy, and bioethical hegemony: the controversy over perinatal AZT trials in Africa. Afr Stud Rev. 2008;51:1–23.

36. Clinical research in resource-limited settings. Council for International Organizations of Medical Sciences 2021. https://cioms.ch/publications/product/clinical-research-in-low-resource-settings/
37. Suba EJ. Response by Eric Suba to Sankaranarayanan et al. Indian J Med Ethics. 2014;11(3):178–80. https://doi.org/10.20529/IJME.2014.044. Epub 2014 June 6.
38. Sankaranarayanan R, Nene BM, Shastri S, Esmy PE, Rajkumar R, Muwonge R, Swaminathan R, Malvi SG, Kane S, Desai S, Kelkar R, Hingmire S, Jayant K. Response to article titled "US-funded measurements of cervical cancer death rates in India: scientific and ethical concerns" by Eric J Suba. Indian J Med Ethics. 2014;11(3):175–8. https://doi.org/10.20529/IJME.2014.045. Epub 2014 June 2.
39. Pérez AC, Smith RN. The revised Declaration of Helsinki: interpreting and implementing ethical principles in biomedical research. Int J Pharm Med. 2001;15:131–43. https://doi.org/10.2165/00124363-200106000-00006.
40. Department of Health and Human Services Food and Drug Administration. 21 CFR Part 312 Human subject protection: foreign clinical studies not conducted under an investigational new drug application. Fed Regist. 2008;73(82):22800–16. https://www.govinfo.gov/ content/pkg/FR-2008-04-28/pdf/E8-9200.pdf
41. International Conference on Harmonisation of Technical Requirements for Registration of Pharmaceuticals for Human Use. ICH harmonised tripartite guideline: Guideline for Good Clinical Practice E6(R1). 10 June 1996.
42. Lurie P, Greco DB. US exceptionalism comes to research ethics. Lancet. 2005;365(9465):1117–9.
43. Trials on trial. Nature 2008;453(7194):427–8. doi:https://doi.org/10.1038/453427b.
44. Anderson M. FDA abandons Declaration of Helsinki for international clinical trials. Social Medicine Portal 2008.
45. Integrity in Science. FDA scraps Helsinki Declaration on protecting human subjects. 2008. https://web.archive.org/web/20081022213154/http://www.cspinet.org/integrity/watch/200805051.html#2
46. International Conference on Harmonisation of Technical Requirements for Registration of Pharmaceuticals for Human Use. ICH harmonised tripartite guideline: choice of control group and related issues in clinical trials E10. 2000. https://database.ich.org/sites/default/ files/E10_Guideline.pdf
47. Corina Bontempo Duca de Freitas and Bruno R. Schlemper Jr. Progress and challenges of clinical research with new medications in Brazil. In: Homedes N, Ugalde A, editors. Clinical trials in Latin America: where ethics and business clash. Cham: Springer; 2014. p. 151–71.
48. Macklin R. Against relativism: cultural diversity and the search for ethical universals in medicine. Oxford: Oxford University Press; 1999.
49. Macklin R. Double standards in medical research in developing countries. Cambridge law, medicine and ethics. Cambridge: Cambridge University Press; 2004.
50. COVID-19 and bioethics part 3: pandemic and research ethics-democracy, placebo and post-trial access. Clin Eval. 2021; 49(Sup 38):9–109. https://cont.o.oo7.jp/49sup38/49sup38contents_e.html.
51. Kurihara C, Greco D, Dhai A, Matsuyama K, Baroutsou V. Vulnerability, social value and the equitable sharing of benefits from research: beyond the placebo and access debates. Front Med. 2024;11:1432267. https://doi.org/10.3389/fmed.2024.1432267.
52. Freedman B. Equipoise and the ethics of clinical research. NEJM. 1987;317:141–5.
53. Rothman KJ. Placebo mania. Br Med J. 1996;313:3–4.
54. Sackett DL. CMAJ. 2000;162:1311–4, CMAJ. 2001;164:1831–2.
55. Miller FG, Brody H. What makes placebo-controlled trials unethical? Am J Bioeth. 2002;2(2):3–9. https://doi.org/10.1162/152651602317533523.
56. Miller FG, Brody H. A critique of clinical equipoise. Therapeutic misconception in the ethics of clinical trials. Hast Cent Rep. 2003;33(3):19–28.

57. Miller FG, Brody H. Clinical equipoise and the incoherence of research ethics. J Med Philos. 2007;32(2):151–65. https://doi.org/10.1080/03605310701255750.
58. The Health and Safety Executive. The tolerability of risks from nuclear power stations. 1988, Revised 1992. https://www.onr.org.uk/media/v1vi3v21/tolerability.pdf.
59. Menikoff J. Protecting participants is not the top priority in clinical research. JAMA. 2024; https://doi.org/10.1001/jama.2024.7677. Epub ahead of print.
60. Saio T. Noblesse oblige of the WMA in peril: save its honor at the diamond anniversary. Clin Eval. 2025;52(3):475–81. https://cont.o.oo7.jp/52pop/W40-W45.pdf.
61. Adebamowo C, Bah-Sow O, Binka F, Bruzzone R, Caplan A, Delfraissy JF, Heymann D, Horby P, Kaleebu P, Tamfum JJ, Olliaro P, Piot P, Tejan-Cole A, Tomori O, Toure A, Torreele E, Whitehead J. Randomised controlled trials for Ebola: practical and ethical issues. Lancet. 2014;384(9952):1423–4. https://doi.org/10.1016/S0140-6736(14)61734-7. Epub 2014 Oct 13.
62. Presidential Commission for the Study of Bioethical Issues. Ethics and Ebola: Public health planning and response. February 2015.
63. World Health Organization. Ethical considerations for use of unregistered interventions for Ebola viral disease: report of an advisory panel to WHO. WHO/HIS/KER/GHE/141 2014. https://apps.who.int/iris/bitstream/handle/10665/130997/WHO_HIS_KER_GHE_14.1_eng.pdf?sequence=1
64. World Health Organization. Emergency use of unproven clinical interventions outside clinical trials: ethical considerations. 2022 Apr 12. https://www.who.int/publications/i/item/9789240041745
65. BBC. Coronavirus: Africa will not be testing ground for vaccine, says WHO. 7 April 2020. https://www.bbc.com/news/world-africa-52192184
66. Dhai A. Equitable access to COVID-19 vaccines, vaccine research, and vaccine apartheid on African continent: challenges and recommendations. In: Kurihara C, Greco D, Dhai A, editors. Ethical innovation for global health: pandemic, democracy and ethics in research. Berlin: Springer; 2023. p. 39–55.
67. WHO Ad Hoc Expert Group on the Next Steps for Covid-19 Vaccine Evaluation. Placebo-controlled trials of Covid-19 vaccines – why we still need them. N Engl J Med. 2021;384(2):e2. doi:10.1056/NEJMp2033538.. Epub 2020 Dec 2.
68. Kurihara C, Saio T. Therapeutic misconception as the basis for vaccine nationalism of Japan: a historical reflection and perspectives for global public health. In: Kurihara C, Greco D, Dhai A, editors. Ethical innovation for global health: pandemic, democracy and ethics in research. Singapore: Springer; 2023. p. 105–30.
69. Draft. U.S. Food and Drug Administration. Considerations for the design and conduct of externally controlled trials for drug and biological products. February 2023. https://www.fda.gov/regulatory-information/search-fda-guidance-documents/considerations-design-and-conduct-externally-controlled-trials-drug-and-biological-products
70. European Medicines Agency. Reflection paper on use of real-world data in non-interventional studies to generate real-world evidence. End of consultation, 31 August 2024. https://www.ema.europa.eu/en/documents/scientific-guideline/reflection-paper-use-real-world-data-non-interventional-studies-generate-real-world-evidence_en.pdf
71. The Council of International Organizations of Medical Sciences. Real-world data and real-world evidence in regulatory decision making: Report of the CIOMS Working Group XIII. 2024.
72. International Council for Harmonisation of Technical Requirements for Pharmaceuticals for Human Use. ICH Harmonised Guideline for Good Clinical Practice E6(R3) Annex 2 Draft version Endorsed on 6 November 2024 Currently under public consultation. https://database.ich.org/sites/default/files/ICH_E6%28R3%29_Annex%202_Step2_DraftGuideline_2024_1024_0.pdf
73. Maeda H, Ng DB. Regulatory approval with real-world data from regulatory science perspective in Japan. Front Med (Lausanne). 2022;9:864960. https://doi.org/10.3389/fmed.2022.864960. PMID: 35492312; PMCID: PMC9051225.

74. WHO Solidarity Trial Consortium, Pan H, Peto R, Henao-Restrepo AM, Preziosi MP, Sathiyamoorthy V, Abdool Karim Q, Alejandria MM, Hernández García C, Kieny MP, Malekzadeh R, Murthy S, Reddy KS, Roses Periago M, Abi Hanna P, Ader F, Al-Bader AM, Alhasawi A, Allum E, Alotaibi A, Alvarez-Moreno CA, Appadoo S, Asiri A, Aukrust P, Barratt-Due A, Bellani S, Branca M, Cappel-Porter HBC, Cerrato N, Chow TS, Como N, Eustace J, García PJ, Godbole S, Gotuzzo E, Griskevicius L, Hamra R, Hassan M, Hassany M, Hutton D, Irmansyah I, Jancoriene L, Kirwan J, Kumar S, Lennon P, Lopardo G, Lydon P, Magrini N, Maguire T, Manevska S, Manuel O, McGinty S, Medina MT, Mesa Rubio ML, Miranda-Montoya MC, Nel J, Nunes EP, Perola M, Portolés A, Rasmin MR, Raza A, Rees H, Reges PPS, Rogers CA, Salami K, Salvadori MI, Sinani N, Sterne JAC, Stevanovikj M, Tacconelli E, Tikkinen KAO, Trelle S, Zaid H, Røttingen JA, Swaminathan S. Repurposed antiviral drugs for Covid-19 – interim WHO solidarity trial results. N Engl J Med. 2021;384(6):497–511. https://doi.org/10.1056/NEJMoa2023184. Epub 2020 Dec 2.
75. Maves RC. Making sense of contradictory evidence in Covid-19 trials. Clin Infect Dis. 2021;78:ciab012. https://doi.org/10.1093/cid/ciab012. Epub ahead of print.
76. The World Medical Association. The Declaration of Taipei on ethical considerations regarding Health Databases and Biobanks. 2002, last revised 2016. https://www.wma.net/policies-post/wma-declaration-of-taipei-on-ethical-considerations-regarding-health-databases-and-biobanks/
77. The Japanese Association of Pharmaceutical Medicine. At the forefront of using Real World Data: insights from the case of the US drug approval landscape in neuroblastoma. February 14, 2025.
78. Kurihara C, Inoue K, Kai H, Suzuki K, Saeki H, Funabashi Y, Kishi N, Kuge A, Murakami T, Saito Y, Uchida E, Tsutsumi N, Imamura K. Our "WMA Declaration of Helsinki": opinions and proposals from patient and public for research ethics. In: Kurihara C, Greco D, Dhai A, editors. Ethical innovation for global health: pandemic, democracy and ethics in research. Singapore: Springer; 2023. p. 243–69.
79. Kurihara C, Saito Y, Kai H, Funabashi Y, Inoue K, Kishi N, Kuge A, Murakami T, Suzuki K, Takahashi H, Uchida E, Imamura K. Patient Public Declaration of Research Ethics (1st edition): research ethics of the people, by the people, for the people-expanding the impact of the 2024 revision of the declaration of Helsinki. Clin Eval. 2024;52(3):W28–39. http://cont.o.oo7.jp/52pop/52pop_contents_e.html
80. Helsinki Statement Stakeholders. Helsinki statement 2024. Clin Eval. 2025;52(3) http://cont.o.oo7.jp/52pop/HelsinkiStatement_18Oct24_final.pdf
81. Clin Eval. 2024; 52(3). http://cont.o.oo7.jp/52pop/52pop_contents_e.html

Chapter 11
Post-trial Access as a Fundamental Right to Health: What Has Changed in the Implementation Principles?

Chieko Kurihara, Dirceu Greco, Ames Dhai, and Takeo Saio

Abstract Post-trial access as the norm of ethics in health research was established by the World Medical Association (WMA) in the 2000 version of the Declaration of Helsinki (DoH), as the right of study participants to continued access to the best-proven interventions if needed after the completion of trial participation. However, due to pressure from regulators, industries, and researchers who argued difficulties with implementation, this principle in the DoH has been downgraded and it became the merely item in the protocol and informed consent form. In its 2024 revision, small improvements were made, reflecting discussions at WMA's meetings as the voices of resource-limited regions were heard but there was a failure to reinstate the past achievement of 2000.

This chapter carefully examines the small improvements made in the post-trial provisions and other relevant paragraphs of the 2024 DoH, to implement post-trial access as health is one of the fundamental human rights. In conclusion, post-trial access must be guaranteed to (1) study participants, (2) study host community, and (3) all those in need in the global community.

Keywords Post-trial access · Right to health · Justice · Research Ethics Committee · Global public good · Sustainable Development Goals (SDGs)

C. Kurihara (✉)
Kanagawa Dental University, Yokosuka, Kanagawa, Japan
e-mail: chieko.kurihara@nifty.ne.jp

D. Greco
School of Medicine, Federal University of Minas Gerais, Belo Horizonte, Brazil

A. Dhai
School of Clinical Medicine, University of the Witwatersrand, Johannesburg, Gauteng, South Africa

T. Saio
Department of Internal Medicine and Psychiatry, Fuji Toranomon Orthopedic Hospital, Shizuoka, Japan

C. Kurihara et al. (eds.), *The 2024 Declaration of Helsinki*,
https://doi.org/10.1007/978-981-96-9294-1_11

1 Introduction: What Was Changed and How Is It Interpreted?

1.1 Downgrading History and Small Improvements in the Paragraph of Post-trial Access

The right to "post-trial access" is an ethical research standard first established internationally as a research ethics norm [1] in the 2000 version of the World Medical Association (WMA)'s Declaration of Helsinki (DoH) [2] ("para." means "paragraph"):

> *At the conclusion of the study, every patient entered into the study should be* ***assured of access to*** *the best proven prophylactic, diagnostic and therapeutic methods identified by the study.* (para. 30, 2000 DoH)

This norm aims to avoid exploitation and abandonment of participants after the conclusion of the study, depriving them of the necessary care that was provided during the study period. This is as if we were using a person as a means to an end just for another purpose, to generate study results. This goes against the Kantian theory [3, 4] and against the physician's obligation to provide care [5]. In addition, interventions proven effective may become available only to wealthier communities, while participants in resource-limited regions may not have access to the proven interventions due to limited healthcare resources in the study host community. This is also clearly contrary to the principle of justice [6, 7].

The fundamental obligation of the physician–researcher to care for the patient was defined in the first DoH in 1964 [8], based on the 1948 Declaration of Geneva [9] and the 1949 International Code of Medical Ethics [10]. Both of which have been cited in the DoH since the first version. In the 1975 DoH [11], this norm was extended to include the guarantee of "best proven" care "including that of a control group," and then, in 2000 to the care of participants after the completion of the study. Meanwhile, the patient's right to care was defined in the 1991 WMA Declaration of Lisbon [12]. All of these are based on the World Health Organization (WHO)'s definition of people's right to health in 1946 [13] as well as United Nation's Declaration of Human Rights in 1948 [14] and Human Rights Covenants in 1966 [15, 16]. The post-trial access norm in the 2000 DoH clarified these international agreements, with consideration specific to the context after the completion of study.

However, soon after the 2000 DoH, objections were raised [17], especially by the US agencies and researchers. They argued the difficulty of providing a proven but not yet authorized intervention, due to the gap of the time until the regulatory approval and inclusion in public health coverage; and that guaranteeing is too hard a task for one individual researcher and that it should be the responsibility of the state or sponsor. For these reasons, the principle of post-trial access established in 2000 was downgraded by the 2004 Note of Clarification being incorporated in the main text in 2008 and modified in the 2013 DoH [1]. Through this process, this ethical norm was then paraphrased as the items of the plan for post-trial access described in the study protocol and informed consent form (ICF). The evaluation would be left to the Research Ethics Committee (REC) and to each study participant, losing the guarantee supported by an agreed international principle.

The 2024 revision [18] made small improvements over the downgraded 2013 version, the impact of which needs to be expanded in practice. Without guaranteeing post-trial access, it becomes just a superficial procedure to change the term "subject" to "participant" and to promote engagement of the participant and their community in all stages of research, calling them partners of cocreation [19].

This chapter describes the DoH's revision process on post-trial access and provides how this prerequisite norm should be applied in research practice and research review, leading up to the next DoH revision. Finally, it clarifies our commitment to strengthen the post-trial access mechanisms, which must be guaranteed to those who need it, at three levels [1]: (1) study participants who still need it (which is in the scope of the DoH text), (2) study host community, and (3) all those who need the developed intervention in the global community (Table 11.1).

1.2 The Contents of the Revision and the Motion from the Uruguayan Medical Union

In the final stage of the 2024 DoH revision process, at the WMA Medical Ethics Committee (October 16, 2024) during the WMA General Assembly in Helsinki, the Uruguayan Medical Union's motion [20] to restore the post-trial access guarantee, once established in the DoH 2000 version (Table 11.2), was discussed together with another motion on the placebo clause (to restore the 2000 principle that allows placebo-controlled trials only when there is no proven intervention: the details are described in the previous chapter).

During the discussion at the WMA Medical Ethics Committee, this motion was supported by Portugal, Argentina, and Spain Medical Associations, but

Table 11.1 Three levels of "post-trial access" and responding levels of discussions

Three levels of "post-trial access"	The levels of discussions
1. Study participants who still need it after the completion (or termination) of the study	A point of discussion in the text of the DoH. Established in the 2000 version but downgraded afterward, with a small improvement in 2024
2. Study host community	Concerns raised in global communities
3. All those in need in the global community, especially the most vulnerable people clearly in need	Authors' argument from the time of the COVID-19 pandemic

The three levels were mentioned in our previous work [1] but are presented here with a more in-depth analysis

Table 11.2 Changes in the post-trial access in the 2024 revision of the Declaration of Helsinki and the motion from the Uruguayan Medical Union

Para. 34 of the 2013 version of the DoH	
"In advance of a clinical trial, sponsors researchers and host country governments should make provisions for post-trial access for all participants who still need an intervention identified as beneficial in the trial. This information must also be disclosed to participants during the informed consent process"	
Changes in the 2024 revision	Proposed changes in the motion from the Uruguayan Medical Union
Authors' note: strikethrough and underline indicate (proposed) updates from 2013	
"In advance of a clinical trial, ~~sponsors, researchers and host country governments should make~~ **post-trial** provisions ~~for post-trial access~~ **must be arranged by sponsors and researchers to be provided by themselves, healthcare systems, or governments** for all participants who still need an intervention identified as beneficial **and reasonably safe** in the trial. **Exceptions to this requirement must be approved by a research ethics committee.** ~~This~~ **Specific** information **about post-trial provisions** must ~~also~~ be disclosed to participants **as part of** ~~during the~~ informed consent. ~~process~~"	"In advance of a clinical trial, post-trial **access** ~~provisions~~ must be **guaranteed** ~~arranged~~ by sponsors and researchers to be provided by themselves, healthcare systems, or governments for all participants who still need an intervention identified as beneficial and reasonably safe in the trial. ~~Exceptions to this requirement must be approved by a research ethics committee~~. Specific information about post-trial **guarantee** ~~provisions~~ must be disclosed to participants as part of informed consent"
Summary of the reasoning in the public consultation document provided by the WMA	Reasoning in the motion from the Uruguayan Medical Union
Responding to concerns on research in low-resource settings were raised at the Vatican regional meeting, it should be stated that post -trial provisions must be arranged for study participants who need access to the trial intervention	*"The Sindicato Médico del Uruguay[a] maintains that post-trial access must be guaranteed for all those who still need an intervention that has been identified as beneficial and reasonably safe in the trial"*
Para. 34 of the 2024 version of the DoH	
"In advance of a clinical trial, post-trial provisions must be arranged by sponsors and researchers to be provided by themselves, healthcare systems, or governments for all participants who still need an intervention identified as beneficial and reasonably safe in the trial. Exceptions to this requirement must be approved by a research ethics committee. Specific information about post-trial provisions must be disclosed to participants as part of informed consent"	

Authors' note: The texts in italics are quotations from the 2013 and 2024 versions of the DoH and from the motion of the Uruguayan Medical Union (document Number: MEC 227/DoH/Oct2024). The citation of the motion is authorized by the Uruguayan Medical Union

[a]Sindicato Médico del Uruguay = Uruguayan Medical Union

members of two Medical Associations objected. Ultimately, the motion was voted down.

Latin American countries have so far maintained a consensus on the 2000 DoH policy on placebo-controlled trials and on post-trial access, in the legal framework [21–23] and in a series of declarations [24–29]. Strengthening post-trial access was also discussed at the WMA's meeting in the Vatican City State, which set the agenda for research in low-resource settings, one of the eight of regional meetings worldwide during the 2 years revision process. This was described as the WMA's reasoning of the revision in the public consultation document (Table 11.2).

According to the press release of the Uruguayan Medical Union [30] published on October 16, 2024, not only Latin American and Caribbean regions but also countries of Southwestern Europe (Spain, Portugal, France, and Italy) have the same position. On the same date, the Confederación Médica Latinoamericana y del Caribe (CONFEMEL), "medical organizations in Ibero-America and the Caribbean, which represents around 500,000 doctors grouped in national associations," expressed a favorable opinion of Uruguay regarding both the placebo-controlled trial and post-trial access [31]. This position of the CONFEMEL had already been expressed in March 2023 [32].

The final requirements adopted in the 2024 DoH [18] are insufficient and unclear, and can be paraphrased in the following three points:

1. Physician–researcher and sponsor **must** arrange the post-trial provision* for all participants who still need an intervention identified as beneficial and reasonably safe in the trial.
2. This provision* must be provided by themselves (researcher and sponsor), healthcare systems, or government.
3. Exception* to this requirement must be approved by a REC.

   Authors' note: * "Provision" here means "a descriptive statement of what to do" but does not necessarily mean to "provide the proven intervention." The second item is not clear whether the researcher/sponsor, healthcare systems, or government must provide a proven intervention or just a description of what to do. Thus, "exception" is not clear for which meaning of duty. However, in any case, we should read this paragraph in the sense that post-trial access to proven intervention must be granted to participants who need it, and only ethically justifiable exception could be approved by REC.

2 Maximizing the Impact of Small Changes

To strengthen the impact of this small improvements, it is not enough to consider the ambiguous paragraph 34 only, but other notable changes in related paragraphs: (1) REC accountability and functioning; (2) community engagement; (3) vulnerability; and (4) accountability of teams and organizations, as follows:

2.1 REC's Responsibility and Functioning to Strengthen Access (Para. 23)

REC's responsibility and independence to examine what kind of "exception" could be approved is of fundamental importance. RECs must not be used as a means to justify the termination of a study intervention necessary for the continued maintenance of a participant's health status (Level 1 in Table 11.1). Some types of intervention may only be necessary for a limited study period. This could be approved and is not an "exception," but an out of the scope of the "requirement" of post-trial access. What kind of "exception" can be approved is not clear from the texts and interpretation of the entire DoH. To clarify the WMA's position, continued deliberations by the global community are needed to avoid exploitation, inequity, and disparity at the local, regional, or global level.

Post-trial provision must be explicitly described, without exception, in each study protocol, and scrutinized by the REC before an informed consent form (ICF) is provided to each participant. As suggested in the 2000 DoH, post-trial access must be assured if the study participants need it continuously, according to the fundamental principle of the DoH to prioritize participants' rights over the goal of research (para. 7). Furthermore, a clear description in the ICF must be required to allow the study participant not only to make an autonomous decision to participate but also to exercise their right of access at the completion of the study. RECs must act in favor of the rights and best interests of the "participant," who is never a mere "subject" of experimentation, and must always be considered as a partner in the research "cocreation." REC members must exercise their responsibility and expand their opportunities, albeit minimally described in the DoH. It is the role of the REC to request post-trial access from the sponsor for participants in need as this is the human and corporate social responsibility of the sponsor.

The REC's functions in paragraph 23 have been strengthened particularly in the following item related to post-trial access [18].

> *The committee must have <u>sufficient familiarity with local circumstances and context</u>, and include at least one member of the <u>general public</u>.*

This requires that the REC respect community values and cultural background including in the perspective of the general public. Furthermore, it would be a prerequisite to clarify how study results could contribute to post-trial access and benefit sharing in the community, with an appropriate implementation strategy (Level 2 in Table 11.1). The committee member(s) representing the local community should have responsibility for evaluating the specific description of the post-trial provision to be able to guarantee access for those in need in the community.

2.2 Strengthened Community Engagement (Para. 6)

For the first time, the WMA included the principle of community engagement [18]:

> *… Since medical research takes place in the context of various structural inequities, researchers should carefully consider how the benefits, risks, and burdens are distributed.*
>
> *Meaningful engagement with potential and enrolled participants and their communities should occur before, during, and following medical research. Researchers should enable potential and enrolled participants and their communities to share their priorities and values; to participate in research design, implementation, and other relevant activities; and to engage in understanding and disseminating results.* (para. 6, underlined by authors)

This concept has been established in the Council of International Organizations of Medical Sciences (CIOMS) International Ethical Guidelines for Health-related Research Involving Humans [5], Guideline 7. This community engagement strategy within CIOMS is the result of long-standing debates aimed at avoiding exploitation that vulnerable communities are exposed to the risk of studies and that results of studies are only available in the wealthiest regions. This issue has not yet been resolved [33–36]. All the interested parties should be engaged in every relevant stage of research in order to guarantee proven intervention not only for study participants but also for all in need in study host communities (Level 2 in Table 11.1).

In the revised DoH, the term "disseminating results" is weak, as it seems to simply mean dissemination of information. Some authors of this chapter expressed the opinion that "implementation (of results)" should be added. Community engagement should be accompanied by a strategic plan to guarantee post-trial access in the study host community.

2.3 The Vulnerability That Should Not Be Exploited (Para. 19, 20)

The following sentence was changed from "*should stands to benefit*" in 2013 to "*only justified if…stands to benefit.*" (para. 20) [18]:

> *Medical research with individuals, groups, or communities in situations of particular vulnerability is only justified if it is responsive to their health needs and priorities and the individual, group, or community stands to benefit from the resulting knowledge, practices, or interventions.*

This change would be due to promoting the inclusion of vulnerable people, with the recognition that "*exclusion from medical research can potentially perpetuate or exacerbate their disparities*" (para. 19). For vulnerable people to participate in research, it is a prerequisite to provide not only adequate safeguards, but also to guarantee post-trial access [37]. Without post-trial access, promoting inclusion risks exploitation, both at the level of each individual participant (Level 1 in Table 11.1) and the level of study host community (Level 2 in Table 11.1).

One regrettable point of the 2024 revision was the removal of the term "social value," although it was presented as the ultimate goal of research in the first public consultation version. "Social value" is included as Guideline 1 of the 2016 CIOMS Guidelines [5]. Guaranteeing post-trial access and benefit sharing, especially for vulnerable people in resource-limited regions, is a necessary item of social value associated with the precondition of scientific value, which can never take precedence over individual rights [5]. Reconceptualizing these notions [37] would be one of the challenges for the next revision of the DoH.

2.4 Responsibility of Teams and Organizations (Para. 2)

Post-trial access may be difficult for an individual researcher both in terms of providing adequate care to participants and in terms of negotiating with the sponsor company. It is noteworthy that one of the highlights of the 2024 revision was the inclusion of the new terms in paragraph 2: "*these principles should be upheld by all individuals, teams, and organizations involved in medical research*" [18]. Based on this, not only the researcher, but also teams and organizations should take responsibility and make every effort to obtain post-trial access for participants who need to continue the intervention (Level 1 in Table 11.1, or Level 2 in the case of community-based organization).

3 History of Injustice and Inequity

3.1 Global Efforts for Post-trial Access (1): HIV/AIDS Crisis

The principle of post-trial access is one of the outcomes of the international controversies that raised questions of "ethical imperialism" in the 1980s [38] (recognized as "structural inequity" in the 2024 DoH), and especially triggered by studies of prevention of mother-to-child transmission of HIV/AIDS in the late 1990s. After effective interventions were established in developed countries [39], 15 placebo-controlled trials, sponsored by institutions of the United States and France were conducted in developing countries to compare less expensive (but not affordable for the most of host communities [40]) against placebo [41]. After interim analysis of some studies, the intervention that had shown to be effective was provided to participants in the placebo arm (Level 1 in Table 11.1). However, for most people in developing countries, even this less expensive intervention was too expensive to be offered in the public health systems (Level 2 or 3 in Table 11.1).

Responding to the HIV/AIDs crises, an international consensus was reached that public health must be prioritized over the protection of intellectual property rights, which was proclaimed in the Doha Declaration in 2001 [42] as an addendum to the

World Trade Organization (WTO)'s Agreement on Trade-Related Aspects of Intellectual Property Rights (TRIPS Agreement). It opened the possibility that essential drugs could be produced by implementing "compulsory licensing" (a State authorizes a non-patent holder to use patent-protected technology without the patent holder's permission) and some countries with generic drugs manufacturing capacity (e.g., Brazil, India) could produce and export such medicines to affected and resource-limited countries most in need (Level 3 in Table 11.1). The WTO Doha Declaration states that "*the TRIPS Agreement does not and should not prevent members from taking measures to protect public health. Accordingly, while reiterating our commitment to the TRIPS Agreement, we affirm that the Agreement can and should be interpreted and implemented in a manner supportive of WTO members' right to protect public health and, in particular, to promote access to medicines for all*" [42].

3.2 *Global Efforts for Post-trial Access (2): COVID-19 Pandemic*

When two COVID-19 vaccines were proven to be safe and effective by interim analyses in late 2020 [43–45], and as the research results were based on global collaboration, especially with the participation of the most affected, including vulnerable communities, the sponsoring companies changed their study protocols to allow study participants in the placebo group to receive vaccines at their own discretion when vaccines became available in their region [1]. This strategy was defined by an agreement of the regulatory authorities and companies to assume the ethical responsibility of ensuring post-trial access. It also appears to be a part of the marketing strategy of sponsoring companies. In addition, there were global efforts by the COVAX Initiative to develop a pooled purchasing system, as well as to ensure access to vaccines, even in the most affected low-income countries, through donations and Official Development Assistance (ODA) (Level 3 in Table 11.1). However, this initiative was hampered by "vaccine nationalism" [46] of wealthier countries that sought to purchase vaccines for their nations through bilateral negotiation with the companies.

The important milestone was the proposal by South Africa and India to the WTO for a waiver from the application of the TRIPS Agreement to COVID-19-related intellectual property rights [47], called the "TRIPS Waiver." The proposal, supported by many countries, was associated with a series of initiatives for the local development of manufacturing capacity [48, 49]. Unfortunately, it was not until June 2022 that a waiver agreement was reached at WTO and was limited to COVID vaccines, "authorizing the use of the subject matter of a patent required for the production and supply of COVID-19 vaccines without the consent of the right holder to the extent necessary to address the COVID-19 pandemic." [50] Global consensus was formed to promote international solidarity and that a COVID-19 vaccine is a

Global Public Good [51–54]. In addition, the COVAX Initiative and the TRIPS Waiver were regarded as a step toward global vaccine equity and solidarity to ensure the equal fundamental right to health [13–16], as a part of Sustainable Development Goals (SDGs) [55].

These global experiences were acknowledged in the text of the DoH as the recognition of structural inequity and disparity. However, the language of the 2024 DoH on post-trial access is insufficient to overcome the future possibility of emerging health crises. This must be overcome by both local and global efforts (Levels 2 and 3 in Table 11.1) to expand the small revision of the 2024 DoH, highlighting the right of each community, region, or country to establish more protective rules with strategies for implementation.

4 Discussion

4.1 Clarifications of Responsibilities of Each Interested Party

Post-trial access is a prima facie ethical norm, based on human rights to avoid exploitation of study participants and ensure the physician's duty of care. Our previous paper [37] highlighted the responsibility of each researcher to care for participants and negotiate with the sponsor (Level 1 in Table 11.1); the sponsor must take primary responsibility as one of the corporate social responsibilities not only for study participants (Level 1 in Table 11.1) but also for study host communities (Level 2 in Table 11.1) and globally, toward SDGs with research governance policies (Levels 2 and 3 in Table 11.1); and that of the government in establishing a regulatory pathway for immediate post-trial access (Level 1 in Table 11.1) and also the responsibility for including in the public health system with ongoing health technology assessment (HTA) (Level 2 in Table 11.1).

The 2024 DoH mentioned that some of these parties should take some responsibility, without explicitly recommending that post-trial access must be guaranteed. Most counterarguments are not based on ethics, but on regulatory barriers, financial limitation, and capacity-building challenges, for which further deliberations are needed. The discussion points are summarized below.

4.2 Need for a Regulatory Framework

The regulatory arguments are that providing unapproved drugs to people who need them may violate the law. In many jurisdictions, due to regulatory frameworks or company policy, schemes already exist to provide unapproved drugs to people who need them through safety studies, extension studies, or expanded access programs (Level 1 in Table 11.1). To implement post-trial access, governments, industry, and research institutes must establish such schemes with rigorous frameworks to maintain participant protection and the integrity of data generated by post-trial treatment, in collaboration with interested parties including participant representatives and the public.

A law is an ethical minimum [56] and ethics is sustained by the internal incentive of autonomous individuals [3]. Ethics suggests the direction to laws because ethics rises above the law, but if the DoH cannot provide clear ethical recommendations (which is a weakness of the 2024 DoH), physician–researchers, sponsor, REC, participants, and the community must act according to their internal incentive for the benefit of not only each participant but also the community (Level 2 in Table 11.1), at the time of each decision-making, as well as for the establishment of the appropriate legal pathways.

4.3 To Overcome Financial Limitation

Financially motivated claims are often made that the cost of medicines will increase if the sponsoring companies are required to prepare to cover the cost of post-trial access. However, the costs required for post-trial access are small, compared to the costs used in post-trial marketing [57]. Another type of claim is that investigator-initiated trials are sometimes conducted by purchasing drugs from companies that are not motivated to develop them, and therefore the costs of providing the drug after the trial cannot be covered within the limited research budget. Such cases are rarer and exceptional cases must be carefully evaluated by the REC. RECs should work to promote partnerships between researchers and manufacturer companies. Furthermore, the relevant community and society promote that industries to be motivated to conduct studies with social value rather than studies motivated by profit.

Meanwhile, the principle of post-trial access is sometimes used as a part of a marketing strategy. Where the notion of post-trial access is used as a means of a "seeding trial" [58] to create a dominant marketing jurisdiction, involving participants who do not need them, this would be most inappropriate example of a "seeding trial" just for marketing purpose and lacking "social value" as explained in the CIOMS Guidelines [5].

On the other hand, if sponsors fulfill their ethical responsibilities and deliver the essential medicines to those who need it, and where leads to better sales results, then this could be considered a win–win relationship between ethics and commercial activities. The implication of community engagement sought by the CIOMS Guidelines [5] and the CIOMS report for patient involvement [59] would be that all interested parties will work together to plan a strategy not only for post-trial access but also for post-trial studies and health technology assessment to identify the needs for access to developed drugs both in domestic and on global scales (Levels 2 and 3 in Table 11.1). A justifiable and equitable scheme for public resource allocation is a future challenge to achieve sustainable post-trial access for all who need it.

4.4 Intellectual Property Right Management and Capacity Development

To expand global post-trial access along with an appropriate marketing strategy, it is necessary to develop an intellectual property right (IPR) management scheme that follows the principles of the Doha Declaration [42] and the TRIPS Waiver [47], taking into account global experiences with HIV/AIDS and COVID-19 (Level 3 in Table 11.1). This requires technology transfer strategies and community capacity building, both in product manufacturing and procurement, as well as in the sociocultural environment.

The Convention on Biological Diversity [60, 61] defined the requirements for benefit sharing for areas holding biological resources, as well as detailed requirements and commitments by the interested parties to achieve this. This benefit-sharing policy is already incorporated in the UNESCO Universal Declaration of Bioethics and Human Rights [62, 63]. This could be applied as guiding principles to expand the impact of the DoH.

The globalization of research is expanding, and data-driven research is being promoted, along with the promotion of the use of real-world evidence for marketing authorization. Data from people suffering from the disease being studied, which may be spread around the world, will form part of the total body of evidence [63]. Those who contribute to such data accumulation at the global scale are entitled/stand to engagement in all stages of research and development, along with capacity development, leading to the benefit-sharing resulting from the research (Level 3 in Table 11.1). In addition, artificial intelligence (AI) tools, especially those developed with data collected from studies in low- and middle-income countries (LMICs), must be considered in post-trial access, which must not be limited to medicines and devices, but include the wider concept of digital health. This would guarantee more equitable access to these tools, promoting justice in global health research [64].

5 Conclusion: Post-trial Access for All

Discussions about post-trial access should not be seen as a technical procedure or question of study cost. This is a fundamental ethical standard for equal human rights to health. The standard of post-trial access for all who need it is a pathway to achieving one of the SDGs. The objectives and steps are clear and obvious, and what is needed is that it is implemented in every trial process, including protocol development; negotiation and the contract between researcher, industry, and relevant parties; research ethics review; study conduct; and post-trial arrangement in terms of regulatory frameworks; IPR management, technology transfer, manufacturing, procurement, and social capacity building; and health technology assessment, including adequate post-marketing study designs.

All stages of post-trial access must be ensured for each participant, for the study host community and from the view of a global scale. To achieve this, radical systems reformation is needed to recognize health as a right and study outcomes as a global common good. As we continue to work toward this goal, the 2024 DoH can be one of the driving forces along with the 2016 CIOMS Guidelines [5] and the UNESCO Declaration on Bioethics and Human Rights [62], for the future development of global ethical standards for research involving humans.

Acknowledgments CK appreciates the International Federation of Associations of Pharmaceutical Physicians and Pharmaceutical Medicine (IFAPP) for granting her travel expenditure for her in-person participation in the WMA Regional Meeting in Tel Aviv (December 2022) and the WMA for granting her travel expenditure for her in-person participation and presentation as a panelist in WMA Regional Meeting in Washington, DC (August 2024), and waiver of registration fee from WMA as a journalist to attend the General Assembly in Helsinki (October 2024). She participated in all the other WMA Regional Meetings at her own expense or online.

DG was officially engaged in discussions for the revisions of the Declaration of Helsinki from 2000 to 2008 as a member of the Brazilian Medical Association and Brazilian Medical Council teams; he also participated in the CIOMS 2016 guidelines as a member of working group; he was invited to the regional meeting in Sao Paulo by the WMA to discuss the ethics of placebo-controlled trials; and he participated, with partial funding from CIOMS, in the 2024 WMA General Assembly that adopted the 2024 version of the DoH.

Conflict of Interest AD was a Board member of the South African Medical Association, which was a member in the Workgroup of the World Medical Association for the 2024 revision of the Declaration of Helsinki. The authors declare no other competing interests relevant to the content of this chapter.

References

1. Kurihara C, Greco D, Dhai A. Post-trial access: historical analysis considering the experience of COVID-19 pandemic. In: Kurihara C, Greco D, Dhai A, editors. Ethical innovation for global health: pandemic, democracy and ethics in research. Singapore: Springer; 2023. p. 225–41.
2. World Medical Association. Declaration of Helsinki: ethical principles for medical research involving human subjects. 2000. Rescinded. https://www.wma.net/wp-content/uploads/2018/07/DoH-Oct2000.pdf
3. Kant I. Groundwork of the metaphysics of morals. 1785.
4. van der Graaf R, van Delden JJ. On using people merely as a means in clinical research. Bioethics. 2012;26(2):76–83. https://doi.org/10.1111/j.1467-8519.2010.01820.x. Epub 2010 Apr 5. PMID: 20377571.
5. Council for International Organizations of Medical Sciences. International ethical guidelines for health-related research involving humans. 2016. https://cioms.ch/publications/product/international-ethical-guidelines-for-health-related-research-involving-humans/
6. The National Commission for the Protection of Human Subjects of Biomedical and Behavioral Research. The Belmont Report: ethical principles and guidelines for the protection of human subjects of research. 1979. https://www.hhs.gov/ohrp/regulations-and-policy/belmont-report/read-the-belmont-report/index.html
7. Rawls J. A theory of justice. Cambridge: Belknap Press; 1971.

8. World Medical Association. Declaration of Helsinki: recommendations guiding doctors in clinical research. June 1964. Rescinded. https://www.wma.net/wp-content/uploads/2018/07/DoH-Jun1964.pdf
9. World Medical Association. WMA Declaration of Geneva. Adopted in September 1948, last amended in October 2017. https://www.wma.net/policies-post/wma-declaration-of-geneva/
10. World Medical Association. WMA International Code of Medical Ethics. Adopted in October 1949, last amended in October 2022. https://www.wma.net/policies-post/wma-international-code-of-medical-ethics/
11. World Medical Association. Declaration of Helsinki: recommendations guiding medical doctors in biomedical research involving human subjects. 1975. Rescinded. https://www.wma.net/wp-content/uploads/2018/07/DoH-Oct1975.pdf
12. The World Medical Association. WMA Declaration of Lisbon on the Rights of the Patient. Adopted in September/October 1981, last reaffirmed in April 2015. https://www.wma.net/policies-post/wma-declaration-of-lisbon-on-the-rights-of-the-patient/
13. World Health Organization. Constitution of the World Health Organization. 1946 July 22. https://www.who.int/about/governance/constitution
14. United Nations. The universal declaration of human rights. Adopted 10 December 1948. https://www.un.org/en/about-us/universal-declaration-of-human-rights
15. United Nations. International Covenant on Economic, Social and Cultural Rights. Adopted by General Assembly resolution 2200A (XXI) of 16 December 1966, entry into force 3 January 1976. https://www.ohchr.org/en/instruments-mechanisms/instruments/international-covenant-economic-social-and-cultural-rights
16. United Nations. International Covenant on Civil and Political Rights. Adopted by General Assembly resolution 2200A (XXI) of 16 December 1966, entry into force 23 March 1976. https://www.ohchr.org/en/instruments-mechanisms/instruments/international-covenant-civil-and-political-rights
17. Pérez AC, Smith RN. The revised Declaration of Helsinki: interpreting and implementing ethical principles in biomedical research. Int J Pharm Med. 2001;15:131–43. https://doi.org/10.2165/00124363-200106000-00006.
18. The World Medical Association. WMA Declaration of Helsinki—ethical principles for medical research involving human participants. First adopted in 1964, last amended in 2024. https://www.wma.net/policies-post/wma-declaration-of-helsinki/
19. Kurihara C, Matsuyama K, Baroutsou V. World Medical Association's Declaration of Helsinki, 2024 revision: celebrating the 60th anniversary, at Helsinki. Clin Eval. 2025;52(3):W1–27. http://cont.o.oo7.jp/52pop/52pop_contents_e.html
20. Sindicato Médico del Uruguay. World Medical Association Submission of a Motion to Amend (on Paragraph 34, submitted to) 227th Council Session and General Assembly, Helsinki, 8 Oct 2024.
21. Greco D, Shimoda K, Watanabe H, Organizer. The past, present, and future of ethics of international health research: research as a stepping-stone to universal public health care access. Clin Eval. 2020;48(1):207–31. http://cont.o.oo7.jp/48_1/w29-w53.pdf
22. Brazilian National Health Council (CNS), Brazilian National Research Ethics Committee (CONEP/CNS/MS). Guidance manual: frequent pending issues in clinical research protocols. Version 1.0. 2015. https://abracro.org.br/images/legislacao/guidance_manual_frequent_pending_issues_in_clinical_research_protocols_CONEP_V1_2015.pdf
23. Brazilian Research Ethics Commission Resolution 466/2012, which succeeded Resolution 404/2008.
24. Homedes N, Ugalde A, editors. Clinical trials in Latin America: where ethics and business clash. Cham: Springer; 2014.
25. The Buenos Aires Declaration on Ethics and Clinical Trials. Buenos Aires, May 13, 2008. In: Homedes N, Ugalde A, editors. Clinical trials in Latin America: where ethics and business clash. Cham: Springer; 2014.

26. Red Latinoamericana y Del Caribe de Bioetica—Redbioetica/UNESCO. Carta de Co'rdoba sobre e'tica en investigaciones con seres humanos. Córdoba, Argentina, 14 de noviembre de 2008. https://www.psi.uba.ar/academica/carrerasdegrado/psicologia/sitios_catedras/obligatorias/723_etica2/material/casuistica/carta_de_cordoba_08.pdf
27. Solbakk JH. National ethics advisory bodies and committees in Norway: history, lessons learnt, and common challenges ahead. National Bioethics Committees in Action. United Nations Educational, Scientific and Cultural Organization; 2010. https://unesdoc.unesco.org/ark:/48223/pf0000189548
28. Confederation Medica latinoamericana Y del Caribe (Latin American and Caribbean Medical Confederation). Declaration of Pachuca 2013.
29. Greco DB. Ethical dilemmas, research, and access to treatment. In: International encyclopedia of the social & behavioral science, vol. 8. 2nd ed. Amsterdam: Elsevier; 2015. p. 31–6.
30. Sindicato Medico del Uruguay. SMU propone cambios en la Declaración de Helsinki. 16 October 2024. https://www.smu.org.uy/el-smu-en-la-asamblea-de-la-asociacion-medica-mundial-propone-cambios-en-la-declaracion-de-helsinki/
31. CONFEMEL. Intervención de la Dra. Zaida Arteta en la Asamblea General de la AMM. 16 octubre 2024. https://www.confemel.com/intervencion-de-la-dra-zaida-arteta-en-la--asamblea-general-de-la-amm/
32. CONFEMEL. CONFEMEL considera adecuado y oportuno el proceso de Revisión de la DoH llevado adelante por la AMM. 24 April 2023. https://www.confemel.com/confemel-considera-adecuado-y-oportuno-el-proceso-de-revision-de-la-doh-llevado-adelante-por-la-amm/
33. Miller JE, Mello MM, Wallach JD, Gudbranson EM, Bohlig B, Ross JS, Gross CP, Bach PB. Evaluation of drug trials in high-, middle-, and low-income countries and local commercial availability of newly approved drugs. JAMA Netw Open. 2021;4(5):e217075. https://doi.org/10.1001/jamanetworkopen.2021.7075.
34. Gyawali B, Carson LM, Berry S, Moraes FY. Challenges of globalization of cancer drug trials – recruitment in LMICs, approval in HICs. Lancet Reg Health Am. 2021;7:100157. https://doi.org/10.1016/j.lana.2021.100157.
35. Rubagumya F, Hopman WM, Gyawali B, Mukherji D, Hammad N, Pramesh CS, Zubaryev M, Eniu A, Tsunoda AT, Kutluk T, Aggarwal A, Sullivan R, Booth CM. Participation of lower and upper middle-income countries in clinical trials led by high-income countries. JAMA Netw Open. 2022;5(8):e2227252. https://doi.org/10.1001/jamanetworkopen.2022.27252. Erratum in: JAMA Netw Open. (2022) Sept 1;5(9):e2235554. https://doi.org/10.1001/jamanetworkopen.2022.35554.
36. Council for International Organizations of Medical Sciences. Clinical research in resource-limited settings. 2021. https://cioms.ch/publications/product/clinical-research-in-low-resource-settings/
37. Kurihara C, Greco D, Dhai A, Matsuyama K, Baroutsou V. Vulnerability, social value and the equitable sharing of benefits from research: beyond the placebo and access debates. Front Med. 2024;11:1432267. https://doi.org/10.3389/fmed.2024.1432267.
38. Angell M. Ethical imperialism? Ethics in international collaborative clinical research. N Engl J Med. 1988;319(16):1081–3. https://doi.org/10.1056/NEJM198810203191608. PMID: 3173435.
39. Connor EM, Sperling RS, Gelber R, et al. Reduction of maternal-infant transmission of human immunodeficiency virus type 1 with zidovudine treatment. Pediatric AIDS Clinical Trials Group protocol 076 study group. N Engl J Med. 1994;331(18):1173–80.
40. Varmus H, Satcher D. Ethical complexities of conducting research in developing countries. N Engl J Med. 1997;337(14):1003–5. https://doi.org/10.1056/NEJM199710023371411.
41. Lurie P, Wolfe SM. Unethical trials of interventions to reduce perinatal transmission of the human immunodeficiency virus in developing countries. N Engl J Med. 1997;337(12):853–6.
42. World Trade Organization. Doha Declaration on the TRIPS Agreement and Public Health. Adopted by the WTO Ministerial Conference of 2001 in Doha on 14 Nov 2001. https://www.wto.org/english/thewto_e/minist_e/min01_e/mindecl_trips_e.htm

43. Polack FP, Thomas SJ, Kitchin N, Absalon J, Gurtman A, Lockhart S, Perez JL, Pérez Marc G, Moreira ED, Zerbini C, Bailey R, Swanson KA, Roychoudhury S, Koury K, Li P, Kalina WV, Cooper D, Frenck RW Jr, Hammitt LL, Türeci Ö, Nell H, Schaefer A, Ünal S, Tresnan DB, Mather S, Dormitzer PR, Şahin U, Jansen KU, Gruber WC, C4591001 Clinical Trial Group. Safety and efficacy of the BNT162b2 mRNA Covid-19 vaccine. N Engl J Med. 2020; https://doi.org/10.1056/NEJMoa2034577. Epub ahead of print. PMID: 33301246.
44. Baden LR, El Sahly HM, Essink B, Kotloff K, Frey S, Novak R, Diemert D, Spector SA, Rouphael N, Creech CB, McGettigan J, Kehtan S, Segall N, Solis J, Brosz A, Fierro C, Schwartz H, Neuzil K, Corey L, Gilbert P, Janes H, Follmann D, Marovich M, Mascola J, Polakowski L, Ledgerwood J, Graham BS, Bennett H, Pajon R, Knightly C, Leav B, Deng W, Zhou H, Han S, Ivarsson M, Miller J, Zaks T, COVE Study Group. Efficacy and safety of the mRNA-1273 SARS-CoV-2 vaccine. N Engl J Med. 2020; https://doi.org/10.1056/NEJMoa2035389. Epub ahead of print. PMID: 33378609.
45. Voysey M, Clemens SAC, Madhi SA, Weckx LY, Folegatti PM, Aley PK, Angus B, Baillie VL, Barnabas SL, Bhorat QE, Bibi S, Briner C, Cicconi P, Collins AM, Colin-Jones R, Cutland CL, Darton TC, Dheda K, Duncan CJA, KRW E, Ewer KJ, Fairlie L, Faust SN, Feng S, Ferreira DM, Finn A, Goodman AL, Green CM, Green CA, Heath PT, Hill C, Hill H, Hirsch I, Hodgson SHC, Izu A, Jackson S, Jenkin D, Joe CCD, Kerridge S, Koen A, Kwatra G, Lazarus R, Lawrie AM, Lelliott A, Libri V, Lillie PJ, Mallory R, Mendes AVA, Milan EP, Minassian AM, McGregor A, Morrison H, Mujadidi YF, Nana A, O'Reilly PJ, Padayachee SD, Pittella A, Plested E, Pollock KM, Ramasamy MN, Rhead S, Schwarzbold AV, Singh N, Smith A, Song R, Snape MD, Sprinz E, Sutherland RK, Tarrant R, Thomson EC, Török ME, Toshner M, Turner DPJ, Vekemans J, Villafana TL, Watson MEE, Williams CJ, Douglas AD, AVS H, Lambe T, Gilbert SC, Pollard AJ, Oxford COVID Vaccine Trial Group. Safety and efficacy of the ChAdOx1 nCoV-19 vaccine (AZD1222) against SARS-CoV-2: an interim analysis of four randomised controlled trials in Brazil, South Africa, and the UK. Lancet. 2020;397(10269):99–111. https://doi.org/10.1016/S0140-6736(20)32661-1. Epub ahead of print. PMID: 33306989; PMCID: PMC7723445.
46. Duke Global Health Innovation Center. New study shows rich country shopping spree for COVID-19 vaccines could mean fewer vaccinations for billions in low-income countries. 2 Nov 2020. https://dukeghic.org/wp-content/uploads/sites/20/2020/11/COVID19-Vax-Press-Release__28Oct2020-1.pdf
47. Council for Trade-Related Aspects of Intellectual Property Rights. Waiver from certain provisions of the TRIPS Agreement for the prevention, containment and treatment of COVID-19: Communication from India and South Africa. 2 Oct 2020; IP/C/ W/669. https://docs.wto.org/dol2fe/Pages/SS/directdoc.aspx?filename=q:/IP/C/W669.pdf&Open=True
48. Dhai A. Equitable access to COVID-19 vaccines, vaccine research, and vaccine apartheid on African continent: challenges and recommendations. In: Kurihara C, Greco D, Dhai A, editors. Ethical innovation for global health: pandemic, democracy and ethics in research. Singapore: Springer; 2023. p. 39–55.
49. Baroutsou V. Medicines development for global health: learning from COVID-19 vaccines R&D. In: Kurihara C, Greco D, Dhai A, editors. Ethical innovation for global health: pandemic, democracy and ethics in research. Singapore: Springer; 2023. p. 273–82.
50. World Trade Organization, Ministerial Conference Twelfth Session Geneva, 12–15 June 2022. Ministerial Decision on the TRIPS Agreement. Adopted on 17 June 2022. 22 June 2022; WT/MIN(22)/30 WT/L/1141 https://docs.wto.org/dol2fe/Pages/SS/directdoc.aspx?filename=q:/WT/MIN22/30.pdf&Open=True
51. The Lancet COVID-19 Commission. Enhancing global cooperation to end the COVID-19 pandemic. February 2021. https://covid19commission.org/enhancing-global-cooperation
52. Corporate author: UNESCO International Bioethics Committee, World Commission on the Ethics of Scientific Knowledge and Technology. Statement on COVID-19: ethical considerations from a global perspective. April 2020. https://unesdoc.unesco.org/ark:/48223/pf0000373115

53. Corporate author: International Bioethics Committee, World Commission on the Ethics of Scientific Knowledge and Technology. UNESCO's ethics commissions' call for global vaccines equity and solidarity: joint Statement by the UNESCO International Bioethics Committee (IBC) and the UNESCO World Commission on the Ethics of Scientific Knowledge and Technology (COMEST). February 2021. https://unesdoc.unesco.org/ark:/48223/pf0000375608
54. Corporate author: International Bioethics Committee, World Commission on the Ethics of Scientific Knowledge and Technology. Joint statement of UNESCO's ethics commissions on ensuring equal access for all to vaccines and therapeutics developed to confront COVID-19: joint statement of the UNESCO International Bioethics Committee (IBC) and UNESCO World Commission for the Ethics of Science and Technology (COMEST). September 2021. https://unesdoc.unesco.org/ark:/48223/pf0000379042
55. United Nations. Resolution adopted by the General Assembly on 25 September 2015 – Transforming our world: the 2030 Agenda for Sustainable Development. 21 Oct 2015. https://www.un.org/en/development/desa/population/migration/generalassembly/docs/globalcompact/A_RES_70_1_E.pdf
56. Jellinek G. Die sozialethische Bedeutung von Recht, Unrecht und Strafe. Wien: Alfred Hölder; 1878. https://archive.org/details/diesocialethisc00jellgoog
57. Greco D, Invited lecturer. Kimura R, Special guest. Victoria Perottino M, Guest Discussant. Saio T, Kurihara C, Organizers & Discussants. Ethics of international collaborative research: perspectives from Brazil: part 1 selected notes on Paulo Freire: part 2 access, compulsory license, case study. Clin Eval. 2020;48(1):273–301. http://cont.o.oo7.jp/48_1/w95-w123.pdf
58. Sox HC, Rennie D. Seeding trials: just say "no". Ann Intern Med. 2008;149(4):279–80. https://doi.org/10.7326/0003-4819-149-4-200808190-00012. Erratum in: Ann Intern Med 2008 Sept 16;149(6):439.
59. Council for International Organizations of Medical Sciences (CIOMS). Patient involvement in the development, regulation and safe use of medicines. CIOMS Working Group report. Geneva, Switzerland. 2022. https://doi.org/10.56759/iiew8982
60. United Nations. Convention on biological diversity. 1992. https://www.cbd.int/doc/legal/cbd-en.pdf
61. Secretariat of the Convention on Biological Diversity, United Nations. Nagoya protocol on access to genetic resources and the fair and equitable sharing of benefits arising from their utilization to the Convention on Biological Diversity: text and annex/Secretariat of the Convention on Biological Diversity. 2011. Available at: https://www.cbd.int/abs/doc/protocol/nagoya-protocol-en.pdf. Available at: https://www.cbd.int/abs/
62. United Nations Educational, Scientific and Cultural Organization. Universal Declaration on Bioethics and Human Rights. 19 Oct 2005. https://www.unesco.org/en/legal-affairs/universal-declaration-bioethics-and-human-rights?hub=66535
63. Matsuyama K, Kurihara C, Crawley FP, Kerpel-Fronius S. Utilization of genetic information for medicines development and equitable benefit sharing. Front Genet. 2023;14:1085864. https://doi.org/10.3389/fgene.2023.1085864.
64. Haidar H, Gooshki ES. Trial-centrism in the declaration of Helsinki: a challenge for post-study access and preventing data colonialism in AIDriven Health Research. Can J Bioeth/Rev Can Bioeth. 2024;7(4):138–9. https://www.erudit.org/en/journals/bioethics/2024-v7-n4-bioethics09717/1114971ar.pdf

Chapter 12
Universities as Guardians of Ethics and Equity in Medical Research: Protecting the Rights of Vulnerable Populations, Ensuring Post-trial Access, and Advancing Transparency in Clinical Trials

Sarai Mirjam Keestra, Yara de Koning, Maurice Luca Remy, Molly Pugh-Jones, Lauren Paremoer, Till Bruckner, and Toby Pepperrell

Abstract Universities have historically been regarded as independent institutions for knowledge creation in service of the public good. However, increasing financial reliance on private sector funding has reshaped their role in medical research. This

S. M. Keestra (✉)
Department of Epidemiology and Data Science, Amsterdam University Medical Centre, University of Amsterdam, Amsterdam, The Netherlands

Universities Allied for Essential Medicines Europe, Berlin, Germany

Y. de Koning
Universities Allied for Essential Medicines Europe, Berlin, Germany

Pharmaceutical Accountability Foundation, Bergeijk, The Netherlands

M. L. Remy
Universities Allied for Essential Medicines Europe, Berlin, Germany

Athena Institute, Faculty of Science, Vrije Universiteit Amsterdam, Amsterdam, The Netherlands

M. Pugh-Jones
Universities Allied for Essential Medicines Europe, Berlin, Germany

L. Paremoer
Department of Political Studies, University of Cape Town, Cape Town, South Africa

T. Bruckner
Department of Clinical Neuroscience, Karolinska Institutet, Stockholm, Sweden

UiT The Arctic University of Norway, Tromsø, Norway

TranspariMED, Stockholm, Sweden

T. Pepperrell
Department of Biological and Environmental Sciences, University of Stirling, Stirling, UK

C. Kurihara et al. (eds.), *The 2024 Declaration of Helsinki*,
https://doi.org/10.1007/978-981-96-9294-1_12

chapter examines universities' responsibility to uphold research integrity, transparency, and equitable access to health technologies, particularly in the context of contracted research with the private sector.

We argue that universities must actively serve as guardians of ethical standards in medical research by ensuring inclusive decision-making, mandating equitable post-trial access to interventions, and enforcing transparency in research design, clinical trial agreements, and reporting. Ethical review committees should broaden their mandate beyond procedural oversight to include proactive risk mitigation strategies, particularly in protecting vulnerable populations in low-resource settings. The latest revisions of the Declaration of Helsinki (DoH) must be leveraged to institutionalise stronger ethical safeguards in research partnerships, empowering universities to negotiate fairer agreements with third parties.

By embedding principles of accountability, justice, and equity into their research frameworks, universities can counterbalance private interests and reinforce the fundamental ethical commitments of the DoH. This chapter calls on universities to take a leadership role in shaping medical research practices that prioritise public health over commercial gain.

Keywords Universities · Public-private partnerships · Inclusive decision-making · Post-trial access · Clinical trial transparency

1 The Declaration of Helsinki and the University Context

In an ideal world, universities should function as global institutions that foster independent thought and innovation, pursuing knowledge creation for the good of the public [1, 2] and are held to higher ethical standards than other actors in medical research and development (R&D) [3]. In a system of medical R&D centred on global health equity, universities would be committed to addressing the global burden of disease based on public health needs, incorporating transparency in research practices and ensuring the accessibility and affordability of new health technologies in all clinical trials [4]. As most universities are historically publicly funded, they are widely seen as responsible for producing research that benefits the public, free of profit-driven motives [5]. However, public universities are increasingly facing underfunding or defunding, compelling them to seek funding from the private sector [6–9]. Universities' knowledge creation and research infrastructure, including access to the patients needed for clinical trials, makes them an appealing partner for contract research funded by the private sector [10, 11]. The difficulty of maintaining ethical standards in contracted research lies in balancing public and private interests, which can undermine universities' ability to fully serve public needs in innovation and to defend the rights and welfare of clinical trial participants [5, 8, 12].

The Declaration of Helsinki (DoH) by the World Medical Association sets out benchmarks for ethical principles in medical research involving human participants [13]. Adopted in 1964 and revised periodically, it provides guidance to safeguard the rights and welfare of participants throughout the scientific process. In this chapter, we discuss the role that universities play in upholding these ethical principles. We explore the responsibility of universities as guardians of public interest to serve the global public good by protecting the rights of vulnerable communities, ensuring post-trial access to interventions and advancing transparency in medical research. We argue that universities must foster equity and transparency in contracted research by representing the public in negotiations and contracts with the private sector. Moreover, universities set ethical examples for upcoming researchers and professionals, thereby shaping the future of socially impactful science [2, 5].

The 2024 revision of the DoH provides an opportunity to set a bold agenda for the next decade [13]. The DoH has historically adapted reactively to developments in the medical research landscape. It is an iterative document that has incorporated relevant additions over time, such as the establishment of ethics committees (e.g., Institutional Review Boards), post-trial provisions, and most recently, timely results reporting. Also in 2024, the World Health Organisation (WHO) published its 'Guidance for best practices for clinical trials', which places a particular focus on operational considerations [14]. The wider systems of governance universities operate within, such as national laws, policies and budgets, or the mandates and capacities of multilateral organisations, are susceptible to sometimes abrupt political and economic changes [15–18]. We argue that universities can and should advocate for clear, actionable ethical frameworks and policies that ensure that their research priorities and partnerships reflect public over private interest, such as the Affordable Access Plan by the Medicines Patent Pool [19]. Through supplying an expert consensus-driven framework, the DoH can provide universities with a standard of reference in negotiating contracted research from a public viewpoint, and a solid foundation for pushing back against wider policies that threaten to undermine the well-being and safety of people participating in medical research.

2 Balancing Public and Private Interests

> ***Article 7 of the 2024 DoH*** [13]***: The primary purpose of medical research involving human participants is to generate knowledge to understand the causes, development and effects of diseases; improve preventive, diagnostic and therapeutic interventions; and ultimately to advance individual and public health. These purposes can never take precedence over the rights and interests of individual research participants***

While contracted research, particularly in public-private collaborations, is often considered critical to accelerating scientific progress [20, 21], ethical concerns arise when transparency and equity are compromised due to conflicts of interest [7, 22]. Many universities fail to disclose details of their contracts with commercial counterparts such as intellectual property agreements, profit-sharing, research aims, and research results [23, 24]. Private interests, such as commercial confidentiality and the profit motive, are then put above public needs [25, 26]. However, as Article 7 of the DoH states, the purposes of clinical research with human participants can never take precedence over the rights and interests of participants [13]. There is a risk that public-private collaborations prioritise profits over equitable access to the resulting health technologies; opaque agreements can limit the public's ability to hold the parties involved accountable [12].

Public universities operate within a framework of public trust and societal responsibility [27, 28]. In contrast, the private sector is primarily accountable to their shareholders and furthering their profit-seeking motives [3, 22, 29, 30]. In public-private research partnerships, the private interests of pharmaceutical companies may dictate terms of the collaboration in their pursuit of a different set of objectives, leading universities astray from serving public interests [31, 32]. Encroachment of private interests into the public research sphere is ongoing in high-income countries (HICs). A recent example of this is seen in the opioid crisis, when universities and academic hospitals continued accepting more than $60 million worth of donations and endowments from companies such as Purdue Pharma and the Sackler Family over a 5-year period, undermining public trust, scientific integrity, and academic independence [25].

Meanwhile, in low- and middle-income countries (LMICs), where public investments in universities have historically been low, attracting private financing is one of the few viable options to fund medical research [33, 34]. Between the 1970s and 1990s, nations undergoing neoliberal economic reforms, including Eastern Europe, became major sites for clinical trials as foreign investment capitalised on low-cost infrastructure and weak ethical regulations in these regions [35, 36]. Pharmaceutical companies expanded their operations into these LMICs at a time when public trust in clinical trials was declining in HICs [37]. For economically vulnerable populations, participation in clinical trials was often the only opportunity to access essential healthcare services [36] placing participants at significant risk of exploitation by industry and researchers alike. Simultaneously, clinical trials offer corporations lucrative profit-generating opportunities should they result in bringing products to market, highlighting a tension between public health objectives and profit-driven motives [26].

When participants are exposed to the risks of novel health technologies without adequate safeguards or post-trial access guarantees, this is a failure of defending the rights and welfare of participants over private interests in an increasingly globalised, industrialised, and commercialised world of clinical research [37]. Rather than being isolated incidents, these are symptoms of larger trends. Many of the same injustices, such as a lack of post-trial access to interventions, were first identified during the HIV pandemic and then repeated during the COVID-19 pandemic [38]. The following sections of this chapter will argue that equity and transparency are critical to the biomedical research ecosystem, and that universities should defend these core values across their academic and contracted research activities, culminating in recommendations for accountability mechanisms. By leading efforts in three critical areas of the evolving global medical research landscape—protecting vulnerable populations, ensuring post-trial access and benefit-sharing, and advancing transparency—academic institutions and healthcare professionals can and should ensure that medical research upholds the highest ethical standards and serves the public good.

3 Supporting and Protecting the Rights of Vulnerable Populations

> ***Article 19 of the 2024 DoH*** [13]***: Some individuals, groups, and communities are in a situation of more vulnerability as research participants due to factors that may be fixed or contextual and dynamic, and thus are at greater risk of being wronged or incurring harm. When such individuals, groups, and communities have distinctive health needs, their exclusion from medical research can potentially perpetuate or exacerbate their disparities. Therefore, the harms of exclusion must be considered and weighed against the harms of inclusion. In order to be fairly and responsibly included in research, they should receive specifically considered support and protections.***

The role carved out for medical professionals and the academic community in protecting participants and their (human) rights in research is rooted in the history of the DoH. The Doctors' Trial at Nuremberg (1946–1947) exposed Nazi physicians' inhumane medical experiments in concentration camps during World War II, including by university researchers [39]. The trial culminated in the Nuremberg Code (1947), detailing ethical standards for medical research [40–42]. The Nuremberg Code prioritised the autonomy and general welfare of participants above all else, marking an important shift from the Hippocratic Oath, which lacks these concepts [40]. It emphasised voluntary consent, scientific necessity, minimising harm, and participants' rights to withdraw at any point as basic premises for conducting medical research [43].

The Declaration of Helsinki (1964) emerged approximately two decades later, re-emphasising the physician's ethical duty to protect research subjects, returning focus to the responsibility of the medical profession [40]. In a process driven by universities and civil society organisations [44–47], the understanding of the role of patients in research has shifted from passive subjects requiring protection to active participants with agency and expertise [39, 40]. Today, patients are commonly described as active participants in the research process, contributing to research agendas, decision-making, and results dissemination [48–52], mirroring the general shift in medicine from physician-centred decision-making to a model that values patient autonomy and shared decision-making [53, 54]. However, the growing involvement of the private sector in patient engagement has raised concerns. For example, by funding and working through patient advocacy groups or by directly including patient voices in companies' commercial R&D activities, patient engagement can be tokenistic, manipulative, or reduce patients to mere consumers in a market-driven health system [26, 36, 55–57].

Despite progress, vulnerable populations are still frequently excluded from medical research or subjected to violations of dignity, autonomy, privacy, or confidentiality of information [58, 59]. We consider populations vulnerable, based on intersectional systems of oppression including gender, race, age, education, economic status, and other factors [60, 61]. Furthermore, the new patient-centred, patient-driven approach to medical research has yet to be consistently translated into practice, especially in setting research priorities and adjusting health technologies to local contexts and public health needs [62–66]. For example, during the COVID-19 pandemic, the development of costly mRNA vaccines requiring cold-chain management fitting HIC settings was prioritised over cheaper, easily distributed vaccines that could be kept at room temperature, as well as oral or nasal alternatives more suitable in LMICs. Vaccine development during this period largely ignored the practical and cultural realities of resource-constrained health systems [67].

Within the medical research ecosystem, universities are uniquely positioned to pioneer innovative approaches to research that serves the public interest. In particular, universities' ethical review boards are tasked with ensuring that medical research is conducted in line with DoH provisions. In addition to the principles of fairness and inclusivity as per Article 19, this also includes the need to facilitate inclusive decision-making (Article 6), protect vulnerable populations and enable their equitable participation in research (Article 9), and the provision of sufficient access to participation for underrepresented groups (Article 13). Although efforts by universities and civil society organisations have begun addressing these gaps [65, 66, 68, 69], these initiatives must be institutionalised to ensure sustained impact. In Article 6 of the DoH, it is mentioned that 'meaningful engagement should occur with potential and enrolled participants and their communities should occur before, during, and following medical research', but to prevent oversights and ensure that research outcomes align with the needs of vulnerable communities, it is important that inclusive decision-making is made more explicit as an ethics requirement.

Universities must be at the forefront of this development by establishing frameworks that guarantee the inclusion of patient groups and community representatives in decision-making processes in all medical research. This means adopting enforceable policies that govern collaborations with the private sector and ensuring vulnerable populations are prioritised throughout the research process. Ethical review processes conducted at universities must go beyond procedural assessments, incorporating rigorous risk-benefit analyses that actively address systemic inequities, prioritise inclusive decision-making and protect participant welfare. Without such measures, ethical oversight risks becoming a tokenistic and bureaucratic procedure, shielding researchers and sponsors from accountability and neglecting not only participants' rights but also their priorities and values in medical research. This is especially important when considering efforts by political powers to undermine diversity, equity, and inclusion in a way that endangers the value of medical research, such as the 2025 purge of guidance on clinical trial diversity by the US government [70]. Reference to global benchmarks set out by the DoH may enable universities to push back against this trend where diversity is politicised. As emphasised in the 2024 WHO Guidance for Best Practices for Clinical Trials, more rigorous oversight by local actors, such as university ethical review boards, is urgently needed to enforce international ethical standards and protect vulnerable populations [14], and universities are in an excellent position to be leading this development.

4 Ensuring Equitable Access Beyond Post-trial Provisions for All

> ***Article 34 of the 2024 DoH*** [13]***: In advance of a clinical trial, post-trial provisions must be arranged by sponsors and researchers to be provided by themselves, healthcare systems, or governments for all participants who still need an intervention identified as beneficial and reasonably safe in the trial. Exceptions to this requirement must be approved by a research ethics committee. Specific information about post-trial provisions must be disclosed to participants as part of informed consent.***

The 2000 revision of the DoH attempted to strengthen ethical safeguards to protect vulnerable populations and address systemic inequities in medical R&D [71]. In the United States, the Belmont Report—published following public scrutiny over the 40-year-long, harmful, and racially exploitative Tuskegee Syphilis Study—outlined ethical principles that became foundational for US research regulations and influenced the global medical ethics landscape concerning clinical trials in vulnerable populations [72]. Of particular relevance are the Belmont Report's principles of beneficence and justice, which emphasise that vulnerable populations should not bear the burdens of research without receiving corresponding benefits [72]. Researchers should therefore not only avoid causing harm but also create benefits for participants of medical research to share in. However, translating these principles into enforceable practices remains a significant challenge. Without strong safeguards, medical R&D can replicate structural violence, defined by Farmer as the 'normalised, interacting social and political structures that cause injury, injustice and oppression' in the name of science [73–75].

The Belmont Report's justice approach has evolved into demands for a maximin approach to justice, where the distribution of benefits must favour the worst-off parties in a negotiation, ensuring that access to basic goods like medicines—valued inherently and not merely as commodities—is not obstructed by power imbalances of the negotiating parties [76, 77]. A maximin approach to justice, where the rights of vulnerable populations are prioritised, needs to be strengthened in various parts of the DoH as it moves towards participant centredness and inclusive decision-making. In this context, the ambiguous phrasing of 'post-trial provisions' in Article 34 of the DoH is both hesitant and imprecise, and thereby an obstacle to achieving true post-trial equity. It leaves unclear who bears responsibility—sponsors, principal investigators, governments, and/or healthcare systems—thereby diluting accountability when this principle is violated. When no stakeholder is given ultimate responsibility, inequities seem likely to remain unaddressed and unresolved.

While patients in HICs are commonly viewed as potential consumers of profitable treatments by pharmaceutical companies, less-wealthy populations in LMICs have been described as sources of 'value generation and knowledge production' with little expectation that they may share in the research's benefits [78]. Clinical trials are often framed as a social good: they enable people living in LMICs to access otherwise unavailable treatments, notably during the HIV epidemic [36]. Yet this access is frequently a mirage. In South Africa during the early phase of the HIV pandemic, participants who bore the risks of clinical trials were often unable to access HIV treatments after the trials concluded due to their prohibitive cost [79–81]. More recently, countries like Brazil and South Africa, with large HIV-positive populations, facilitated trials for long-acting antiretrovirals for HIV/AIDS such as cabotegravir and lenacapavir. However, these nations continue to struggle with the scale and cost of integrating these life-saving treatments into public health systems and have been left out of voluntary licensing and patent pooling schemes [82–84].

The COVID-19 pandemic repeated some of these inequities seen in the HIV/AIDS crisis [36], suggesting that little has changed. LMICs hosted numerous COVID-19 vaccine trials, but participants were not guaranteed access to the resulting treatments. For example, South Africa faced high vaccine costs and companies delayed seeking regulatory approval for the distribution and use of the Moderna and Novavax vaccines, preventing them from being fully integrated into its public health programs [85, 86]. More broadly, the principle of post-trial access could be considered to apply to entire host populations rather than just trial participants, especially for preventative interventions such as antiretrovirals or vaccines. For these interventions, a level of community transmission of the disease is required for trials to be successful, as they provide an essential real-world control arm for trials. This blurs the definition of who counts as a 'trial participant'. If the whole population is not considered in post-trial access, a disconnect forms between the contributions of trial participants in LMICs and the equitable distribution of the resulting health benefits, as any benefits gained during the trial phase will be insignificant in a society that lacks access to preventative health technologies.

Universities and academic hospitals play a central role in enabling clinical trials in LMICs by providing critical infrastructure and expertise. However, their efforts to safeguard post-trial access are insufficient. In particular, university ethics review boards often focus narrowly on harm reduction during trials, paying little attention to access and benefit-sharing measures that could promote equitable access to health technologies for vulnerable populations after trials end. Taking such a broader perspective would require universities' ethics review boards to have the mandate and capacity to review and assess contracts negotiated with the private sector. However, this gap is rarely addressed. For example, the Clinical Trials Resolution adopted at the 75th World Health Assembly focused on technical improvements such as data quality and comparability, rather than addressing post-trial access concerns [87]. Similarly, while the Africa CDC's emphasis on ethical standardisation across the continent is promising, it is framed as part of a broader strategy to attract more clinical trials [88]. Not requiring access and benefit-sharing conditions to form part of clinical trial contracts, and not empowering ethics reviews to assess such conditions, perpetuates the status quo.

Contracts should provide clarity on post-trial access provisions before trials begin; universities' technology transfer offices could support such negotiations. Medecins Sans Frontieres has called on universities, amongst others, to ensure that explicit requirements on transparency, pricing, access and use of technologies and know-how, and enhancing local capacity in R&D, manufacturing and supply are embedded in the research design from the start [89]. This includes assessing open license agreements, knowledge and know-how sharing, needs-based production capacity options, and possibilities for pooled procurement or other discount mechanisms for LMICs [90]. In addition, the 2024 WHO Guidance for best practices for clinical trials suggests that equitable post-trial access processes 'can be enabled by governments and other funders attaching concrete and enforceable conditions when they support or fund clinical trials' [14].

5 Making Research Designs, Outcomes, and Costs Transparent

Article 36 of the 2024 DoH [13]***: Researchers, authors, sponsors, editors, and publishers all have ethical obligations with regard to the publication and dissemination of the results of research. Researchers have a duty to make publicly available the results of their research on human participants and are accountable for the timeliness, completeness, and accuracy of their reports. All parties should adhere to accepted guidelines for ethical reporting. Negative and inconclusive as well as positive results must be published or otherwise made publicly available. Sources of funding, institutional affiliations, and conflicts of interest must be declared in the publication. Reports of research not in accordance with the principles of this Declaration should not be accepted for publication.***

Before a study begins, prospective trial registration creates a comprehensive overview of past and present research activities, which support the setting of research priorities, identification of gaps, and prevention of needless duplication of studies. Subsequently, the benefits, harms, and cost-effectiveness of health technologies can only be reliably assessed if all relevant clinical trial results are available [91–94]. In addition, clinical trial transparency builds trust with study participants and the wider public [89]. Mandatory trial pre-registration (Article 35) and reporting (Article 36) were first introduced in the 2008 revision of the DoH [95]. While this marked a shift towards greater transparency in medical research, its introduction lagged behind legal developments in key jurisdictions, including the European Union's Clinical Trials Directive (2001) and the United States' FDA Amendments Act (2007) [96]. Both jurisdictions require some trials to be pre-registered, and their results to be made public by the sponsor within one year of trial completion. In 2017, the WHO Joint statement on public disclosure of results from clinical trials was adopted. While it is not legally binding and was primarily directed towards funders, the Joint Statement sets out specific transparency benchmarks that are equally applicable to universities and other entities that act as the legal sponsors of clinical trials. Notably, the WHO document extended key transparency requirements, including the 1-year timeline for reporting, to all types of interventional trials worldwide [97].

Nonetheless, many clinical trials still fail to report results within a year of the primary completion date, even those legally required to do so [23, 73, 94, 98, 99]. Lack of legal enforcement and concrete sanctions mean that sponsors are rarely held accountable for failures to comply even with mandatory registration and reporting requirements [23, 100, 101]. A recent Cochrane review found that currently 53% of all clinical trials have not reported results, and those that did, typically did so only after a delay of more than 2 years [102]. The 2024 revision of the DoH represents progress in that it now mentions for the first time the need for 'timeliness' of results reporting [103]. However, does not clearly state which stakeholder is responsible for registration and reporting, and 'timeliness' is an inherently elastic term that is arguably open to multiple interpretations.

The DoH clearly states in Article 36 that studies not adhering to its principles should not be accepted for publication by scientific journals. The International Committee of Medical Journal Editors announced in the early 2000s that the results of unregistered clinical trials would henceforth no longer be published, leading to a sharp increase in trial registrations [104, 105]. Nonetheless, reports of unregistered trials or late registrations continue to be published by lower-quality medical journals [105] even though they are in violation of Article 35 of the DoH. Data gathered by WHO show that over 10,000 trials per year are still not being adequately preregistered [106]. On the positive side, recent political attention to clinical trial transparency by parliament and civil society in the United Kingdom has led to significant improvements in the trial reporting performance of British universities [23, 94], demonstrating that universities have the capacity to effectively address this problem in cases where they themselves act as legal sponsors of trials. Similarly, US universities have significantly improved their legal compliance in recent years [107, 108]. In both cases, universities set up effective oversight and compliance mechanisms that could in the future be extended and used additionally to embed other transparency principles in their operations.

However, transparency gains with regard to trial registration and reporting have so far not translated into strong progress in addressing opacity within other areas of the research ecosystem [23]. Between 2019 and 2022, half of all clinical trials were exclusively funded by industry; those trials were three times as likely to report 'positive' outcomes than other trials [109]. The DoH states that researchers should declare 'sources of funding, institutional affiliations and conflicts of interest' in their publications [103], but reporting this information has largely become a tick-box exercise whose real-life impacts remain questionable. Public disclosure of sources of funding and potential conflicts of interest helps to put a study's findings into context. Research funding transparency is necessary because funders dictate capacity, priorities, capabilities, and which technologies are commercialised [73, 110]. Similarly, transparency around contracts between universities and private sector players is essential because it allows trial participants and the wider public to assess whether public interests have been sufficiently taken into account and safeguarded.

An even larger unresolved challenge is the lack of transparency regarding trial costs, which the DoH currently does not address at all. The cost of clinical trials contributes to decision-making regarding product pricing and availability [24]. However, the relative contributions of private versus public sector funders, as well as the absolute R&D costs of developing new health technologies, typically remain shrouded in secrecy. This hinder efforts to determine the overall production costs of healthcare products, making it difficult to compare production costs with the prices charged by pharmaceutical and medical device companies [111, 112]. Greater R&D cost transparency could reduce prices by strengthening governments' and insurers' negotiation positions. Universities can play a leading role in this area by providing the public with detailed breakdowns of various funding sources and the costs of running clinical trials.

6 Recommendations for Protecting Patients and Strengthening Public Health

The increasingly globalised landscape of medical R&D, and clinical trials in particular, demands renewed commitment to equity and transparency in research ethics. R&D must become an equitable and democratic process that actively involves researchers and affected communities in setting priorities and defining the accessibility, acceptability, and quality of novel health technologies [76]. For the DoH to deliver on its promise of providing a strong and meaningful standard that universities and other stakeholders can operationalise to ensure that medical research is conducted in an ethical manner, further work is required on two fronts. First, the scope and content of the DoH itself need to be strengthened. Second, universities must leverage their unique position in medical research to lead efforts to institutionalise equitable research practices.

6.1 Future Iterations of the Declaration of Helsinki

While we acknowledge that the DoH is a norm-setting document, future iterations should reflect and possibly reference more detailed standards set out by the WHO and other authoritative bodies. This would help to bridge the gap between principles and practice. Future iterations of the DoH should define which stakeholder is responsible for meeting any given ethics requirement, and additional accountability and transparency measures should be added. The Declaration of Helsinki should be amended in Article 6 for inclusive decision-making with affected populations to be prioritised throughout the research process as a requirement for equitable and democratic R&D, to platform post-trial access and equitable benefit sharing. Article 34, currently referring only to 'post-trial provisions', should be expanded to 'provisions for post-trial access that employ realistic premeditated mechanisms for access to trial interventions for participants, as well as wider communities in the case of preventive treatments, including through plans to share knowledge and know-how'. Additionally, Article 36 of the DoH should incorporate existing WHO standards by specifying that the legal sponsor of a clinical trial must ensure that a trial's results are made public within 1 year. The DoH should clearly state that ethical review boards have the mandate and responsibility to ensure that inclusive decision-making is pursued, adequate post-trial access provisions are adopted and implemented, and timely reporting is enforced. In sum, future iterations of the DoH should set out clear ethical standards to anchor legally enforceable frameworks that strengthen research integrity and advance global health equity.

6.2 Roles and Responsibilities of Universities

Universities and academics bear responsibility for upholding justice through their decisions and actions. Specifically, universities should prevent clinical research from being dominated by private sector interests that prioritise private profits over public health gains. They can do this by implementing existing DoH provisions and addressing current gaps in the scope and specificity of the DoH. In this context, universities should regard the DoH as setting out minimum ethical obligations for certain aspects of research, rather than defining the boundaries of universities' ethical obligations across the entire spectrum of activities related to biomedical research.

As centres of knowledge creation and in their role as intermediaries of the private and public sector in a globalised clinical trial landscape, universities can and should play a pivotal role in driving a shift away from market-driven models that can exploit vulnerable populations, and towards participant-centred initiatives that prioritise public health equity, inclusive decision-making, transparency, and accountability towards the global public. Going beyond rhetorical commitments to 'decolonise' global health research, universities must forge equitable partnerships that benefit host communities and promote global health [113–115]. In particular, negotiating explicit and binding access, pricing, and capacity-building clauses in clinical trial agreements would help to address long-standing inequities. Universities should commit to transparency as a foundational value to be applied at all stages throughout their medical research. Transparency must be promoted through internal policies and supporting infrastructure, ensuring researchers have the resources to comply with and exceed legal transparency requirements. Universities' end goal should be to become proactive guardians of equity, transparency, and accountability towards the global public in biomedical research.

6.3 Balancing Public and Private Interests

- Universities should adopt legally binding and enforceable policies that govern collaborations with private actors, ensuring that the public interest is prioritised throughout the research process, such as the Affordable Access Plan by the Medicines Patent Pool. By uniting around shared ethics policies anchored in the DoH, universities can strengthen their collective negotiating position.

6.4 Supporting and Protecting the Rights of Vulnerable Populations

- Universities should establish frameworks that guarantee the inclusion of patient groups and community representatives in decision-making processes, as per Article 6.
- Universities should institutionalise policies and processes that ensure diverse and representative populations in clinical trials, as per Articles 9 and 13.
- Universities should prioritise justice and equity in their research practices and protect the rights of (vulnerable) participants in clinical trials, their dignity, autonomy, privacy, and confidentiality of information, as per Article 17.

6.5 Ensuring Equitable Access Beyond Post-trial Provisions for All

- Responsibility for post-trial access should be addressed at project conception, during the contract negotiation phase, with universities insisting on explicit and legally binding safeguards to protect the rights of the populations that bear the risks of research.
- Universities should ensure that meaningful and effective post-trial access mechanisms are put into place. These may include open license agreements, knowledge and know-how sharing, needs-based production capacity possibilities, possibilities for pooled procurement, or other discount mechanisms for LMICs.

6.6 *Making Research Designs, Outcomes, and Costs Transparent*

- Universities should ensure consistent and full compliance with DoH clinical trial registration and reporting requirements, as per Articles 35 and 36 and the provisions of the WHO Joint statement on pre-registering and public disclosure of results from clinical trials.
- Universities should disclose the absolute values and proportions of public and private funding in each clinical trial, in addition to R&D costs. This would clarify the contributions of various stakeholders, helping to ensure that public investments in research translate into equitable health outcomes.

Conflict of Interest The authors have no competing interests to declare that are relevant to the content of this chapter.

References

1. Calhoun C. The university and the public good. Thesis Eleven [Internet]. 2006 [cited 2025 Feb 26];84(1):7–43. https://doi.org/10.1177/0725513606060516
2. Menezes I, Coelho M, Amorim JP. Social and public responsibility, universities. In: Teixeira PN, Shin JC, editors. The International encyclopedia of higher education systems and institutions [Internet]. Dordrecht: Springer; 2020 [cited 2025 Feb 17]. p. 2540–6. https://doi.org/10.1007/978-94-017-8905-9
3. Keestra S. Sofosbuvir: the creation of an [in]valuable medicine for hepatitis C. Glob Health Annu Rev [Internet]. 2021 [cited 2025 Feb 16];1(6). Available from: https://journals.mcmaster.ca/ghar/article/view/2667
4. Gotham D, Meldrum J, Nageshwaran V, Counts C, Kumari N, Martin M, et al. Global health equity in United Kingdom university research: a landscape of current policies and practices. Health Res Policy Syst [Internet]. 2016;14. Available from: https://health-policy-systems.biomedcentral.com/track/pdf/10.1186/s12961-016-0148-6?site=health-policy-systems.biomedcentral.com
5. Owen-Smith J. Research universities and the public good: discovery for an uncertain future [Internet]. Stanford University Press; 2018 [cited 2025 Feb 16]. Available from: https://www.degruyter.com/document/doi/10.1515/9781503607095/html
6. Lewis K. Post-trial access to treatment: How managed access programs could be a solution. Medicine Access @ Point of Care [Internet]. 2018;2:239920261879543–239920261879543. https://doi.org/10.1177/2399202618795431
7. Moynihan R, Bero L, Hill S, Johansson M, Lexchin J, Macdonald H, et al. Pathways to independence: towards producing and using trustworthy evidence. BMJ [Internet]. 2019 [cited 2025 Feb 17];16576. Available from: https://www.bmj.com/lookup/doi/10.1136/bmj.l6576
8. Thornton M. Privatising the public university. London: Routledge; 2011. 264 p.
9. Washburn J. University, Inc: the corporate corruption of American higher education [Internet]. New York: Basic Books; 2004 [cited 2025 Feb 16]. 362 p. Available from: http://archive.org/details/universityinccor0000wash
10. Wirsich A, Kock A, Strumann C, Schultz C. Effects of university–industry collaboration on technological newness of firms. J Prod Innov Manag [Internet]. 2016 [cited 2025

Feb 17];33(6):708–25. Available from: https://onlinelibrary.wiley.com/doi/abs/10.1111/jpim.12342
11. Rossoni AL, de Vasconcellos EPG, de Castilho Rossoni RL. Barriers and facilitators of university-industry collaboration for research, development and innovation: a systematic review. Manag Rev Q [Internet]. 2024 [cited 2025 Feb 17];74(3):1841–77. https://doi.org/10.1007/s11301-023-00349-1
12. Noguera PC. Conflicts of interest in research. In: Valdés E, Lecaros JA, editors. Handbook of bioethical decisions volume II [Internet]. Cham: Springer; 2023 [cited 2025 Feb 17]. p. 65–74. (Collaborative Bioethics; vol. 3). Available from: https://link.springer.com/10.1007/978-3-031-29455-6_6
13. World Medical Association. WMA declaration of Helsinki – ethical principles for medical research involving human participants [Internet]. 2024. Available from: https://www.wma.net/policies-post/wma-declaration-of-helsinki/
14. World Health Organisation. Guidance for best practices for clinical trials. 2024. Available from: https://www.who.int/publications/i/item/9789240097711
15. RFI [Internet]. 2025 [cited 2025 Feb 28]. France's proposed budget cuts set to slash overseas development aid. Available from: https://www.rfi.fr/en/france/20250205-france-proposed-budget-cuts-slash-overseas-development-aid-coordinationo-sud
16. Nolen S. Abandoned in the middle of clinical trials, because of a Trump order. The New York Times [Internet]. 2025 Feb 6 [cited 2025 Feb 28]; Available from: https://www.nytimes.com/2025/02/06/health/usaid-clinical-trials-funding-trump.html
17. Nolen S. U.S. Terminates Funding for Polio, H.I.V., Malaria and Nutrition Programs Around the World. The New York Times [Internet]. 2025 Feb 27 [cited 2025 Feb 28]; Available from: https://www.nytimes.com/2025/02/27/health/usaid-contract-terminations.html
18. Keir Starmer cuts aid to fund increase in defence spending [Internet]. 2025 [cited 2025 Feb 28]. Available from: https://www.bbc.com/news/articles/clyrkkv4gd7o
19. Medicines Patent Pool. Upstream access plan for health technologies in LMICs [Internet]. 2024 [cited 2025 Feb 28]. Available from: https://medicinespatentpool.org/what-we-do/upstream-access
20. Siegel DJ. Social partnership, cross-sector collaboration. In: Teixeira PN, Shin JC, editors. The international encyclopedia of higher education systems and institutions [Internet]. Dordrecht: Springer; 2020 [cited 2025 Feb 17]. p. 2562–6. Available from: https://link.springer.com/10.1007/978-94-017-8905-9
21. Figueiredo NL, Ferreira JJM. More than meets the partner: a systematic review and agenda for University–Industry cooperation. Manag Rev Q [Internet]. 2022 [cited 2025 Feb 17];72(1):231–73. Available from: https://link.springer.com/10.1007/s11301-020-00209-2
22. Buchman DZ, Magel B, Shier R, Davies TE, Sud A, Mahajan S, et al. Canadian cannabis researcher perspectives on the conduct and sponsorship of scientific research by the for-profit cannabis industry. Soc Sci Med [Internet]. 2025 [cited 2025 Feb 17];364:117556. Available from: https://www.sciencedirect.com/science/article/pii/S0277953624010104
23. Keestra S, Rodgers F, Osborne R, Wimmer S. University patenting and licensing practices in the United Kingdom during the first year of the COVID-19 pandemic. Glob Public Health. 2022;17(5):641–51.
24. Cross S, Rho Y, Reddy H, Pepperrell T, Rodgers F, Osborne R, et al. Who funded the research behind the Oxford-AstraZeneca COVID-19 vaccine? BMJ Glob Health [Internet]. 2021;6(12):1–11. https://doi.org/10.1101/2021.04.08.21255103
25. Marks JH. Lessons from corporate influence in the opioid epidemic: toward a norm of separation. Bioeth Inq [Internet]. 2020 [cited 2025 Feb 28];17(2):173–89. https://doi.org/10.1007/s11673-020-09982-x
26. Hill KP, Ross JS, Egilman DS, Krumholz HM. The advantage seeding trial: a review of internal documents. Ann Intern Med [Internet]. 2008 [cited 2025 Feb 28];149(4):251–8. Available from: https://www.acpjournals.org/doi/full/10.7326/0003-4819-149-4-200808190-00006

27. Chase HW. The social responsibility of the state university. Socl Forces [Internet]. 1923 [cited 2025 Feb 17];1(5):517–21. Available from: https://academic.oup.com/sf/article-lookup/doi/10.2307/3005122
28. Godonoga A, Sporn B. The conceptualisation of socially responsible universities in higher education research: a systematic literature review. Stud High Educ [Internet]. 2023 [cited 2025 Feb 17];48(3):445–59. https://doi.org/10.1080/03075079.2022.2145462
29. Montalban M. Shareholder value, political work and globalization in the pharmaceutical industry. In: Jullien B, Smith A, editors. Industries and globalization: the political causality of difference [Internet]. London: Palgrave Macmillan UK; 2008 [cited 2025 Feb 17]. p. 92–128. https://doi.org/10.1057/9780230282155_4
30. UCL Institute for Innovation and Public Purpose. The people's prescription: RE-imagining health innovation to deliver public value [Internet]. London: IIPP, Global Justice Now, Just Treatment, STOPAIDS; 2018 [cited 2025 Feb 17]. Available from: https://www.ucl.ac.uk/bartlett/public-purpose/sites/public-purpose/files/peoples_prescription_report_final_online.pdf
31. 't Hoen E. The responsibility of research universities to promote access to essential medicines. Yale J Health Policy Law Ethics [Internet]. 2003;3(2). Available from: http://hdl.handle.net/20.500.13051/6007
32. Kapczynski A, Chaifetz S, Benkler Y, Katz Z. Addressing global health inequities: an open licensing approach for university innovations [Internet]. Rochester: Social Science Research Network; 2005 [cited 2025 Feb 16]. Available from: https://papers.ssrn.com/abstract=1323527
33. Chikwari CD, Tadesse AW, Shanaube K, Shepherd A, McQuaid CF, Togun TO. Achieving equitable leadership in Global Health partnerships: barriers experienced and strategies to improve grant funding for early- and mid-career researchers. BMC Glob Public Health [Internet]. 2024 [cited 2025 Feb 17];2(1):17. Available from: https://bmcglobalpublichealth.biomedcentral.com/articles/10.1186/s44263-024-00047-4
34. Malavige GN, Medigeshi GR, Sasmono RT, Amaratunga C. Funding and collaboration inequalities in infectious disease research – why does it matter? Trends Microbiol [Internet]. 2024 [cited 2025 Feb 17]; Available from: https://www.cell.com/trends/microbiology/abstract/S0966-842X(24)00290-7
35. Petryna A. Experimentality: on the global mobility and regulation of human subjects research. PoLAR: Polit Legal Anthropol Rev [Internet]. 2007 [cited 2025 Feb 16];30(2):288–304. Available from: https://onlinelibrary.wiley.com/doi/abs/10.1525/pol.2007.30.2.288
36. Petryna A. When experiments travel: clinical trials and the global search for human subjects [Internet]. Princeton University Press; 2009. Available from: https://press.princeton.edu/books/paperback/9780691126579/when-experiments-travel
37. Rettig RA. The industrialization of clinical research: clinical research has become an industry of its own, one that warrants careful scrutiny to protect human research subjects. Health Aff [Internet]. 2000 [cited 2025 Feb 17];19(2):129–46. Available from: http://www.healthaffairs.org/doi/10.1377/hlthaff.19.2.129
38. South African Medical Research Council. SAMRC launches a study to evaluate the effectiveness of the Moderna booster. 2022. Available from: https://www.samrc.ac.za/press-releases/samrc-launches-study-evaluate-effectiveness-moderna-booster
39. Annas GJ, Grodin MA. The Nazi doctors and the nuremberg code: human rights in human experimentation. The Nazi Doctors And The Nuremberg Code [Internet]. 1992 May. Available from: https://academic.oup.com/book/54203
40. Shuster E. The Nuremberg code: Hippocratic ethics and human rights. Lancet [Internet]. 1998;351(9107):974–7. Available from: https://pubmed.ncbi.nlm.nih.gov/9734958/
41. Vollmann J, Winau R. Informed consent in human experimentation before the Nuremberg code. BMJ [Internet]. 1996;313(7070):1445–7. Available from: https://www.bmj.com/content/313/7070/1445

42. Ghooi R. The Nuremberg Code-a critique. Perspect Clin Res [Internet]. 2011;2(2):72. Available from: https://pubmed.ncbi.nlm.nih.gov/21731859/
43. Nuremberg Military Tribunals. Trials of war criminals before the Nuremberg Military Tribunals under control Council Law No. 10 [Internet]. Vol. 2. 1949 [cited 2025 Feb 16]. Available from: http://archive.org/details/TrialsOfWarCriminalsBeforeTheNurembergMilitaryTribunalsUnderControlCouncil
44. Caron-Flinterman JF, Broerse JEW, Bunders JFG. The experiential knowledge of patients: a new resource for biomedical research? Soc Sci Med. 2005;60(11):2575–84.
45. Den Oudendammer WM, Noordhoek J, Abma-Schouten RY, Van Houtum L, Broerse JEW, Dedding CWM. Patient participation in research funding: an overview of when, why and how amongst Dutch health funds. Res Involv Engagem [Internet]. 2019;5(1). Available from: https://pubmed.ncbi.nlm.nih.gov/31720008/
46. Ennis L, Wykes T. Impact of patient involvement in mental health research: longitudinal study. Br J Psychiatry [Internet]. 2013;203(5):381–6. Available from: https://pubmed.ncbi.nlm.nih.gov/24029538/
47. Wiig S, Rutz S, Boyd A, Churruca K, Kleefstra S, Haraldseid-Driftland C, et al. What methods are used to promote patient and family involvement in healthcare regulation? A multiple case study across four countries. BMC Health Serv Res [Internet]. 2020;20(1):1–15. Available from: https://bmchealthservres.biomedcentral.com/articles/10.1186/s12913-020-05471-4
48. Faulkner SD, Pittens CACM, Goedhart NS, Davies EH, Manning E, Diaz-Ponce A, et al. Optimising multi-stakeholder practices in patient engagement: a gap analysis to enable focused evolution of patient engagement in the development and lifecycle management of medicines. Ther Innov Regul Sci [Internet]. 2021;55(6):1165–79. Available from: https://link.springer.com/article/10.1007/s43441-021-00313-9
49. Fox G, Fergusson DA, Daham Z, Youssef M, Foster M, Poole E, et al. Patient engagement in preclinical laboratory research: a scoping review. EBioMedicine [Internet]. 2021;70. Available from: https://pubmed.ncbi.nlm.nih.gov/34280783/
50. Pittens C. Knowledge co-production in health research, policy and care practice. 2013.; Available from: https://research.vu.nl/en/publications/knowledge-co-production-in-health-research-policy-and-care-practi
51. Slattery P, Saeri AK, Bragge P. Research co-design in health: a rapid overview of reviews. Health Res Policy Syst [Internet]. 2020;18(1). Available from: https://pubmed.ncbi.nlm.nih.gov/32046728/
52. van de Bovenkamp HM, Zuiderent-Jerak T. An empirical study of patient participation in guideline development: exploring the potential for articulating patient knowledge in evidence-based epistemic settings. Health Expect [Internet]. 2015;18(5):942–55. Available from: https://onlinelibrary.wiley.com/doi/full/10.1111/hex.12067
53. Whitney SN, McGuire AL, McCullough LB. A typology of shared decision making, informed consent, and simple consent. Ann Intern Med [Internet]. 2004;140(1):54–9. Available from: https://pubmed.ncbi.nlm.nih.gov/14706973/
54. Sacristán JA. Patient-centered medicine and patient-oriented research: Improving health outcomes for individual patients. BMC Med Inform Decis Mak [Internet]. 2013;13(1):1–8. Available from: https://bmcmedinformdecismak.biomedcentral.com/articles/10.1186/1472-6947-13-6
55. Boutin M, Dewulf L, Hoos A, Geissler J, Todaro V, Schneider RF, et al. Culture and process change as a priority for patient engagement in medicines development. Ther Innov Regul Sci [Internet]. 2017;51(1):29–38. Available from: https://pubmed.ncbi.nlm.nih.gov/28232876/
56. Zvonareva O. Patient engagement in drug development: configuring a new resource for generating innovation. Crit Public Health [Internet]. 2023;33(5):506–17. Available from: https://www.tandfonline.com/doi/abs/10.1080/09581596.2023.2188140
57. Batt S. Health Advocacy, Inc.: how pharmaceutical funding changed the breast cancer movement [Internet]. UBC Press; 2019 [cited 2025 Feb 28]. Available from: https://www.ubcpress.ca/health-advocacy-inc

58. Dasgupta S. How Do You Catch Mistakes in a Clinical Trial? An Ethics Expert Explains. – The Wire Science. 2021 [Internet]. 2021; Available from: https://science.thewire.in/health/how-do-you-catch-mistakes-in-a-clinical-trial-an-ethics-expert-explains/
59. Fleming TR, Sharpies K, McCall J, Moore A, Rodgers A, Stewart R. Maintaining confidentiality of interim data to enhance trial integrity and credibility. Clin Trials [Internet]. 2008;5(2):157. Available from: https://pmc.ncbi.nlm.nih.gov/articles/PMC2703711/
60. Crenshaw K. Mapping the margins: intersectionality, identity politics, and violence against women of color. Stanford Law Rev [Internet]. 1991;43(6):1241–99. Available from: http://www.jstor.org/stable/1229039
61. Dahlgren G, Whitehead M. Policies and strategies to promote social equity in health. Background document to WHO – Strategy paper for Europe. Institute for Futures Studies, Arbetsrapport. 1991;14.
62. Clark LT, Watkins L, Piña IL, Elmer M, Akinboboye O, Gorham M, et al. Increasing diversity in clinical trials: overcoming critical barriers. Curr Probl Cardiol. 2019;44(5):148–72.
63. Kelsey MD, Patrick-Lake B, Abdulai R, Broedl UC, Brown A, Cohn E, et al. Inclusion and diversity in clinical trials: actionable steps to drive lasting change. Contemp Clin Trials. 2022;116:106740.
64. Mason S, Hussain-Gambles M, Leese B, Atkin K, Brown J. Representation of South Asian people in randomised clinical trials: analysis of trials' data. BMJ [Internet]. 2003;326(7401):1244–5. Available from: https://www.bmj.com/content/326/7401/1244
65. Shea L, Pesa J, Geonnotti G, Powell V, Kahn C, Peters W. Improving diversity in study participation: Patient perspectives on barriers, racial differences and the role of communities. Health Expect [Internet]. 2022;25(4):1979. Available from: https://pmc.ncbi.nlm.nih.gov/articles/PMC9327876/
66. Witham MD, Anderson E, Carroll C, Dark PM, Down K, Hall AS, et al. Developing a roadmap to improve trial delivery for under-served groups: results from a UK multi-stakeholder process. Trials. 2020;21(1)
67. Baker PJ. Advantages of an oral vaccine to control the COVID-19 Pandemic. Am J Med [Internet]. 2022;135(2):133–4. Available from: https://pubmed.ncbi.nlm.nih.gov/34562412/
68. Bodicoat DH, Routen AC, Willis A, Ekezie W, Gillies C, Lawson C, et al. Promoting inclusion in clinical trials – a rapid review of the literature and recommendations for action. Trials [Internet]. 2021;22(1):1–11. Available from: https://trialsjournal.biomedcentral.com/articles/10.1186/s13063-021-05849-7
69. Farooqi A, Jutlla K, Raghavan R, Wilson A, Uddin MS, Akroyd C, et al. Developing a toolkit for increasing the participation of black, Asian and minority ethnic communities in health and social care research. BMC Med Res Methodol [Internet]. 2022;22(1):1–16. Available from: https://bmcmedresmethodol.biomedcentral.com/articles/10.1186/s12874-021-01489-2
70. Herper M, Lawrence L. FDA purges material on clinical trial diversity from its site, showing stakes of Trump DEI ban. STAT News [Internet]. 2025. Available from: https://www.statnews.com/2025/01/23/fda-purges-pages-clinical-trial-diversity-after-trump-dei-ban/
71. World Medical Association. WMA Declaration of Helsinki – ethical principles for medical research involving human participants [Recinded] [Internet]. Edinburgh: World Medical Association; 2000 p. 5. (Declaration of Helsinki). Report No.: 6. Available from: https://www.wma.net/wp-content/uploads/2018/07/DoH-Oct2000.pdf
72. Ryan KJ, Brady JV, Cooke RE, Height DI, Jonsen AR, King P, et al. The Belmont Report: ethical principles and guidelines for the protection of human subjects of research. United States of America: The National Commission for the Protection of Human Subjects of Biomedical and Behavioral Research, Department of Health, Education, and Welfare; 1979.
73. Keestra S. Structural violence and the biomedical innovation system: what responsibility do universities have in ensuring access to health technologies? BMJ Glob Health. 2021;6(5):e004916.
74. Freire P. Pedagogy of the oppressed. 30th ed. Continuum; 2005.

75. Farmer PE, Nizeye B, Stulac S, Keshavjee S. Structural violence and clinical medicine. PLoS Med [Internet]. 2006 [cited 2025 Feb 26];3(10):e449. Available from: https://dx.plos.org/10.1371/journal.pmed.0030449
76. Ballantyne AJ. How to do research fairly in an unjust world. Am J Bioethics [Internet]. 2010;10(6):26–35. Available from: https://pubmed.ncbi.nlm.nih.gov/20526966/
77. Rawls J. A theory of justice. Rev. ed. 5.-6. printing. Cambridge, MA: Belknap Press of Harvard University Press; 2003. 538 p.
78. Rajan KS. Subjects of speculation: emergent life sciences and market logics in the United States and India. Am Anthropol [Internet]. 2005;107(1):19–30. Available from: https://onlinelibrary.wiley.com/doi/full/10.1525/aa.2005.107.1.019
79. Waterhouse C, Lebogang K. Why doesn't SA use this bargaining chip when it makes deals with drugmakers? 2022 Jul.; Available from: https://www.news24.com/news24/opinions/analysis/analysis-why-doesnt-sa-use-this-bargaining-chip-when-it-makes-deals-with-drugmakers-20220730
80. Bayer R. The debate over maternal-fetal HIV transmission prevention trials in Africa, Asia, and the Caribbean: racist exploitation or exploitation of racism? Am J Public Health [Internet]. 1998;88(4):567–567. Available from: https://pmc.ncbi.nlm.nih.gov/articles/PMC1508416/
81. Annas GJ, Grodin MA. Human rights and maternal-fetal HIV transmission prevention trials in Africa. Am J Public Health [Internet]. 1998;88(4):560–3. Available from: https://pubmed.ncbi.nlm.nih.gov/9550993/
82. Pepperrell T, Cross S, Hill AM. Cabotegravir – global access to long-acting pre-exposure prophylaxis for HIV. Open Forum Infect Dis. 2023;10(1)
83. Hill A, Levi J, Fairhead C, Pilkington V, Wang J, Johnson M, et al. Lenacapavir to prevent HIV infection: current prices versus estimated costs of production. J Antimicrob Chemother [Internet]. 2024;79(11). Available from: https://pubmed.ncbi.nlm.nih.gov/39225016/
84. Medecins Sans Frontieres. Pharmaceutical Corporation Viiv must urgently supply HIV drug CAB-LA according to global public health needs. 2024 Jul. Available from: https://www.msf.org.za/news-and-resources/press-release/pharmaceutical-corporation-viiv-must-urgently-supply-hiv-drug-cab
85. South African Medical Research Council. SAMRC launches a study to evaluate the effectiveness of the Moderna booster. 2022 Jun. Available from: https://www.samrc.ac.za/press-releases/samrc-launches-study-evaluate-effectiveness-moderna-booster
86. Novavax Inc. No.2019nCoV501 Phase 2ab South Africa Clinical Trial Protocol. 2020. Available from: https://www.wits.ac.za/media/wits-university/news-and-events/images/documents/2020/2019nCoV_501_Phase%202ab%20South%20Africa_Protocol.pdf
87. World Health Organisation. Strengthening clinical trials to provide high-quality evidence on health interventions and to improve research quality and coordination. Draft resolution proposed by Argentina, Peru, United Kingdom of Great Britain and Northern Ireland. Seventy-fifth World Health Assembly. WHA75.8. Agenda item 16.2. [Internet]. Geneva; 2022 May. Available from: https://www.who.int/news/item/18-05-2017-joint-statement-on-registration.
88. Ndembi N, Mekonen TT, Folayan MO, Dereje N, Kruger A, Fokam J, et al. Strengthening and expanding capacities in clinical trials: advancing pandemic prevention, preparedness and response in Africa. Nat Commun [Internet]. 2024;15(1):1–6. Available from: https://www.nature.com/articles/s41467-024-53126-3
89. Medecins Sans Frontieres. MSF comments on the clinical trial resolution at the 75th World Health Assembly [Internet]. 2022 May. Available from: https://campaign.msfaccess.org/sites/default/files/2023-05/MSF-AC_CTTransparency_Roundtable%202_Summary_ENG_22.5.2023.pdf
90. Pepperrell T, Koivusalo M, Grant L, McCallum A. Transforming trade for vaccine equity: Policy gaps and barriers. PLOS Global Public Health, 2025;5(6):e0004012. https://doi.org/10.1371/journal.pgph.0004012.
91. Hudson KL, Lauer MS, Collins FS. Toward a new era of trust and transparency in clinical trials. JAMA. 2016;316(13):1353–4.

92. Talebi R, Redberg RF, Ross JS. Consistency of trial reporting between ClinicalTrials.gov and corresponding publications: one decade after FDAAA. Trials. 2020;21(1):675.
93. Rodgers F, Pepperrell T, Keestra S, Pilkington V. Missing clinical trial data: the evidence gap in primary data for potential COVID-19 drugs. Trials [Internet]. 2021 [cited 2025 Feb 27];22(1):59. Available from: https://trialsjournal.biomedcentral.com/articles/10.1186/s13063-021-05024-y
94. DeVito NJ, Morley J, Goldacre B. Barriers and best practices to improving clinical trials transparency at UK public research institutions: a qualitative interview study. Health Policy. 2024;142:104991.
95. World Medical Association. WMA Declaration of Helsinki – ethical principles for medical research involving human participants [Recinded] [Internet]. Seoul:Republic of Korea; 2008. Available from: https://www.wma.net/wp-content/uploads/2018/07/DoH-Oct2008.pdf
96. Viergever RF, Li K. Trends in global clinical trial registration: An analysis of numbers of registered clinical trials in different parts of the world from 2004 to 2013. BMJ Open. 2015;5(9)
97. World Health Organisation. Joint statement on public disclosure of results from clinical trials [Internet]. 2017 [cited 2025 Feb 28]. Available from: https://www.who.int/news/item/18-05-2017-joint-statement-on-registration
98. Riedel N, Wieschowski S, Bruckner T, Holst MR, Kahrass H, Nury E, et al. Results dissemination from completed clinical trials conducted at German university medical centers remained delayed and incomplete. The 2014 –2017 cohort. J Clin Epidemiol. 2022;144:1–7.
99. Nilsonne G, Wieschowski S, DeVito NJ, Salholz-Hillel M, Ahnström L, Bruckner T, et al. Results reporting for clinical trials led by medical universities and university hospitals in the Nordic countries was often missing or delayed. J Clin Epidemiol [Internet]. 2025;111710. Available from: https://linkinghub.elsevier.com/retrieve/pii/S0895435625000435
100. DeVito NJ, Bacon S, Goldacre B. FDAAA TrialsTracker: a live informatics tool to monitor compliance with FDA requirements to report clinical trial results. bioRxiv [Internet]. 2019;266452. Available from: https://www.biorxiv.org/content/10.1101/266452v4
101. De Vito NJ, Bacon S, Goldacre B. Compliance with legal requirement to report clinical trial results on ClinicalTrials.gov: a cohort study. The Lancet [Internet]. 2020 [cited 2025 Feb 26];395(10221):361–9. Available from: https://linkinghub.elsevier.com/retrieve/pii/S0140673619332209
102. Showell MG, Cole S, Clarke MJ, DeVito NJ, Farquhar C, Jordan V. Time to publication for results of clinical trials. Cochrane Database of Syst Rev [Internet]. 2024;2024(11). Available from: https://www.cochranelibrary.com/cdsr/doi/10.1002/14651858.MR000011.pub3/full
103. WMA. WMA Declaration of Helsinki – ethical principles for medical research involving human participants [Internet]. Helsinki; 2024 [cited 2025 Feb 16]. Available from: https://www.wma.net/policies-post/wma-declaration-of-helsinki/
104. Huser V, Cimino JJ. Evaluating adherence to the International Committee of Medical Journal Editors' policy of mandatory, timely clinical trial registration. J Am Med Inform Assoc. 2013;20(E1):e169–74.
105. Loder E, Loder S, Cook S. Characteristics and publication fate of unregistered and retrospectively registered clinical trials submitted to The BMJ over 4 years. BMJ Open [Internet]. 2018;8(2):e020037. Available from: https://bmjopen.bmj.com/content/8/2/e020037
106. World Health Organisation. Percent of clinical trials with retrospective registration [Internet]. 2024 [cited 2025 Feb 28]. Available from: https://www.who.int/observatories/global-observatory-on-health-research-and-development/monitoring/percent-of-trials-with-enrollment-year-preceding-registration-year
107. EBM DataLab. FDAAA TrialsTracker [Internet]. 2018 [cited 2025 Feb 28]. Available from: https://fdaaa.trialstracker.net/trials/
108. UAEM. Universities Allied for Essential Medicines. 2023 [cited 2025 Feb 28]. UAEM's Freedom of Information Act (FOIA) Discovery Investigates Clinical Trial Reporting. Available from: https://www.uaem.org/tools/uaems-foia-discovery

109. Siena LM, Papamanolis L, Siebert MJ, Bellomo RK, Ioannidis JPA. Industry involvement and transparency in the most cited clinical trials, 2019-2022. JAMA Netw Open [Internet]. 2023;6(11):e2343425. Available from: https://jamanetwork.com/journals/jamanetworkopen/fullarticle/2811814
110. Charani E, Abimbola S, Pai M, Adeyi O, Mendelson M, Laxminarayan R, et al. Funders: the missing link in equitable global health research? PLOS Global Public Health [Internet]. 2022;2(6):e0000583. Available from: https://journals.plos.org/globalpublichealth/article?id=10.1371/journal.pgph.0000583
111. Viergever RF. The mismatch between the health research and development (R&D) that is needed and the R&D that is undertaken: an overview of the problem, the causes, and solutions. Glob Health Action [Internet]. 2013;6:22450. Available from: https://www.tandfonline.com/doi/abs/10.3402/gha.v6i0.22450
112. Lezaun J, Montgomery CM. The pharmaceutical commons: sharing and exclusion in global health drug development. Sci Technol Hum Values [Internet]. 2015 [cited 2025 Feb 26];40(1):3–29. Available from: https://journals.sagepub.com/doi/10.1177/0162243914542349
113. Price R, Snijder M, Apgar M. Defining and evaluating equitable partnerships: a rapid review. UK Research and Innovation GCRF Urban Disaster Risk Hub; 2020.
114. Addison J, Browne J, van de Kamp J, Glas-van Dijk L, Gonani C, Kodan L. Overcoming challenges to build more equitable partnerships for international medical electives. MT Bull NVTG. 2022;60(3):18–20.
115. Sekalala S, Forman L, Hodgson T, Mulumba M, Namyalo-Ganafa H, Meier BM. Decolonising human rights: how intellectual property laws result in unequal access to the COVID-19 vaccine. BMJ Glob Health [Internet]. 2021;6(7):e006169. Available from: https://gh.bmj.com/content/6/7/e006169

Chapter 13
Patient Public Declaration of Research Ethics: Opinions and Proposals from Patients and the Public on Research Ethics

Chieko Kurihara, Yoshiko Saito, Hiroto Kai, Yoshikazu Funabashi, Keiko Inoue, Noriko Kishi, Akemi Kuge, Toshie Murakami, Katsura Suzuki, Hiroko Takahashi, Eiko Uchida, and Kyoko Imamura

C. Kurihara (✉)
Kanagawa Dental University, Yokosuka, Kanagawa, Japan
e-mail: chieko.kurihara@nifty.ne.jp

Y. Saito
Breast Cancer Survivor, Tokyo, Japan

H. Kai
Faculty of Education, Aichi University of Education (Science Course, Major in High School Education), Kariya, Aichi, Japan

Y. Funabashi
Cancer Patient and Survivor, Tokyo, Japan

K. Inoue
Association for Medical Malpractice Victims, Tokyo, Japan

N. Kishi
CMT Association Japan, Tokyo, Japan

A. Kuge
Freelance Social Worker, Takarazuka, Hyogo, Japan

T. Murakami
Sagamihara Kyodo Hospital Cancer Patients' Association "hukkiso", Sagamihara, Kanagawa, Japan

K. Suzuki
Person with Parkinson, Kawasaki, Kanagawa, Japan

H. Takahashi
Hereditary Hemorrhagic Telangiectasia (HHT) Patients' Association Japan, Osaka, Japan

E. Uchida
Breast Cancer Patient Support Group BOUGAINVILLEA/Patients-voice Council, Tachikawa, Japan

Y. Imamura
Japanese Institute for Public Engagement (Ji4pe), Yokohama, Japan

C. Kurihara et al. (eds.), *The 2024 Declaration of Helsinki*,
https://doi.org/10.1007/978-981-96-9294-1_13

Abstract We are members of the Bioethics Working Group of the Japanese Institute for Public Engagement (Ji4pe), composed of patients and the public. Since 2020, we have been continuing monthly Web meetings to learn the World Medical Association (WMA)'s Declaration of Helsinki (DoH) and have achieved mainly the following three publications:

- "Our WMA Declaration of Helsinki": a Japanese paper in which each article of the DoH was paraphrased into our plain language together with our opinions
- An English paper consolidating the above opinions
- The Patient Public Declaration of Research Ethics

During the process, we submitted our opinions in response to two periods of public consultation on the revision of the DoH, organised by the WMA.

We hope that the 2024 DoH will be adhered to by all the relevant parties and that our opinions can be included in future global norms. We believe that expressing our opinions will help maximise the impact of the new DoH. This chapter introduces the above process and reproduces the Patient Public Declaration of Research Ethics.

Keywords Patient public involvement · Research ethics · Fundamental human rights · Plain language · Cutting-edge technologies · Future generations

1 Interactions with the World Medical Association

1.1 Our Direct Commitment to the 2024 Revision of the Declaration of Helsinki (DoH)

Since 2020, we have been continuously learning about the Declaration of Helsinki (DoH) by the World Medical Association (WMA) [1], through monthly Web meetings. We are members of the Bioethics Working Group of the Japanese Institute for Public Engagement (Ji4pe). This institute provides systematic programmes for learning about medical systems, including medical research, shared among patients, the public, and experts. Each member of our group started learning for a variety of motivations and voluntarily continued collaborative activities. During the process of learning the DoH, we also learned about especially important international human rights norms [2–6], and other declarations by the WMA [7–9].

Our learning results were dedicated to our chapter contribution [10] to the book *Ethical Innovation for Global Health: Pandemic, Democracy and Ethics in research* (2023 November, Springer). There was also a contribution to this book from the WMA representatives and we expressed our opinions in English on the revision of the DoH, from the perspective of patients and the public. At the webinar to disseminate the launch of this book (2024 December), where the authors from the WMA also participated, our two members (Inoue, Uchida) made a presentation in English [11]. In the first half of 2024, the WMA set two periods of public consultation for the revision of the DoH, and we submitted our opinions both times. It was an exciting experience to bring our views directly to the WMA.

In August 2024, another member (Saito) also presented our opinions on the DoH [12] in a webinar co-organised by the Brazilian Society of Bioethics, the International Federation of Associations of Pharmaceutical Physicians and Pharmaceutical Medicine (IFAPP), and other organisations, being supported by the Japanese Association for Bioethics.

1.2 Discussions at the WMA's Regional Meeting in Washington, DC

During this process, Kurihara was invited as a speaker at the WMA regional meeting in Washington, DC, representing the IFAPP, connecting patients and experts. During the discussion, Kurihara conveyed the following message from our members:

> ***Study results are "gifts" to future patients and the public.***
> ***We believe that protecting the rights of patients means protecting the rights of physicians.***

A researcher in the United States who heard this said that she was almost moved to tears. We think she was impressed by the fact that patients suffering from some illness expressed, in their own words, altruism and gratitude towards physicians. Patients and the public will participate in research and collaborate with researchers because they expect the results of research to be delivered equally to the relevant communities.

After the conference, a letter of gratitude from the organisers, the WMA and the American Medical Association (AMA), stated the following:

> ***We also appreciate your work on research participant communication, which was a helpful reminder for the workgroup to strive for plain language and that outreach efforts beyond the revision itself may be needed.***

Our opinion is that the DoH itself should be written in plain language, but the 2024 revision includes a new requirement that each potential participant must be informed about the study in plain language.

1.3 Paradigm Change: Meaningful Engagement

Three major delightful changes were achieved in the 2024 DoH. It would be the result of the voices from many people around the world, not only ours.

- **The key term "research subject" was changed to "research participant".**
- **Meaningful community engagement at all stages of research is now encouraged.**

- **In Research Ethics Committees, diversity and familiarity with the local context have come to be required, particularly with the specification of the need for a member of the general public.**

These changes are sincerely appreciated. This is a "*paradigm change*", which the Secretary General of the WMA stated as "*more than a matter of courtesy, but recognises that patients play a key role in co-creation rather than being passive subjects*" [13]. It should be also noted that he states that "*care should be taken to avoid manipulation of research data or results*". Patient engagement should also contribute to the credibility and integrity of research.

This paradigm change has not yet been a reality in most research practice; thus, we wish to disseminate this policy from the WMA. We, the patients and the public, wish to take responsibility for this direction to shared decision-making (SDM) in both healthcare and research.

In particular, the implication of the word "meaningful" must be taken seriously: engagement should not be just a formal, administrative procedure. In Japan, the term "patient and public involvement" is used often, rather than "community engagement". This may be because the power of communities is relatively weak in Japan. "Meaningful" may vary up to regional, cultural contexts or levels of healthcare. It is necessary to consider what "meaningful engagement" means in each situation.

Patients and the public are rather than just "observers" or "those whose opinions should be considered". Patients and the public must have an equal role in decision-making, having voting rights with other interested parties. To achieve this, patients and the public also need to deepen their own voluntary and systematic learning.

2 Patient and Public Declaration of Research Ethics

2.1 Research Ethics of the People, by the People, for the People

With the goal of bringing our views to the 2024 DoH, we have been able to produce a number of achievements. The latest one was the following publication in a medical journal [14]. This is reproduced with small updates (from 1.0 to 1.1 edition) in the latter part of this chapter.

Patient Public Declaration of Research Ethics (1.1 edition):
Research Ethics of the People, by the People, for the People
-Expanding the Impact of the 2024 Revision of the Declaration of Helsinki

Our Declaration calls for adherence to the 2024 DoH, but also expresses the hope that the ethical principles not included in the DoH will become a universal code of research ethics for the future. In addition, we would work together with various interested parties to maximise the impact of the new DoH.

During the above-mentioned 2023 webinar on disseminating our book published by Springer, Prof Ames Dhai, one of the book editors, who heard our presentation, acknowledged that the content of our opinions had important significance not found in the DoH and encouraged us to issue a declaration on research ethics by patients and the public from here in Japan. Encouraged by her words, we set about drafting the Declaration.

2.2 *Online Seminar Held on 20 January 2025*

In the online seminar held on 20 January 2025, Kurihara presented in the first half the background and contents of the 2024 revision of the DoH, as an external collaborator, and in the second half, eight members of our group read out the Patient Public Declaration of Research Ethics, describing its contents and the thoughts.

There were 525 registrants, with 357 accessing this seminar (due to free-of-charge participation, the number of accesses was less than that of registrations, as is ordinarily the case). This shows that the Japanese research community is very much interested in the impact of the new DoH.

Participants raised a number of comments and questions. Particularly, it was impressive that several questions were raised by company participants about the DoH's level of acceptable risks in placebo-controlled trials when there is a proven intervention, which is so high. In Japan, it is difficult to include patients in placebo-controlled trials, and researchers are interested in global efforts exploring the use of real-world data as external control.

There were also discussions about the challenges of sharing with patients some study outcomes still in the development phase, in advance of marketing approval. “Meaningful” involvement of patients and the public needs further discussions on such issues shared by patients, the public, and experts.

3 Our Achievements and Goals for the Future

3.1 *Our Achievements*

Our group’s activities started with the Web meetings to learn about the DoH. During the process, we published “Our WMA Declaration of Helsinki” [15] to convey the contents of the DoH in our own plain language (Fig. 13.1). This publication is composed of three parts below:

1. **A reproduction of each paragraph of the DoH, using the Japanese translation by the Japan Medical Association (JMA) (based on the permissions from the WMA and JMA)**

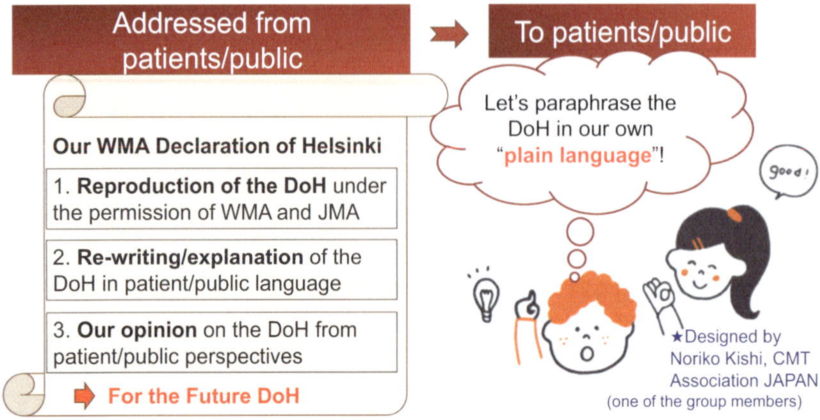

Fig. 13.1 Our WMA Declaration of Helsinki (DoH) [15]

2. **A paraphrase of each paragraph of the DoH in our own plain language**
3. **Our opinions on each paragraph of the DoH**

Because the part "(3) Our opinions on each paragraph of the DoH" includes the ethical principles beyond the scope of the DoH, this part was expanded and became a chapter in the 2023 book from Springer [10] and then "Patient Public Declaration of Research Ethics" [14]. A summary of the opinions is outlined in Table 13.1, some of which were accepted in the 2024 revision.

3.2 *Goals for the Future*

Because the already published "Our WMA Declaration of Helsinki" [15] was based on the previous, 2013, version of the DoH, our next goal is to publish "Our WMA Declaration of Helsinki" based on the 2024 DoH.

We also wish to develop the Patient Public Declaration of Research Ethics into a second edition, based on discussions with expanded domestic and global interested parties.

In addition, we wish to publish a report on Patient, Public, and Community Engagement, oriented by us, the patients, and the public. In order to make community engagement "meaningful", we ourselves need to clearly indicate how we wish to be involved and what we wish to achieve, resulting from our participation.

The DoH is a set of ethical principles for research; thus, we wish to broaden our scope and explore in-depth "bioethics", aiming for the achievement of holistic healthcare, considering the relationship between life, science, and technology, from our own perspective.

Table 13.1 A summary of authors' opinions and proposals, not included in the 2013 DoH, some of which were included in the 2024 DoH and should be discussed for future research ethics principles

• Early evaluation of the research and development plan considering the impact on study target populations, society, global contexts, and future generations
– Avoid discrimination or stigmatization against the target populations of the study
– Aim for the United Nation's Sustainable Development Goals (SDGs)
– Assess the impact on future generations, the environment, and society
• Strengthen patient/public perspectives on the ethical principles beyond the scope of autonomous norms established by and addressed to physicians
– Use plain language, e.g. avoiding "double-negative" phrases.[a]
– The scope of the Declaration needs to adequately address respect for embryos, foetuses, and the deceased person on which studies are performed
– Ensure the participation of patients and the public in all processes of research
– Multidisciplinary collaboration, including patient peer supporters.[b]
– Shared decision-making based on informed consent
– Diversity, fairness, and equity in the composition of the Research Ethics Committee.[c]
– Protect the dignity and rights of an individual who is incapable of consent and has no available proxy
– Develop international research ethics norms that are agreed upon by all the interested parties.

: Items where our suggestions were somewhat accepted in the DoH 2024 revision

[a]Incorporated as using plain language in the informed consent process, although we hope the DoH itself is written in plain language

[b]Multi-professional collaboration was incorporated as an expansion of the DoH recommendations for teams and organisations, but patients' peer support was not considered

[c]Diversity and public involvement in RECs were directly incorporated, but the issue of "fairness" was not sufficiently considered

Note: The table is updated from our previous publication [10] under the permission of the publisher with some additional information after the 2024 revision of the DoH

4 Cutting-Edge Technologies and the Vision for Future Generations

4.1 Expectations and Concerns

One of the reasons for the need for patient and public involvement is that cutting-edge technologies have been rapidly developed. Interests of industries are shifting from the development of drugs for chronic diseases with large markets to genetic diseases, rare and intractable diseases, and new methods using gene technology, such as genome editing. These treatments are normally expensive, reflecting development costs and small market size, and thus are insufficiently accessible to those who need them. We wish to participate in the design of health systems to solve the problem of the disparity in access to healthcare between the rich and poor. Patient and public involvement must be required to achieve equitable access to safe and effective treatments and equitable benefit sharing.

In addition, we have seen progress in the development of technologies that can artificially create human life and mental activity, such as artificial intelligence developed by analysing large amounts of personal data, technology to create and fertilise eggs and sperms from iPS cells, and brain organoids. We will evaluate both the positive and negative impacts of these technologies and discuss and collaborate with interested parties in order to pave the way for technology to contribute to the improvement of the quality of healthcare.

4.2 Together with Future Generations

Our group, including surrounding colleagues, is composed of patients living with a variety of illnesses, including rare incurable diseases, neurodegenerative diseases, cancer patients undergoing treatment, cancer survivors engaged in peer support activities, haematological disorders, mental disorders, patient families, and those who have experienced medical errors.

Patients and their families often do not have enough time to think about research ethics and healthcare policy, and even if they do think about it, they do not have the opportunity to communicate and discuss their opinions. Each of the members of our Working Group has a specific motivation for learning, which has promoted our learning, gathering together. As patients, the public, and patient families, we wish to learn more as advocators and to communicate with a view to the future, out of a sense of solidarity.

5 Conclusion

Protecting health and well-being, living with and overcoming illness is not a problem for individuals or families alone, but for society. What is desired as the outcome of research and development would be the realisation of holistic healthcare and equitable global access to essential medicines. Such objectives need to be shared between healthcare providers, researchers, patients, and the public from the very beginning stage of the research and development.

The results of the research would be meaningless if they could not be implemented in the community. The DoH has raised difficult but extremely valuable challenges, such as "benefit sharing", "post-trial access to essential medicines for all in need globally", and "value of the research for the community and the world". We, the patients and the public, are committed to tackling these challenges. We also must discuss the consequences of research and development, including the potential for harm to individuals and society.

In order to set out our vision for the future and to create the foundations for a new code of ethics for new societal challenges, we have issued this Patient Public Declaration of Research Ethics. This Declaration is the starting point. We hope that through this publication we can discuss and develop this Declaration with all relevant, various interested parties having equitable commitments.

Note

A very small portion of this chapter includes the texts in previous publications which led to the original version of the Patient and Public Declaration of Research Ethics. Such duplications and reproduction with some update of the Declaration are under the agreement of each publisher (Clinical Evaluation and Springer) and in the range of "acceptable duplicated publication" in the recommendations of the International Committee of Medical Journal Editors. Acknowledgement and conflict of interest disclosure below are same with some updates as the ones in the original version of the Patient Public Declaration of Research Ethics reproduced below.

Acknowledgement All the authors appreciate Ms Midori Senoo, Managing Director, Myotonic Dystrophy Patients' Group of Japan (DM-family) and Ms Konomi Wakasa (Patient's Association of Nesidioblastosis) who participated in some part of our series of Web meetings and provided meaningful opinions; Prof. Dirceu Greco (Professor Emeritus of Infectious Diseases and Bioethics at the School of Medicine, Federal University of Minas Gerais, Brazil) and Prof. Ames Dhai (Professor Bioethics, School of Clinical Medicine, University of the Witwatersrand, South Africa), editors who have given significantly valuable suggestions, including the original idea of this Declaration; Ms Nami Saito, who provided the language review on the English version of the Declaration.

CK appreciates the International Federation of Associations of Pharmaceutical Physicians and Pharmaceutical Medicine (IFAPP) for granting her travel expenditure for her in-person participation in the WMA Regional Meeting in Tel Aviv (2022 December) and the WMA for granting her travel expenditure for her in-person participation and presentation as a panellist in WMA Regional Meeting in Washington, DC (2024 August), and waiver of registration fee from WMA as a journalist to attend the General Assembly in Helsinki (2024 October). She participated in all the other WMA Regional Meetings at her own expense or online.

Conflict of Interest The authors declare no competing interests relevant to the content of this chapter.

KS is an employee of ASKA Pharmaceuticals.

Author Contribution All contents are constructed from the opinions expressed by authors including patients and the public: HK, who has made the most critical reconstruction, and YS, YF, KI, NK, AK, TM, KS, and EU. Among them, KS is an expert in drug development. AK practises as a qualified mental health social worker. CK edited the words expressed by them into the terminology used in research ethics. KI consulted as an expert in medicine and drug development. Contributions of all the authors are in accordance with the recommendations of the International Committee of Medical Journal Editors.

Patient Public Declaration of Research Ethics (1.1 Edition)[1]

Purpose and Scope

The Patient and Public Declaration of Research Ethics is a declaration of ethical principles for research involving humans, of the patient and public, by the patient and public, for the patient and public. Our Declaration requires all those who conduct health research involving humans to adhere to the World Medical Association (WMA)'s Declaration of Helsinki (DoH) (Box 13.1).

Our Declaration covers health research involving not only living humans but also research on prenatal embryos, foetuses, deceased persons, and materials or data derived from humans.

On the premise that the DoH is adhered to, we declare additional ethical principles, most of which are not included in the DoH, which was developed primarily by physicians. Our Declaration aspires for its principles to be recognised in the future as world common norms, addressing all patients and the public, which include physicians and other experts.

Box 13.1 Summary of the Principles of Research Ethics Established in the Declaration of Helsinki (Keywords in the 2013 and 2024 Versions of the DoH Are Underlined. Important New Items in the 2024 Version Are in Bold, Except Subheadings)

- **Preamble**
  - The Declaration of Helsinki is a set of ethical principles for medical research involving human participants.
  - **The term "human subjects" was changed to "human participants".**
  - **Recommended not only to physicians but also to teams and organisations.**
  - The Declaration covers research on not only living persons but also individually identifiable human materials and data.

(continued)

[1] This part is a reproduction from the first edition published as a part of a previous publication, reference number 14, with small updates to edition 1.1. Main updates are as follows: In Box 13.1, essential points of revisions in the 2024 DoH were added to a summary of the principles in the 2013 DoH. In the text of the Patent Public Declaration, statements based on the new principles in the 2024 DoH are given with DoH's paragraph numbers in the form of (§1) , and statements based on the principles already in the 2013 version of the DoH are given in the form of (§1, 2013). Other texts are an expression of our opinions not included in the DoH.

Box 13.1 (continued)

- **General principles**
  - The rights and interests of research participants take precedence over the goal of research to generate new knowledge.
  - Research must be conducted only by qualified individuals.
  - Harms resulting from research procedures must be appropriately compensated, and necessary treatment must be provided.
- **Risk and benefit**
  - Risk benefit assessment, risk minimisation, and risk management are continuously required. Research may not be initiated or continued unless the benefits outweigh the risks, and must be discontinued if necessary.
  - **Distribution of risks and benefits should be assessed considering various structural inequities.**
  - **Meaningful engagement with potential and enrolled participants and their communities should occur before, during, and following medical research.**
- **Vulnerability**
  - **Considerations are recommended on fixed, contextual, or dynamic vulnerability and the possibility that their exclusion from research may exacerbate their disparities, which must be weighed against the harms of inclusion.**
- **Research protocol and Research Ethics Committee approval**
  - Research may not be initiated or continued unless a research protocol describing the design and performance of the research has been approved by a research ethics committee.
  - **Research Ethics Committees must have diversity and sufficient familiarity with local circumstances and include a member of the general public.**
- **Informed consent**
  - Participation of individuals giving informed consent must be voluntary after adequate necessary information is provided. For a potential participant incapable of giving consent, informed consent from an authorised representative is required. In this case, assent from this individual must be sought.
  - **Values and preferences of participants incapable of giving informed consent must be considered.**

(continued)

Box 13.1 (continued)

- **Placebo use and post-trial provisions**
  - The benefits and risks of a new intervention must be tested against those of the best proven intervention, if there is. Exceptional comparison with a placebo or no intervention may be acceptable if there is no additional risk of serious or irreversible harm*[1] resulting from not receiving the existing best proven intervention.
  - Provisions for post-trial access for participants who still need the tested interventions **must** be made in advance of starting the clinical trial*[2]. **Exceptional arrangement of post-trial provisions must be approved by a research ethics committee.***[3]

    *1: Acceptable risk in this case described in the CIOMS guideline [16] is "no more than a minor increase above minimal risk".

    *2: There are opinions that post-trial access should be ensured to trial participants, not only to make provision in advance. In addition, access should be for the host community as well as those in need globally. **In the 2024 revision, "should" for making provision was changed to "must".**

    *3: **This should not be deemed to be an excuse; rather, the responsibility of the committee should be strictly recognised.**
- **Research registration and publication**
  - Research must be registered in a publicly accessible database before the recruitment of the first participant. Results must be accurate and must be published even if they are not what was expected.
- **Unproven interventions in clinical practice**
  - An unproven intervention may be used by a physician for the treatment of a patient, under some limited conditions, including seeking expert advice and obtaining informed consent from a patient or representative. This intervention should subsequently be made the object of research to evaluate its safety and efficacy.

Source: Summarised from the World Medical Association's Declaration of Helsinki: Ethical principles for medical research involving human participants, 2024 version.

Fundamental Principles

Research Ethics Principles in Plain Language

Research ethics principles must be described in plain language that can be understood by all.

Ethical Principles Based on Universal Human Rights Norms

This Declaration is based on internationally recognised norms of fundamental human rights [2–6], patient rights [7], and principles of medical ethics [8, 9].

Both researchers and research participants should be aware of these basic norms.

Research Ethics Principle Based on Responsibility and Awareness as "Research Participants"

The rights of research participants (§1 and others)must be respected and assured.

On this premise, research participants will take on the role of a collaborative partner to co-create research, with appropriate responsibilities and awareness.

Furthermore, we wish to establish research ethics principles covering the ethics of patient-initiated research.

Promotion of Patient and Public Involvement

We will participate proactively in the research and development of medical products and healthcare policy.

Additionally, we will advocate and support the participation of those who are unable to proactively express their wishes and will deliberate together with them on how their meaningful participation (§6)should take place.

Respect for Human Rights and Welfare of Research Participants

The core principle of the DoH stating that the rights and interests of research participants take precedence over the goal of research(§7, 2013) must be upheld at all times.

Realising the Value of Research

The ultimate goal of medical research is to realise holistic healthcare through recognising the fundamental rights of patients, considering diversity, and respecting the values of the communities of the research participants.

Everyone involved in research should share this goal as a core "value" of research.

Prevention of Discrimination and Stigmatisation, and Consideration of the Impact on the Environment, Society, and Future Generations

Discrimination and stigmatisation of individuals due to the conduct or results of the research must be prevented.

All those who are engaged in research must consider the impact of the research on human society, the human mind, future generations, and the sustainability of the ecosystem.

For this reason, patient and public involvement from the early stage of the research and development program(§6) is required.

Advocate the Rights and Interests of the Vulnerable and Promote Their Participation in Research

The participation of vulnerable people in research provides opportunities for their social inclusion. Their contribution to the improvement of the well-being of their community and beyond creates valuable research for society as a whole.

People who have particular difficulties in expressing and making decisions due to their physical or social conditions are more likely to be put at risk of research. They are also likely to be disadvantaged in the sharing of benefits from research. Because of these reasons, they must receive additional protection (§6, 19, 20, 2013).

Any form of exploitation in the name of research must be prohibited.

Self-advocacy of vulnerable people to express their views and make their own decisions, as well as advocacy on their behalf, must be promoted.

Ensuring the Scientific Validity of Research and Animal Welfare

Scientific validity and research integrity must be assured (§21, 22, 2013), avoiding research waste (§21) and strictly prohibiting research misconduct (§12).

Conduct of animal experimentation must be in accordance with the principles of the 3Rs (Replacement, Reduction, Refinement). Animal welfare (§21, 2013) must be respected.

Research Ethics Committees

Diversity of Research Ethics Committees and Patient Public Involvement

Diversity of Research Ethics Committee members must be assured and participation of members representing patients and the public must be promoted (§23).

Members representing the patients and the public must respect the perspectives of those who are unable to express their opinions, ask appropriate questions, express opinions, and participate in deliberations and voting, taking the given responsibilities.

Open Recruitment for Fairness

Research Ethics Committees must ensure transparency, independence (§23, 2013), and fairness in their establishment, membership, and operation.

The selection of committee members must not be arbitrary. There must be a system whereby motivated members who have achieved relevant systematic education programmes are assigned through open recruitment. They are given training opportunities and eventually replaced after serving a certain period of time, so that they can be followed by well-trained successors.

Assessment of Advanced Technology and Impact on Society and Future Generations

It is desirable that Research Ethics Committees have the function of deepening the discussions on the impact of advanced medical technologies on society and future generations that emerge from the review of individual research protocols, of presenting views as ancillary opinions, and of encouraging a broader conversation within society.

Informed Consent

Informed Consent and Shared Decision-Making

A person's decision to participate in research must be based on a fully informed understanding and free and voluntary consent (§25, 2013).

Explanations about the research must be given in a way that is understandable (§26) to the candidate participant and that correctly conveys the nature of the research.

Research participants must be guaranteed the right to disagree, discontinue participation, or withdraw consent without detriment (§26).

In addition to the assurance of individual informed consent, patient-centred, shared decision-making should be promoted, collaborating with the patient's family members, supporters, physicians, and a multidisciplinary research team. They can work together on a treatment plan, including whether or not to participate in research, and continuously review the patient's participation in research as circumstances change.

Advocating Decision-Making for People Without Capacity to Consent

For a person who is incapable of giving informed consent; due to immaturity or physical or mental situations, the representative must advocate for the patient's will, including previously expressed ones (§28), and the best interest of the candidate participant.

In addition, advocacy and support are required so that such a person can express an assent (§29, 2013), an expression of willingness to participate in research, after receiving sufficient information adjusted to the person's comprehension level.

Dignity and Rights of Persons Without Capacity to Consent and Without Relatives

The representation of decision-making of persons who do not have the capacity to consent, have no relative, and are unable to give their own or proxy consent even at a later date has not been adequately considered.[2]

We wish to discuss from the perspective of patients and the public the rights and conditions for participation in research of such persons, who would be the most vulnerable.

Broad Informed Consent and Dynamic Consent

The Declaration of Taipei for Health Databases and Biobanks [17] must be adhered to where data or materials obtained in research may be subject to secondary use (§32).

When a research participant gives consent to the secondary use of materials or data, the consent should be "broad informed consent" [16] based on sufficient understanding of the information given as far as possible about the purpose, who

[2] In Japan, relevant guidelines have been published as below in relation to medical care but this kind of consideration has not yet been made in the area of research. Ministry of Health, Labour and Welfare. Guidelines and case studies for supporting hospitalisation and medical care for those with difficulties in decision-making, without relative.

https://www.mhlw.go.jp/stf/seisakunitsuite/bunya/kenkou_iryou/iryou/miyorinonaihitoheno-taiou.html

could use it, and the governance arrangement of possible future secondary uses. We request that instructions on how to access the secondary use information be disclosed and the right of withdrawal of consent be assured.

We wish to promote an emerging concept of "dynamic consent", which means giving continuing consent for the use of materials or data, being informed about new secondary uses, the progress of the research, and any other relevant information generated from research, including how to access the secondary use information and the right to refusal or withdrawal [18]. We hope for the development of social consensus for the methods and procedures for such type of consent and an environment where participants can make an adequate decision.

Right to Know and Right Not to Know

A research participant's "right to know" and "right not to know" must be respected. Any meaningful information for an individual participant, including secondary findings obtained from research, must be explained according to their wishes, based on due consideration of its scientific validity and its clinical significance in the treatment program.

Principles of Controlled Clinical Trials and Post-trial Access

Controlled Clinical Trials

Clinical trials comparing treatments(§33, 2013) are conducted when it is not known which method is better.

We will be able to make a decision whether or not to participate in a placebo-controlled trial only when the significance and the compelling reasons are explained.

When there is a safe and effective treatment, it is essential to compare a new treatment with existing safe and effective treatment. Nevertheless, if the placebo-controlled trial is conducted in such a setting, the placebo-related risk must be minimal.

We wish to be involved in the discussion on acceptable conditions of the placebo use, from the perspectives of patients and the public.

Post-trial Access

Research participants (patients) still in need of safe and effective treatment proven in the research (§34, 2013) should be provided with the treatment after the completion of the trial.

Research is co-creation involving patients and the public and the results of research are to be delivered to future society. To ensure that the results from the research can be accessed by those who need them, we will work together with those involved in research to develop the infrastructure to realise post-trial access.

Publication of Research Results

Database Registration of Research

All information on research and its findings must be free and accessible to all and registered in a publicly accessible database (§35, 2013).

The results of the research should be released with a lay summary written in plain language understandable by the general public.

Journal Publication of Research

For the purpose of promoting patient public involvement, "open access" of the report in peer-reviewed journals should be facilitated so that everyone can read the full text through fair and equitable cost-sharing.

"Open science" should be also promoted so that the data supporting the research results can be disclosed to the public or shared conditionally with those who need them. In addition, "citizen science" where patients and the public can make use of such data and participate in the research process should be promoted.

Unproven Treatments

The DoH states that physicians may use some unproven treatment for patient care under some conditions but recommends that it should be made the objective of research to evaluate its safety and efficacy (§37, 2013).

When an unproven intervention is used, relevant data should be accumulated and safety and efficacy should be monitored.

The Wishes of Patients and the Public

Medical care and research depend on trust between the physicians/researchers and the patients and the public.

The goal of medicine is not to pursue profit, but to move towards holistic care and realise the physical, mental, and social well-being of humans.

Research is altruistic in nature and is supported by the altruism of all involved. Study results are "gifts" to future patients and the public.

We believe that protecting the rights of patients means protecting the rights of physicians.

We must always be aware that healthcare is based on research supported by such an ethical foundation.

In order to create new ethical norms responding to emerging social issues, with a vision for the future, we proclaim this Patient and Public Declaration of Research Ethics.

References

1. The World Medical Association. WMA Declaration of Helsinki – ethical principles for medical research involving human participants. First adopted in 1964, last amended in 2024. https://www.wma.net/policies-post/wma-declaration-of-helsinki/
2. World Health Organization. Constitution of the World Health Organization. 1946 July 22. https://www.who.int/about/governance/constitution. Accessed 30 April 2025.
3. United Nations. The Universal Declaration of Human Rights. Adopted December 10, 1948. https://www.un.org/en/about-us/universal-declaration-of-human-rights
4. United Nations. International Covenant on Economic, Social and Cultural Rights. Adopted by General Assembly resolution 2200A (XXI) of 16 December 1966, entry into force 3 January 1976. https://www.ohchr.org/en/instruments-mechanisms/instruments/international-covenant-economic-social-and-cultural-rights
5. United Nations. International Covenant on Civil and Political Rights. Adopted by General Assembly resolution 2200A (XXI) of 16 December 1966, entry into force 23 March 1976. https://www.ohchr.org/en/instruments-mechanisms/instruments/international-covenant-civil-and-political-rights
6. United Nations Educational, Scientific and Cultural Organization. Universal Declaration on Bioethics and Human Rights. 19 October 2005. https://www.unesco.org/en/legal-affairs/universal-declaration-bioethics-and-human-rights?hub=66535
7. The World Medical Association. Declaration of Lisbon on the Rights of the Patient. 1981, last reaffirmed in 2015. https://www.wma.net/policies-post/wma-declaration-of-lisbon-on-the-rights-of-the-patient/
8. The World Medical Association. The Declaration of Geneva. 1948, last amended 2017. https://www.wma.net/policies-post/wma-declaration-of-geneva/
9. The World Medical Association. The international code of medical ethics. 1949, last revised in 2022. https://www.wma.net/policies-post/wma-international-code-of-medical-ethics/
10. Kurihara C, Inoue K, Kai H, Suzuki K, Saeki H, Funabashi Y, Kishi N, Kuge A, Murakami T, Saito Y, Uchida E, Tsutsumi N, Imamura K. Our "WMA Declaration of Helsinki": opinions and proposals from patient and public for research ethics. In: Kurihara C, Greco D, Dhai A, editors. Ethical innovation for global health: pandemic, democracy and ethics in research. Springer; 2023. p. 243–69.
11. Inoue K, Uchida E. Our "WMA Declaration of Helsinki": opinions and proposals from patient and public for research ethics. In: Co-organized by Clinical Evaluation, Pharmaceutical Study Group. Ethical Innovation for Global Health. 2023 Dec 6; Online. http://cont.o.oo7.jp/sympo/eigh.html. Accessed 30 April 2025.
12. Saito Y. Patient and public opinions to the revision of the Declaration of Helsinki: our proposals to the WMA. In: the 2024 Declaration of Helsinki: taking forward bioethics and human rights. Clin Eval. 2025; 52(3): 394-97. To be published on Feb 25, 2025. http://cont.o.oo7.jp/52pop/52pop_contents_e.html

13. Kurihara C, Matsuyama K, Baroutsou V. World Medical Association's Declaration of Helsinki, 2024 revision: celebrating the 60th anniversary, at Helsinki. Clin Eval. 52(3):W1–W27. http://cont.o.oo7.jp/52pop/52pop_contents_e.html
14. Kurihara C, Saito Y, Kai H, Funabashi Y, Inoue K, Kishi N, Kuge A, Murakami T, Suzuki K, Takahashi H, Uchida E, Imamura K. Patient public declaration of research ethics (1st edition): research ethics of the people, by the people, for the people-expanding the impact of the 2024 revision of the Declaration of Helsinki. Clin Eval. 2024;52(3):W28–39. http://cont.o.oo7.jp/52pop/52pop_contents_e.html
15. Kurihara C, Inoue K, Kai H, Suzuki K, Saeki H, Funabashi Y, Kishi N, Kuge A, Murakami T, Saito Y, Uchida E, Tsutsumi N, Imamura K. Our "WMA Declaration of Helsinki"– Opinions and proposals from patient and public–. Clin Eval. 2024; 52(1): 91–131. Japanese. http://cont.o.oo7.jp/52_1/p91-131.pdf
16. Council for International Organizations of Medical Sciences. International ethical guidelines for health-related research involving humans. 2016. https://cioms.ch/publications/product/international-ethical-guidelines-for-health-related-research-involving-humans/
17. The World Medical Association. The Declaration of Taipei on ethical considerations regarding Health Databases and Biobanks. 2002, last revised 2016. https://www.wma.net/policies-post/wma-declaration-of-taipei-on-ethical-considerations-regarding-health-databases-and-biobanks/
18. Council for International Organizations of Medical Sciences (CIOMS). Patient involvement in the development, regulation and safe use of medicines. CIOMS Working Group report. Geneva; 2022. https://doi.org/10.56759/iiew8982

Part III
Alternative Frameworks for Ethical Innovation Towards Global Health

Chapter 14
Medicines Development for Global Health: The Evolution of IFAPP's Ethical Framework in the Ecosystem of the Life Sciences in a New World Order for Health

Varvara Baroutsou

Abstract The Faculty of Pharmaceutical Medicine of the Royal Colleges of Physicians of the United Kingdom defines Pharmaceutical Medicine as the medical science discipline concerned with the discovery, development, evaluation, registration, supervision and medical aspects of the marketing of medicinal products for the benefit of patients and the health of the community.

The International Federation of Associations of Pharmaceutical Physicians and Pharmaceutical Medicine (IFAPP) is dedicated to promoting the ethical practice of pharmaceutical medicine (synonymous with drug development) in a rapidly changing scientific and medical landscape due to emerging innovations and accelerating technologies. IFAPP emphasises scientific integrity, ethics, values and culture and fosters a sense of belonging and purpose among its members.

Since its founding in 1975, IFAPP has worked with global stakeholders in medical and research ethics, including the recent revision of the World Medical Association's Declaration of Helsinki.

IFAPP's initiative embraces innovation and new technologies, enabling members, health professionals and scientists to keep abreast of developments and enhance their skills through continuous, high-quality education and training in science and ethics.

This chapter describes IFAPP's strategy to integrate ethical skills into practice, enabling pharmaceutical medicine professionals to navigate complex ethical dilemmas in research and patient care using the IFAPP Ethics Framework.

Keywords Ethics · Global health · Innovation · Medicines development · New technologies · Patients' rights · Pharmaceutical medicine

---

V. Baroutsou (✉)
International Federation of Associations of Pharmaceutical Physicians and Pharmaceutical Medicine (IFAPP), Woerden, The Netherlands

Consultant in Internal Medicine and Pharmaceutical Medicine, Athens, Greece
e-mail: varbara.baroutsou@ifapp.org; secreteriat@ifapp.org; baroutsb@otenet.gr

C. Kurihara et al. (eds.), *The 2024 Declaration of Helsinki*,
https://doi.org/10.1007/978-981-96-9294-1_14

1 Background

The context and impact of the ethical perspective of the International Federation of Associations of Pharmaceutical Physicians and Pharmaceutical Medicine (IFAPP) are linked to the global community of the development of medicines, including several world organisations, stakeholders and institutions. These collaborations will be addressed to highlight the unified approach to ethical challenges in health research.

The Faculty of Pharmaceutical Medicine of the Royal Colleges of Physicians of the United Kingdom defined Pharmaceutical Medicine as the medical scientific discipline concerned with the discovery, development, evaluation, registration, monitoring and medical aspects of marketing of medicines for the benefit of patients and the health of the community [1].

The foundations of IFAPP were laid in 1970 by three British Pharmaceutical Physicians, Bill Burland, Jan Jouhar and Richard Rondel, who conceived the idea of holding an international meeting of Medical Advisers (as pharmaceutical physicians were then known), which would bring together physicians and scientists from the pharmaceutical industry with those working in research institutes and academic medicine. Their objective was to internationalise the advancement of knowledge of therapeutic agents and their actions, and also to foster the relationship between industry and academia [2].

The IFAPP has developed the International Ethical Framework (IEF) to guide pharmaceutical physicians and medicines development scientists in their professional activities by publishing the first edition of its International Code of Ethical Conduct in April 2003 and updating it in a revised edition named IEF of 2018 [3].

In a recent paper published by IFAPP Ethics Working Group members, the professional significance and application of a collection of ethical advice from the IFAPP Ethical Framework have been summarised by the senior experts and founders of the IFAPP IEF, Sander Kerpel-Fronius and Alexander Becker [4].

The IFAPP Ethics Framework reflects the IFAPP's vision: to be a leading global organisation in advancing the ethical practice in the field of pharmaceutical medicine; and mission: to promote the ethical conduct of pharmaceutical medicine through education, advocacy and collaboration with stakeholders worldwide.

The IFAPP's Ethics Framework [3] is based on the core values of:

- Integrity: Ensuring honesty and transparency in all professional activities.
- Respect for Persons: Upholding the dignity, rights and autonomy of all persons involved in pharmaceutical research and development.
- Beneficence and non-maleficence: Striving to do good and avoid harm in all professional activities.

The primary areas of responsibility defined in the IFAPP's Ethics Framework [3] refer to:

1. Duty of Clinical Care: Adhering to the Declaration of Helsinki and Good Clinical Practice (GCP) guidelines to ensure the safety and well-being of patients participating in medical research
2. Scientific Competence and Diligence: Maintaining high standards of professional competence and diligence in all research and professional activities
3. Impartiality in Communication: Ensuring evidence-based, unbiased and accurate communication of scientific information and data, particularly in relation to the use of medicinal products
4. Probity: Exhibiting high moral standards
5. Integrity and Accountability: Demonstrating trustworthy behaviour

In addition, collaboration with IFAPP stakeholders' joint responsibility areas has been defined focusing on:

- Studies of Clinical Pharmacology and Therapeutic Use: Conducting post-marketing surveillance and studies to monitor the safety and efficacy of medicines in real-world settings
- Special and Vulnerable Patient Groups: Paying special attention to the needs and protection of vulnerable populations in clinical research
- Awareness of Innovation and Technology: Staying informed about new developments and innovations in pharmaceutical medicine and incorporating them responsibly

Furthermore, the framework includes detailed guidance on important ethical issues about:

- Use of Placebos: Ethical considerations for using placebos in clinical trials
- Post-trial Access: Ensuring patients have access to beneficial treatments after the conclusion of clinical trials
- Secondary Data-Driven Research: Addressing the ethical use of secondary data in research according to the Declaration of Taipei (DoT) [3, 5]

There has never been a more important time to stand up for patients' rights in global health.

2 Build Continuous Improvement for Best Practices

2.1 Challenges and Opportunities for Innovation in Adverse Circumstances

A new global perspective on research and development (R&D) in the academic, regulatory, public and private life sciences ecosystem is needed to ensure better evidence, more transparency and co-creation with patients for medical research that leads to more targeted, effective, safer and affordable medicines and vaccines to benefit patients, society and embed evidential practice resulting in equitable health outcomes.

The need for new antibiotics is well-recognised, and on 26 September 2024, the United Nations General Assembly discussed political actions in a high-level meeting on antimicrobial resistance, or 'AMR'. The AMR Accelerator—a public–private partnership involving 9 European projects and 98 organisations—urges government leaders and private actors to invest in the development of antibiotics and research on antimicrobial resistance [6].

The call to action, published in Nature Reviews Drug Discovery, emphasises the need for coordinated action and commitments to meet the threat of antimicrobial resistance and secure a sustainable future for European antibiotic development. The return on investment is low, and many large pharmaceutical companies have abandoned the field. According to the authors, collaboration and risk-sharing can help keep companies in anti-infective drug development [7].

Of course, the AMR is one European programme in the right direction that may motivate more relevant efforts that are much needed due to the current climate crisis, geopolitical tensions and regional wars. However, we continue to see governments, policymakers and industry retreat from funding medical research for real unmet patient needs, creating therapeutic challenges and inequalities for vulnerable populations, mostly in low-income countries.

In the current context, ethical frameworks need to be updated to meet emerging needs and reduce the burden on patients and society through innovative solutions.

The IFAPP Ethics Working Group proactively engages with international stakeholders to advocate and ensure that ethical standards evolve in a way that remains relevant and effective [2]. For example, the IFAPP team has been actively involved in the recent tenth revision of the Declaration of Helsinki [8] and has contributed comments to the VoiREthics initiative for Phase I healthy volunteers [9], the WHO Guidance for Best Practices for Clinical Trials [10] and the forthcoming implementation of ICH E6 R3 [11] through public consultation submissions to ICH and the European Medicines Agency (EMA). As a result, the IFAPP Ethics Framework is in the process of being revised to incorporate all advances as soon as the latest ICH E6 R3 Annex 2 text is finalised, which is expected in July 2025 at the time of writing.

2.2 *IFAPP's Ethics and Innovation Strategic Plan*

The IFAPP's strategic plan for ethics and innovation comprises three pillars, which are described below [2]:

Education and training that cultivates professional identity by providing continuous professional development opportunities and interdisciplinary education for biomedical scientists and pharmaceutical physicians for their career planning in modern research and clinical development with new technologies and multiomics-based biomarkers in the emerging field of precision medicine and digital therapeutics, among others.

Advocacy with regulatory agencies (e.g. European Union—EU Register, European Medicines Agency—EMA) and other international stakeholders, such as the World Medical Association (WMA), Council of International Organisations of Medical Sciences (CIOMS), International Council of Harmonisation (ICH), World Health Organisation (WHO), academia and patient advocacy groups, to promote human rights, ethics, scientific integrity and the ethical use of new technologies—artificial intelligence (AI).

Collaborating and partnering with international organisations to pursue best practices in pharmaceutical medicine along the life cycle of medicines and medical devices that are future-proof for better patient outcomes.

2.3 IFAPP Collaboration with WMA and its Contributions to the Tenth Revision of the Declaration of Helsinki

IFAPP and WMA have a long-standing collaboration, based on a memorandum of understanding (MoU) since 2017 and joint efforts on projects of common interest. The WMA's revision of the Declaration of Helsinki (DoH) was a top-priority project for both organisations. IFAPP organised sessions on the DoH with WMA regarding needed changes during its International Conference of Pharmaceutical Medicine in Athens, Greece, in October 2022 and at its European Regional Meeting in Amsterdam in June 2023.

In parallel, IFAPP was invited to the WMA sessions for the revision of the DoH that started in December 2022 and ended in October 2024 with the adoption of the tenth revision of the Declaration.

Furthermore, IFAPP's Ethics Working Group members submitted several articles relevant to suggested changes in the IFAPP TODAY Journal, in peer-reviewed publications and textbooks.

The details of the relevant actions of IFAPP and WMA (meetings, sessions, articles, peer reviewed publications and book chapters) in relation to the DoH mentioned above are depicted in Table 14.1 and were published in Clin.Eval.52 (3) 2025 [12].

Table 14.1 IFAPP members' participation in WMA Regional Meetings and General Assembly and contributions in journals

Meetings		
Sessions with WMA at IFAPP meetings		
Year, Month, date Meeting venue	Title of the session	Speakers invited from the WMA
2022 Oct 21 ICPM Athens, Greece	IFAPP Workshop on the future revision of the Declaration of Helsinki: Dialogue with the WMA	Hybrid, Jack Resneck, Workgroup chair for DoH, Otmar Kloiber, Secretary General of the WMA, online invited
2023 Jun 29 IFAPP Regional Meeting Amsterdam, Netherland	IFAPP & WMA Workshop on the Revision of the Declaration of Helsinki focusing on Data-Driven Research	Hybrid, Daniel Fu-Chang Tsai, WG member for DoH, Jeppe Berggreen Høj, advisor of the WMA, online invited
Sessions with IFAPP at WMA Regional Meetings, Public Consultations, General Assembly		
Year, month, date Meeting venue	Topics	Participants from the IFAPP
2022 Dec 9-11 Tel-Aviv, Israel	General Data-driven research	Chieko Kurihara, EWG of the IFAPP, in-person participation
2023 Feb 24, 25 Sao Paulo, Brazil	Placebo-controlled trial	Chieko Kurihara, EWG of the IFAPP, in-person participation
2023 Sep 21, 22 Copenhagen, Denmark	Emerging trial design	Varvara Baroutsou, President of the IFAPP, invited as a speaker, in-person participation
2023 Nov 30, Dec 1 Tokyo, Japan	Disaster settings	Chieko Kurihara, EWG of the IFAPP, in-person participation
2024 Jan 13 – Feb 7	WMA Public Consultation (phase 1)	IFAPP Members submitted opinions
2024 Jan 18, 19 Vatican City State	Resource-poor settings Vulnerable population; post-trial access	Chieko Kurihara, EWG of the IFAPP, online observation
2024 Feb 18, 19 Johannesburg, South Africa	Vulnerability	Chieko Kurihara, EWG of the IFAPP, online observation
2024 May 14, 15 Munich, Germany	Research with vulnerable people	Chieko Kurihara, EWG of the IFAPP, online observation
2024 Jun 3 – 24	WMA Public Consultation (phase 2)	IFAPP Members submitted opinions
2024 Aug 15, 16 Washington DC, United States	General Maximizing impact of the DoH: communications, advocacy and implementation	Chieko Kurihara, EWG of the IFAPP, invited as a speaker, in-person participation
2024 Oct 16-19 General Assembly Helsinki, Finland	Adoption of the 2024 revision of the DoH	Varvara Baroutsou, Chieko Kurihara, and Kotone Matsuyama, IFAPP, in-person participation

(continued)

Table 14.1 (continued)

IFAPP TODAY articles	
Year, Month, No. pages	Title, - Topics
2021 Jul/Aug No. 16: 4-7	Webinar on COVID-19 and Bioethics: Pandemic and Research Ethics: Democracy, Placebo, and Post-Trial Access.
2021 Nov/Dec No. 19: 13-15	A proposal for the Revision of the Declaration of Helsinki to promote data-driven science and strengthening human subject protection.
2022 Jan No. 20: 3-7	IFAPP Recommendations for the Revision of the Declaration of Helsinki, Version 2013 - General, whole the construction of the DoH
2022 Nov/Dec No. 29: 5-7	Future Revision of the Declaration of Helsinki: Dialogue with the WMA in Athens. - General, whole the construction of the DoH
2023 May No. 34: 8-12	Next Revision of the Declaration of Helsinki: Meetings of the WMA and IFAPP. IFAPP TODAY. - General, whole the construction of the DoH (Tel Aviv) - Placebo-controlled trial (Sao Paulo)
2023 Sep No 37: 5-8	Discussion in Amsterdam on Data-driven Research and the WMA Declaration of Helsinki.
2023 Nov/Dec No 39, 15-17	Exploring New and Emerging Trial Designs Considering the Revision of the Declaration of Helsinki
Book, peer-reviewed papers	
Year, Month	Title, - topics
2020 Jul	Linking the Declarations of Helsinki and of Taipei: Critical Challenges of Future- Oriented Research Ethics. doi: 10.3389/fphar.2020.579714.
2023 Nov	Ethical innovation for global health: pandemic, democracy and ethics in research. Springer (Several number of chapters are contributed from the members of the IFAPP) https://link.springer.com/book/10.1007/978-981-99-6163-4
2024 Apr	Declaration of Helsinki: ethical norm in pursuit of common global goals - Data-driven research; Broad consent, Dynamic consent; Participants' rights, Individual patient data sharing; Open science; Social value; Risk minimization in placebo-controlled trials; Post-trial access with best proven intervention; Co-creation with stakeholders in future-oriented research frameworks. doi: 10.3389/fmed.2024.1360653.
2024 Sep	Vulnerability, social value and the equitable sharing of benefits from research: beyond the placebo and access debates. doi: 10.3389/fmed.2024.1432267.

2.4 *Impact of the Tenth Revision of the Declaration of Helsinki*

As the result of the above-mentioned collaboration, the impact of the tenth revision of the Declaration of Helsinki could be summarised as follows, introducing: (1) the "Helsinki Statement," of which the author is one of the contributors, (2) evaluations from other experts, and (3) the author's individual perspective at this moment:

1. Helsinki Statement

   As a contributor to the Helsinki Statement on the tenth revision of the Declaration of Helsinki, I join the signatories in congratulating the World Medical Association (WMA) on the 60th anniversary of the Declaration of Helsinki (DoH), since its first adoption in 1964.

   The significant revisions to the 2024 DoH are not fully described here, but can be found on the WMA website at:
   https://www.wma.net/policiespost/wma-declaration-of-helsinki/

It is of the utmost importance that all those involved in research involving human participants should carefully consider the newly agreed ethical principles, particularly those set out below, in order to implement and adhere to them in all research and research review settings:

(Reprint from the Helsinki Statement [13], including footnotes)

1. Throughout the Declaration, the words "research subjects" are replaced with "research participants".
2. Although the recommendations are directed to physicians, their scope should be also applied to non-physicians, research teams and organizations. (§2*[1])
3. In recognition of "various structural inequalities" in research, considerations on benefits, risks and burdens must be reinforced. (§6)
4. Meaningful engagement with participants and their communities is fundamental and should be required at all stages of the research. (§6)
5. The principles of the Declaration must also be upheld during public health crises. (§8)
6. The contextual and dynamic nature of vulnerability must be recognized and it is important to emphasise that the exclusion of vulnerable can exacerbate their disparities. Therefore, their inclusion in research aiming at their benefit must be promoted, with adequate protection. (§19, 20)
7. The functions of research ethics committees have been strengthened. And the committee must be familiar with the local context and involve at least one member of the general public. (§23)
8. The preferences and values of incapable research participants shall be considered during the process of deciding to participate in research. (§28, 29)
9. Data or specimens from research shall be handled in accordance with the "Declaration of Taipei" on Health Databases and Biobanks. (§32)
10. The clinical use of unproven interventions must never be used to circumvent the protections set out by the (DoH), and must be the subject of research to evaluate their safety and efficacy. (§37)

However, we express our concern that the items below were rejected following discussions during the revision process. We very much hope that these ethical principles will be agreed upon by relevant stakeholders, their institutions or organizations and jurisdictions, implemented in research and research review settings and consequently reflected in future revisions of the DoH:

1. All texts must be written in plain language that is understandable to research participants.
2. The "social value" of research has already been incorporated into policy in several jurisdictions and guidelines. Although it was included in the first public consultation version, it was deleted in the approved 2024 DoH. It goes together with the notion that the purpose of research "*never takes precedence over the*

[1] The numbers with §mean the numbers of paragraphs in the 2024 DoH: https://www.wma.net/policies-post/wma-declaration-of-helsinki/

rights and interests of individual participants". Thus, social value must be reinforced in order to implement relevant research and also to ensure that benefits are shared through co-creation with community.

3. The use of placebo in clinical trials can only be acceptable when there is no proven effective and safe comparator. This was stated in the 2000 DoH,[2] in accordance with DoH's fundamental principle that the purpose of research "*never takes precedence over the rights and interests of individual participants*". Some of the signers are in favour of the CIOMS 2016 language: that is, placebo "*in exceptional circumstances could be used even when there is an active comparator if delaying or withholding the established effective intervention will result in no more than a minor increase above minimal risk to the participant and risks are minimized, including through the use of effective mitigation procedures*". Nevertheless, since the 2008 Declaration of Helsinki (DoH), the accepted level of risk to participants has been much higher than that set out by the Council for International Organizations of Medical Sciences (CIOMS), and this remains the case in the 2024 DoH.[3]
4. At the end of the trial, research participants who still need interventions that have been shown to be safe and effective must have access to them ensured by the sponsor(s). The 2024 DoH is a lot weaker than the 2000 DoH in relation to participant right. This is because instead of ensuring access, it just says "it must be arranged" (§ 34).[4] In addition, to tackle health disparities, scientifically sound and ethical strategies are needed to ensure access in public health system in the host community, and most importantly, for all in need globally.
5. The timely dissemination of research results and technology transfer of the products of research involving humans must be recognised as public goods, so that transparency is ensured and everyone can have fair and equitable access to the research results and benefits.[5]

In closing:

> The 2000 DoH presented the highest level of protection to research participants in terms of placebo use and post-trial access. As the DoH has long been a driving force and respected

---

[2] 2000 DoH (Rescind, archived version): https://www.wma.net/wp-content/uploads/2018/07/DoH-Oct2000.pdf

[3] 2024 DoH in §33, placebo can be used, even if there is proven intervention, when the participant "*will not be subject to additional risks of serious or irreversible harm as a result of not receiving the best proven intervention*".

[4] Compared to 2013 version which stated that "…***should*** *make provisions for post-trial access…*", the text in 2024 was strengthened to "*…post-trial provisions* ***must*** *be arranged…*" but remains weaker compared to 2000 version which stated that "*…should be* ***assured of access***". In addition, in the 2024 DoH, any exception must be explicitly approved by a research ethics committee. Such excuse is not relevant to a high-level ethical principle; however, in practice, the strengthened responsibility of REC must be recognised.

[5] Regarding the research results, new texts for avoiding research misconduct (§21) and research waste (§12) suggest improvement, however, considering that "social value" was once proposed and then deleted, and insufficient transparency and benefit sharing measures in result disclosure policy, this topic was not included in ten celebrated items.

document for ethics in human research, WMA should aim for the ideal of the highest ethical standards being applied.

We will collaborate to strengthen research ethics principles with the equal participation of all relevant stakeholders, including civil society, and in relation to the 2024 DoH, we present the current proposal aimed at maximizing the ethical impact of the DoH on research practice, research review process and the protection of participants globally.

(End of Reprint)

The above is a reproduction from the original publication accessed below, together with Arabic and Japanese translation. Greek translation is under preparation.

http://cont.o.oo7.jp/52pop/52pop_contents_e.html

The background to the development of the Statement and the Copyright arrangement is described in the Statement as below:

(Reprint from the note in Helsinki Statement [13].)

This statement was discussed (by e-mails and two times web meetings on 15 and 19–20 October, connecting from Helsinki with global participants) and agreed upon by the authors/signatories, and published online in 18 October; and was updated on 21 October with clarifications reflecting the 2024 Declaration of Helsinki adopted by the World Medical Association on 19 October 2024. Additional signatories were accepted until November 30; authors updated it on 16 December 2024.

- Background information: Kurihara C, Matsuyama K, Baroutsou V. World Medical Association's Declaration of Helsinki, 2024 revision: Celebrating the 60th anniversary, at Helsinki. Clin Eval. 52(3); Other related articles:
- http://cont.o.oo7.jp/52pop/52pop_contents_e.html
- Contact: chieko.kurihara@nifty.ne.jp
- Copyright: This statement may be freely used, disseminated and reproduced including publication in the original or translated language (for which authors cannot take responsibility of accuracy).

(End of Reprint)

2. Evaluations from other experts

As noted above, the Tenth revision of the Declaration of Helsinki has received both praise and criticism from the bioethics community's published views. There are also substantial number of evaluations from other experts have been published, including the following two analyses. Well-recognised common perspectives can be found among these evaluations.

The substantial changes in the 2024 DoH are classified into two areas [14]:

1. *Participant-centred inclusion and co-creation, respect and protection, including recognition of participant vulnerability, calls for community engagement, pursuit of global justice, obtaining free informed consent, with use of participant-centred language.*
2. *Research beneficence and value, including the pursuit of "individual and public health", upholding scientific rigor and integrity, and considered distribution of benefits, risks and burdens.*

Another evaluation provided in editorial in Medical Science Monitor can be summarised as follows [15]:

1. Terminology been changed from "subject" to "participant," and gender-neutral language has been adopted.
2. Research inclusivity of interdisciplinary teams and organisations has been addressed.

3. Meaningful engagement with research participants at all stages of research has been emphasised.
4. Fair and responsible inclusion of vulnerable populations has been advised, taking into account long-term and global benefits, risks and safety.
5. Informed consent for the collection, storage and secondary use of data and biological material for secondary use has been written, with cross-references to the WMA Declaration of Taipei.
6. It has been clarified that the ethical principles of the DoH must be upheld even during public health emergencies.
7. It has been clarified that unproven interventions should not bypass ethical safeguards or evade evaluation through controlled clinical trials.
8. The commitment to environmental sustainability and rigorous trial design to prevent wasteful research has been strengthened.
9. Adequate resources for research ethics committees have been recommended.
10. Participants' values and preferences should be considered, including when a proxy gives consent.

3. Author's perspective

   Considering the above evaluation, the author of this chapter regards the main concerns as referring to eight areas [13, 15, 16]:

   (a) Placebo use: The ongoing debate about the ethical use of placebos in clinical trials for conditions with proven treatments remains unresolved. Critics argue that using placebos in such cases can be unethical, as it may deny participants effective treatment.
   (b) Post-trial access: Ensuring post-trial access to interventions for participants who need them continues to be a contentious issue. Commentators highlight that the Declaration is not clear on how to guarantee this access.
   (c) Global justice: While the revision emphasises global justice, experts argue that it lacks specific mechanisms to address disparities in research benefits between high-income and low-income countries.
   (d) Ethical diversity: The revision attempts to address ethical diversity, but the declaration does not sufficiently tackle these complex issues.
   (e) In practical terms, the principles of the DoH should be further supported by guidelines possibly drafted by CIOMS and/or ICH E6 R(3) in order to be applied in areas such as:
   (f) AI and Digital Technologies in medical research: Ethical concerns and guidance on Data Privacy and Security.
   (g) Environmental impact: Although the revision urges researchers to consider environmental sustainability, it does not provide standards for minimising environmental harm.
   (h) Vulnerable populations: The Declaration calls for the inclusion of vulnerable populations but does not offer specific strategies to ensure their protection and fair treatment in research.

(i) The overall implementation of the updated version of DoH requires education, practical guidance and research institutional governance on how to apply ethical principles effectively in diverse research settings.

3 Global Research Unmet Priorities for Global Health

In the year 2025, we need to focus global research efforts on the remaining unmet medical priorities, since we lack medicines in numerous and complex therapeutic areas that often vary by region and population. Despite progress, current evidence indicates that diseases like malaria, tuberculosis and HIV/AIDS continue to pose challenges in low-income countries plus the residual risk of harm from adverse drug reactions, emerging infectious diseases, prevailing antimicrobial resistance and lack of new antibiotics are critical public health underserved needs [17]. Furthermore, the 2023 World Health Assembly resolution provides an opportunity for WHO to review the availability of data on unmet needs globally and to provide recommendations for data collection and analysis that can feed into monitoring of universal health coverage gaps and access barriers. Systematic measurement of unmet healthcare needs enhances global health monitoring and comparability, informing strategies to address health system gaps, mitigate disease burden, and improve population health outcomes [18].

If the COVID-19 crisis was not a catalyst for establishing global pathogen surveillance and preparedness to save lives, we should now address the increasingly challenging and accumulating threats posed by (a) climate change and natural disasters and (b) ongoing wars in Ukraine and the Middle East as well as other conflicts.

From a global perspective, the imbalance between research needs and research efforts lies with sponsors in high-income countries, while researchers in middle-income countries prefer researching diseases in high-income countries to receive more citations instead of researching prevalent diseases in their own countries. In contrast, low-income countries have not adequate infrastructure, resources, governance and funding to research and manage public health issues, endemic diseases and emergencies. However, evidence synthesis is needed to improve health systems performance and health outcomes.

Global research under the umbrella of WHO can improve in several ways to address these unmet needs [19–21]:

- Equitable research funding for prevalent diseases in low- and middle-income countries
- Collaborative international efforts by pooling resources and expertise to tackle global health issues more effectively
- Focus on implementation of scientific protocols to effectively implement therapeutic schemes in the diverse settings

- Addressing the social determinants of health with research that considers social, economic, environmental factors, behavioural and genetic aspects that impact health

3.1 Potential Scientific and Technological Research Opportunities in the Horizon for Global Health

Promising science and technologies for global health and research are on the horizon with the potential to improve global public health, if the International Health Organisation WHO and or other institutions have the funding needed.

The growing field of digital health, including telemedicine and the responsible use of AI in healthcare, has great potential to support underserved patient communities in middle- and low-income countries by supplementing their healthcare services.

Genomics and Precision Medicine can be realised to predict and treat on time chronic and serious diseases, reduce disease burden, improve patients' quality of life, extend lifespan and reduce health expenditure.

Infectious diseases and antimicrobial resistance represent a huge global health issue that could be addressed through coordinated efforts by the scientific community under the umbrella of collaborative efforts with International Organisations like WHO and other Institutions.

Climate change and its impact on health presents a serious global threat to citizens and public health that require global political consensus and international scientific commitment.

Health equity and social determinants of health are critical for individual and public heath that require appropriate health literacy, public policies and governmental actions with concerted universal commitment.

The complete withdrawal of the United States from the WHO in January 2025 and the abrupt cuts in funding for the US National Institutes of Health (NIH) are not grounds for optimism about medical research worldwide and global health in particular. Large-scale research initiatives funded by the NIH and the Wellcome Trust annually have led to major advances in vaccines for infectious diseases, neurological disorders, and cancer. However, recent policy changes by the NIH are ending foreign subawards, jeopardizing thousands of international research collaborations in these fields [22].

Reversing this trajectory will require rethinking investment in R&D, reliable and strong partnerships with academia, governments, non-governmental organisations (NGOs), the health sector ecosystem and global health diplomacy.

Cutting spending on biomedical research leads to reduced research and development of innovative therapies, which will be detrimental to societal progress and improvements in public health [23].

3.2 Current Challenges for Global Health

Global health challenges, including neglected diseases, frequent epidemics, pandemics, and climate-related health crises require international cooperation. The withdrawal of the United States from the WHO following an executive order by the US President on 20 January 2025 has paused funding and scientific support by recalling US personnel working with the WHO in any capacity [24].

Historically, the United States influenced global health through funding, governance and international collaborations.

Global health and research will be globally impacted by the new administration in the White House in the United States, at a time when infectious disease outbreaks are increasing in frequency and severity. The WHO has declared seven global infectious disease emergencies since the 2009 H1N1 pandemic averaging one every 2 years [25]. Emerging global infectious diseases during the period 2002–2022 can be summarised as follows:

- Severe acute respiratory syndrome (SARS) (2002–2003): Emerged in China and spread globally, causing significant morbidity and mortality.
- H1N1 Influenza (2009): Pandemic originated in Mexico and spread worldwide, leading to widespread illness and deaths.
- Middle East respiratory syndrome (MERS) (2012): First identified in Saudi Arabia and caused severe respiratory illness.
- Ebola (2014–2016): Outbreak in West Africa resulted in thousands of deaths and highlighted the need for improved global health responses.
- Zika virus (2015–2016): Spread rapidly in the Americas, causing birth defects and neurological complications.
- COVID-19 (2019-present): Caused by SARS-CoV-2 and has had a profound impact on global health, economies and daily life.
- Monkeypox (2022): Outbreaks were reported in multiple countries, raising concerns about its spread and impact.

The WHO's comprehensive approach to managing health crises works through preparedness, prevention, detection and execution of emergency preparedness, emergency response plans, risk assessment, contingency plans, emergency risk communication and collaboration with stakeholders.

What the withdrawal of United States means for the WHO isn't entirely clear at the moment. The WHO is a United Nations agency whose mandate includes strengthening global infectious disease surveillance and preparing for and coordinating global responses to dangerous epidemics and pandemics, with mission to promote and protect global health.

The United States accounted for more than $1 billion of the WHO's $6.7 billion budget in 2022–2023. If those funds aren't available through other revenue sources, the impact of America's withdrawal will be substantial, especially if the Centers for Disease Control and Prevention (CDC) remains forbidden from working with the WHO in any capacity. The CDC-WHO partnership has been a positive one for improving health and reducing disease risk globally; its dissolution would be a huge setback [26].

In theory, major funders such as the Bill and Melinda Gates Foundation, Germany and the UK could step in to fill the WHO's budget gaps [27].

The WHO and global health groups should work together to address the urgent gaps in the fight against diseases like malaria, HIV, sexual health, polio eradication and health emergencies by sharing countries' resources and emergency financing to prevent shortages of tests, treatments and trained healthcare professionals according to Reuters on 11 March 2025 [28].

Following the United States' withdrawal from global health initiatives, China now has the opportunity to further expand its own global health agenda as the largest single contributor. Brazil will hold the next BRICS presidency, representing Brazil, Russia, India, China and South Africa, as well as Iran, Egypt, Ethiopia and the United Arab Emirates. South Africa and Canada are set to hold the next G20 and G7 presidencies, respectively [27, 29].

The broader humanitarian agenda led by UN agencies, including UNHCR, UNRWA, and UNFPA as well as organizations like Médecins Sans Frontières (MSF), continue to face mounting challenges amid record levels of displacement and conflict. With rising global needs and potential funding shortfalls, these organizations are increasingly looking to Europe and other donor countries for sustained support [27].

The New US White House Administration may potentially challenge FDA to lower scrutiny and accelerate new drug authorisation timelines. These potential decisions by the FDA will have an impact on Project Orbis, a US-led global collaboration to promote faster authorisation and access to cancer drugs [30]. Regulatory agencies participating in Project Orbis will have to decide whether it is in their interest to continue to align with the United States through Project Orbis.

The WHO European Region comprises 53 countries in Europe and Central Asia, where preliminary data suggest that there have been 38 measles deaths in 2024. However, according to WHO, more than half of the reported measles cases required hospitalisation. Globally, 359,521 cases of measles were reported in 2024, making the European total just over one-third (35.4%) of those worldwide and Europe's highest in 25 years. The WHO blamed falling vaccination rates and said that children younger than 5 years accounted for more than 40% of reported cases [31].

At the upcoming World Health Assembly in May 2025, the WHO decision-making body must reach a pandemic agreement, re-establish a fundamental commitment to WHO and set a roadmap for the essential reforms to secure its future [32].

At this time, a parallel effort is needed to rebuild trust in medical science and research which was eroded during the COVID-19 crisis. The application of new strategies for improved science communication, research methods and data transparency that support openness for better public understanding, engage communities to share their concerns and perspectives, and promote participatory research would lead to improved fairness, equity, health and research literacy based on evidence information.

4 Conclusion

It is the purpose of the IFAPP to continue to strive for the highest respect for human rights and to make research more ethical, safe, equitable and efficient for the benefit of healthy volunteers, patients and vulnerable participants, respecting them all as co-creators. Researchers, experts, institutions, sponsors, academics, ethics committees, and regulatory authorities should support the implementation of the tenth revision of the Declaration of Helsinki by undertaking relevant actions and trainings to ensure research integrity for global health.

New technologies and AI in research and development would be required to streamlining the development of next generation medicines and vaccines to the extent needed for better therapeutics by safeguarding ethics and the well-being of research participants according to the recent version of the DoH of October 2024. The COVID-19 pandemic has shown that public funding and government stewardship can reduce costs and risks, make affordability and transparency more feasible, as well as share knowledge that is more grounded. A combination of regulations, incentives and governmental policies can support private and public interest in converging for the common good.

Collaboration among society, industry, government, academia and funding agencies is essential to establishing ethical standards in global biomedical research and supporting initiatives for low- and middle-income countries. This collective effort aims to ensure equitable access to healthcare worldwide, in line with the Universal Declaration of Human Rights and the Tenth World Health Assembly. With the world more fragmented than ever, and the United States' withdrawal from the WHO, the impact on global health may be critical.

There is a pressing need to foster global public health goodwill through multilateral engagement in diplomacy. This approach should tackle significant global health issues such as pandemic preparedness and response, non-communicable diseases, sustainable development and climate change.

In this new world health order, the IFAPP is committed to continue prioritising ethics in pharmaceutical medicine and pursuing continuous improvement of its ethical framework and lifelong education of its members. It is important to emphasise the contemporary risks associated with new technologies and unprecedented innovation in R&D and public health, through evidence generation, communication and collaboration with international scientific, academic and patient communities. These initiatives highlight IFAPP's commitment to advancing pharmaceutical medicine through education, ethical standards and international collaboration networks for global health.

Acknowledgement The author appreciates IFAPP for granting her expenses for her personal attendance as a speaker at the WMA Regional Meeting in Copenhagen in September, 2023, and at the WMA General Assembly (GA) in Helsinki in October, 2024, and a free registration waiver by the WMA for her attendance at the GA in Helsinki in 2024.

Conflict of Interest The author has no competing interests to declare that are relevant to the content of this chapter.

References

1. The Faculty of Pharmaceutical Medicine of the Royal Colleges of Physicians of the United Kingdom. https://www.fpm.org.uk/ Accessed 7 April 2025.
2. The International Federation of Pharmaceutical Physicians and Pharmaceutical Medicine (IFAPP). https://ifapp.org/ Accessed 7 April 2025.
3. International Federation of Associations of Pharmaceutical Physicians & Pharmaceutical Medicine. IFAPP Ethics Framework for Pharmaceutical Physicians and Medicines Development Scientists 2018. https://ifapp.org/wp-content/uploads/2021/07/2nd_ed_2018_Ethics_Framework.pdf Accessed 7 April 2025.
4. Kerpel-Fronius S, Becker AL, Members of the IFAPP Ethics Working Group. The value and importance of a professional ethical code for medicines development: IFAPP international ethics framework. Pharmaceut Med. 2025; https://doi.org/10.1007/s40290-024-00547-6. Epub ahead of print
5. The Declaration of Taipei: Research on Health Databases, Big Data and Biobanks. Adopted by the 53rd WMA General Assembly, Washington, DC, USA, October 2002, and revised by the 67th WMA General Assembly, Taipei, Taiwan, Oct 2016. https://www.wma.net/what-we-do/medical-ethics/declaration-of-taipei/. Accessed 7 April 2025.
6. Innovative Health Initiative. How to fight AMR: public-private partnerships lead the way. 26 Sept 2024. https://www.ihi.europa.eu/news-events/newsroom/how-fight-amr-public-private-partnerships-lead-way. Accessed 7 April 2025.
7. Fernow J, Olliver M, Couet W, Lagrange S, Lamers MH, Olesen OF, Orrling K, Pieren M, Sloan DJ, Vaquero JJ, Miles TJ, Karlén A. The AMR Accelerator: from individual organizations to efficient antibiotic development partnerships. Nat Rev Drug Discov. 2025;24(1):1–2. https://doi.org/10.1038/d41573-024-00138-9.
8. World Medical Association. WMA Declaration of Helsinki – ethical principles for medical research involving human participants. First adopted in 1964, last amended in 2024. https://www.wma.net/policies-post/wma-declaration-of-helsinki/
9. VolREthics, Inserm. A global ethics charter for the protection of healthy volunteers in clinical trials. June 2024. https://www.inserm.fr/wp-content/uploads/Inserm_VolREthics_EthicsCharterFinalVersion_July24.pdf
10. World Health Organization. Guidance for best practices for clinical trials. 25 Sept 2024. https://www.who.int/publications/i/item/9789240097711
11. International Council for Harmonisation of Technical Requirements for Pharmaceuticals for Human Use. ICH harmonised guideline: guideline for good clinical practice E6(R3) Final version Adopted on 06 January 2025. https://database.ich.org/sites/default/files/ICH_E6%28R3%29_Step4_FinalGuideline_2025_0106.pdf
12. Baroutsou V. Collaboration with the WMA and the IFAPP's perspective. Clin Eval. 2025;52(3):420–7. http://cont.o.oo7.jp/52pop/52pop_contents_e.html
13. Helsinki Statement from an Independent Stakeholders' Group to Expand the Impact of the 2024 Revision of the WMA Declaration of Helsinki. Clin Eval. 2025; 52(3). http://cont.o.oo7.jp/52pop/HelsinkiStatement_18Oct24_final.pdf
14. Bioethics Observatory Institute of life sciences of the Catholic University of Valencia. New version of the Declaration of Helsinki, the ethical framework for medical research involving human participants. 30 Oct 2024. https://bioethicsobservatory.org/2024/10/new-version-of-the-declaration-of-helsinki-the-ethical-framework-for-medical-research-involving-human-participants/47209/. Accessed 7 April 2025.

15. Parums DV. Editorial: The 2024 Revision of the Declaration of Helsinki and its continued role as a code of ethics to guide medical research. Med Sci Monit. 2024;(30):e947428. https://doi.org/10.12659/MSM.947428.
16. The Council of International Organizations of Medical Sciences. International guidelines for good governance practice for research institutions. Geneva. 2023. https://cioms.ch/working_groups/principles-of-goodgovernance-for-research-institutions/
17. Swaminathan S, Pécoul B, Abdullah H, Christou C, Gray G, IJsselmuiden C, Kieny MP, Mazzucato M, von Messling V, Ogutu B, Reeder J, Røttingen JA, Swarup R, Tanner M, Trindade Lima N, Childs M, Harris A, Torreele E, Moon S. Reboot biomedical R&D in the global public interest. Nature. 2022;602(7896):207–10. https://doi.org/10.1038/d41586-022-00324-y.
18. Rosenberg M, Kowal P, Rahman MM, Okamoto S, Barber SL, Tangcharoensathien V. Better data on unmet healthcare need can strengthen global monitoring of universal health coverage. BMJ. 2023;382:e075476. https://doi.org/10.1136/bmj-2023-075476.
19. Yegros-Yegros A, van de Klippe W, Abad-Garcia MF, Rafols I. Exploring why global health needs are unmet by research efforts: the potential influences of geography, industry and publication incentives. Health Res Policy Syst. 2020;18(1):47. https://doi.org/10.1186/s12961-020-00560-6.
20. World Health Organization. Understanding global health through data collection. https://www.who.int/activities/understanding-global-health-through-data-collection. Accessed 7 April 2025.
21. Our World in Data. Our World in Data is a project of Global Change Data Lab, a nonprofit based in the UK (Reg. Charity No. 11864330) https://ourworldindata.org/
22. Gaind N. How the NIH dominates the world's health research – in charts. Nature. 2025;639(8055):554–5. https://doi.org/10.1038/d41586-025-00754-4.
23. Galea S, Bibbins-Domingo K. The value of academic health research. JAMA Health Forum. 2025;6(2):e250509. https://doi.org/10.1001/jamahealthforum.2025.0509.
24. Yamey G, Titanji BK. Withdrawal of the United States from the WHO – How President Trump is weakening public health. N Engl J Med. 2025;392(15):1457–60. https://doi.org/10.1056/NEJMp2501790. Epub 2025 Mar 5
25. bluedot. https://bluedot.global/. Accessed 7 April 2025.
26. World Health Organization. Executive summary: progress on the triple billion targets and the sustainable development goals. https://www.who.int/about/accountability/results/who-results-report-2022-2023/executive-summary. Accessed 7 April 2025.
27. Kings College London. Sullivan R. COMMENT: what a Trump presidency means for global health. 12 Nov 2024.https://www.kcl.ac.uk/news/comment-what-trump-presidency-means-global-health. Accessed 7 April 2025.
28. Reuters. Countries, Global health groups band together as US Aid Gaps threaten lives. 11 Mar 2025. https://www.usnews.com/news/world/articles/2025-03-11/countries-global-health-groups-band-together-as-us-aid-gaps-threaten-lives. Accessed 7 April 2025.
29. Kickbusch I; founder. Trump, the rise of the Global South, and a new world order for health. BMJ. 2024;387:q2708. https://doi.org/10.1136/bmj.q2708.
30. U.S. Food and Drug Administration. Project Orbis https://www.fda.gov/about-fda/oncology-center-excellence/project-orbis. Accessed 7 April 2025.
31. Medscape. Meredith S. WHO 'Wake-up call': Measles in Europe highest in 25 years. 14 Mar, 2025. https://www.medscape.com/viewarticle/who-wake-call-measles-europe-highest-25-years-2025a100067m?form=fpf. Accessed 7 April 2025.
32. Focus 2030. Global health timeline 2025. https://focus2030.org/Global-health-timeline-2025. Accessed 7 April 2025.

Chapter 15
Ethical Principles for the Sustainable Development of Drugs for Neglected Tropical Diseases in Neglected Populations: Facilitating Meaningful Participation of Vulnerable People and Communities

Chieko Kurihara, Takeo Saio, Dirceu Greco, and Ames Dhai

Abstract The changes in the 2024 Declaration of Helsinki (DoH) to promote the engagement of vulnerable people and communities will certainly contribute to innovative research and development (R&D) of interventions for neglected tropical diseases (NTDs).

Firstly, this presupposes the establishment of robust community participation in all stages of the research to overcome structural inequalities in healthcare around the world. Secondly, counteracting previous cautionary measures that systematically excluded vulnerable populations from research will avoid perpetuating the vicious cycle of health disparities. Thirdly, with these changes in the 2024 DoH, research with vulnerable participants could be facilitated for their benefit. Above all, to protect against exploitation, post-trial benefit-sharing must be guaranteed.

This chapter reviews the improvement of the renewed DoH from the viewpoint of technological innovation from R&D for NTDs in neglected populations (twice neglected) to achieve equitable and sustainable goals of global health.

Keywords Neglected tropical disease (NTD) · Vulnerability · Health inequity · Community engagement · Research and development (R&D) · Sustainable Development Goals (SDGs)

C. Kurihara (✉)
Kanagawa Dental University, Yokosuka, Kanagawa, Japan
e-mail: chieko.kurihara@nifty.ne.jp

T. Saio
Department of Internal Medicine and Psychiatry, Fuji Toranomon Orthopedic Hospital, Shizuoka, Japan

D. Greco
School of Medicine, Federal University of Minas Gerais, Belo Horizonte, Brazil

A. Dhai
School of Clinical Medicine, University of the Witwatersrand, Johannesburg, Gauteng, South Africa

C. Kurihara et al. (eds.), *The 2024 Declaration of Helsinki*,
https://doi.org/10.1007/978-981-96-9294-1_15

1 Introduction: Unequivocal Changes in the Renewed DoH

The 2024 revision of the Declaration of Helsinki (DoH) by the World Medical Association (WMA) [1] achieved several improvements, especially in three paragraphs (Paragraphs 6, 19, and 20), promising to facilitate research and development (R&D) for neglected tropical diseases (NTDs). NTDs are infectious diseases with the outcome of serious, irreparable damage and suffering to affected people and are endemic mainly in low- and middle-income countries (LMIC), especially among large numbers of the most vulnerable populations. Thus, historically these diseases have not been sufficiently studied for the development of preventive and therapeutic interventions due to the lack of potential commercial benefit for pharmaceutical companies. The World Health Organization (WHO) has been working towards 20 diseases falling under their definition of NTDs, setting a number of milestones and targets, ultimately aiming to eliminate them [2, 3].

The 2024 revision of the DoH added new principles to support the R&D for NTDs. Firstly, Paragraph 6 added new text highlighting "various structural inequities" recommending meaningful community participation in all stages of research as described below:

> Paragraph 6 of the DoH [1] (bold, underlined by authors): "*.....Since medical research takes place in the context of* ***<u>various structural inequities</u>****, researchers should carefully consider how the benefits, risks, and burdens are distributed.*
>
> ***<u>Meaningful engagement with potential and enrolled participants and their communities should occur before, during, and following medical research</u>****. Researchers should enable potential and enrolled participants and their communities to share their priorities and values; to participate in research* ***<u>design, implementation, and other relevant activities; and to engage in understanding and disseminating results</u>***".

Secondly, Paragraph 19 underscores the nature of the vulnerability of study participants not only as fixed categories (e.g. children, incapable adults, women, marginalized or economically disadvantaged populations), but also as being in "contextual and dynamic" situations. In this manner, the new DoH points out that the exclusion of these people from research may "*perpetuate or exacerbate their disparities*" and that their inclusion in research is justified only when the research is responsive to their health needs and priorities. Thirdly, and most importantly, the studied community must stand to benefit from the results, as stated in Paragraph 20:

> Paragraph 19 of the DoH [1]: "*Some individuals, groups, and communities are in a situation of more vulnerability as research participants due to factors that may be fixed or contextual and dynamic, and thus are at greater risk of being wronged or incurring harm.* ***<u>When such individuals, groups, and communities have distinctive health needs, their exclusion from medical research can potentially perpetuate or exacerbate their disparities.</u>*** *Therefore, the harms of exclusion must be considered and weighed against the harms of inclusion. In order to be fairly and responsibly included in research, they should receive* ***<u>specifically considered support and protections</u>***".
>
> Paragraph 20 of the DoH [1]: "*Medical research with individuals, groups, or communities in situations of particular vulnerability is* ***<u>only justified if it is responsive to their health needs and priorities</u>*** *and the individual, group, or community* ***<u>stands to benefit from the resulting knowledge, practices, or interventions</u>****. Researchers should only include those in situations of particular vulnerability when the research cannot be carried out in a less*

vulnerable group or community, or when excluding them would perpetuate or exacerbate their disparities".

2 Alternative R&D Strategy for Vulnerability, Community Engagement, and Benefit Sharing

These pivotal changes might build global awareness to "neglected" tropical diseases (NTDs) prevalent in "neglected" populations. As the WHO defined, NTDs are diseases of neglected populations which is often to say "twice neglected" [4]. We also point out that this is "thrice neglected" because neglected animal diseases (NADs) should be thoroughly researched to conquer or eradicate NTDs from "one health" approach. This means an integrated interdisciplinary collaborative strategy recognizing links between the health of people, animals and ecosystems [5].

To protect vulnerable people involved as study participants, meaningful community participation strategies are needed to ensure both post-trial access and benefit sharing for study host communities [6], while avoiding harmful impact of the study outcome, e.g. unintentionally leading to discrimination and stigmatization from research findings. For this purpose, community participation from the early stages of R&D is crucially important.

Furthermore, the developmental strategy must include all logistics of product licensing management, manufacturing capacity development, procurement and delivery strategies, as well as health technology assessment for effective inclusion in the publicly funded healthcare system. The experiences gained from product development partnership (PDP) have assisted in coming up with an alternative, sustainable scheme of knowledge management and capacity development [7, 8].

At the critical stage of the research ethics review, a new additional text in the DoH requires that the research ethics committee "*must have sufficient familiarity with local circumstances and context*" (Paragraph 23).

With these constitutive changes in mind, the focus of research involving humans must be shifted from for-profit blockbusters to interventions for unequivocal health problems affecting billions of vulnerable people around the world.

3 What Are the Neglected Tropical Diseases (NTDs)?

The World Health Organization (WHO) defines the NTDs [9] as a "*diverse group of conditions caused by a variety of pathogens (including viruses, bacteria, parasites, fungi, and toxins) and associated with devastating health, social, and economic consequences*". NTDs include Buruli ulcer; Chagas disease; dengue and chikungunya; dracunculiasis; echinococcosis; foodborne trematodiases; human African trypanosomiasis; leishmaniasis; leprosy; lymphatic filariasis; mycetoma, chromoblastomycosis and other deep mycoses; noma; onchocerciasis; rabies; scabies and other ectoparasitoses; schistosomiasis; soil-transmitted helminthiases; snakebite envenoming; taeniasis/cysticercosis; trachoma; and yaws.

The WHO also states that [10] "*NTDs are diseases of neglected populations that perpetuate a cycle of poor educational outcomes and limited professional opportunities; in addition, they are associated with stigma and social exclusion*".

According to WHO estimates, the number of people affected by NTDs is more than 1 billion; and including this, 1.5 billion require preventive or curative interventions. Overcoming NTDs is one of the crucial factors for achieving the United Nations' Sustainable Development Goals (SDGs) [11].

Over the past two decades or more, public and private investments in R&D for preventive or curative interventions for NTDs have gradually and then significantly increased. The WHO's first roadmap, drawn up in 2012 setting the goal of eliminating 10 NTDs by 2020, was endorsed by the "London Declaration" [12] issued in the same year as a joint statement by 13 of the world's leading pharmaceutical companies, DNDi, Bill & Melinda Gates Foundation, USAID and the World Bank, adopted by the governments of United Kingdom and United States, as well as ministries of health in NTD-endemic countries, to collaborate on the provision of free medicines, and R&Ds [9, 10]. This situation has significantly triggered further R&D on NTDs.

4 Progress and Road Map Towards Elimination of NTDs

As mentioned above, according to the WHO, the number of people who need preventive or curative interventions is still extremely high (1.5 billion, as of 2023), but this represents a relative drop by 31% [13] compared to the 2.19 billion estimated in 2010. As of February 2025, at least one NTD had already been eliminated in 55 countries [14]. In addition, the WHO has acknowledged several countries for eliminating 2, 3, or 4 conditions [13]. The reported number of cases of Human African Trypanosomiasis decreased by 98% from 1999 to 2023 [14]. In total, the WHO had completed 75 acknowledgement processes [13].

The WHO has also provided a "roadmap" to prevent, control, eliminate, and eradicate 20 NTDs, establishing targets and milestones from 2021 to 2030 [2, 3]. The four levels of disease-specific classified targets are assigned for 20 diseases. A specifically described indicator is defined for each disease. The following is a summary extracting some examples of diseases with targeted changes from 2020 to 2030 as outcome measures (the order here is from the ultimate goal to surrogate outcome, following the table in the roadmap).

1. "Eradiation": For Dracunculiasis and Yaws, both with indicators of "free of transmission": the former from 98% in 2020 to 100% in 2030, whereas the latter from 1% in 2020 to 100% in 2030.
2. "Elimination (interruption of transmission)": For Leprosy, with the indicator of "zero new autochthonous", 26% in 2020 to 62% in 2030; for Onchocerciasis, with the indicator of "interruption of transmission" from 1% in 2020 to 31% in 2030.
3. "Elimination as a public health problem": For Chagas disease, Human African trypanosomiasis, Leishmaniasis, Lymphatic filariasis, Rabies, Schistosomiasis, Soil-transmitted helminthiases, Trachoma. A detailed indicator is described for

each disease. The expected change in the indicators for most of the diseases is more than 80%, whereas for Chagas disease and Human African trypanosomiasis, expected changes are 37% and 61%, respectively.

4. "Control": for the remaining nine diseases. For example, for Scabies and other ectoparasites, the indicator is "incorporated scabies management in the universal health coverage package of care" and the expected change is from 0% to 100%. On the other hand, for Echinococcosis, Mycetoma, chromoblastomycosis, and other deep mycoses, the indicator is the number of countries with some specified control programme. In 2020, both were present only in one country. Target numbers in 2030 are 17 and 15, respectively and the latter is only 50%.

A description here is just to show some of the characteristics of the design of the roadmap. It also illustrates that there are some promising areas, while others face difficulties and challenges. The actual contents of this targeted strategy can be seen in WHO's full report and executive summary [2, 3].

Analysis of the first three years has been reported as presenting several successful progress, while disruptions due to COVID-19 are also presented in the WHO report in 2023 [15].

5 The R&D Portfolio of the Drug for Neglected Diseases Initiative (DNDi)

In the above report by the WHO [15], the activities of the "Drug for Neglected Diseases initiative" (DNDi) are introduced as the representative example of a main driver of innovation for providing new treatment approaches for NTDs. DNDi was launched under the leadership of the Médecins Sans Frontières (MSF). DNDi has been dedicated to developing drugs for NTDs and, to date (First Semester 2025), has achieved success in developing treatments for three kinetoplastid diseases—human African trypanosomiasis (HAT, also known as sleeping sickness), visceral leishmaniasis and Chagas disease. Thus far, disease targeting has been expanded and 13 treatments have been developed, as shown in the portfolio (Fig. 15.1).

Their activities, forming "Product Development Partnerships (PDPs)", have been introduced in our previous publication "*Ethical Innovation for Global Health: Pandemic, Democracy and Ethics in Research*" [8], as one prominent example in the part of "Alternative Frameworks for Innovation and Drug Development Strategies". PDPs are *not-for-profit organizations that develop new medicines, vaccines, and diagnostics by building partnerships between government, private sector, academia, and philanthropic organizations"* [8]. The project brings together people from different parts of the world who share common objectives. This presents an alternative framework transformed from traditional company-initiated "closed innovation" aimed at commercial profit-making. An alternative approach could be called "open innovation" for a global common objective leading to public health, respecting the social value of the affected communities.

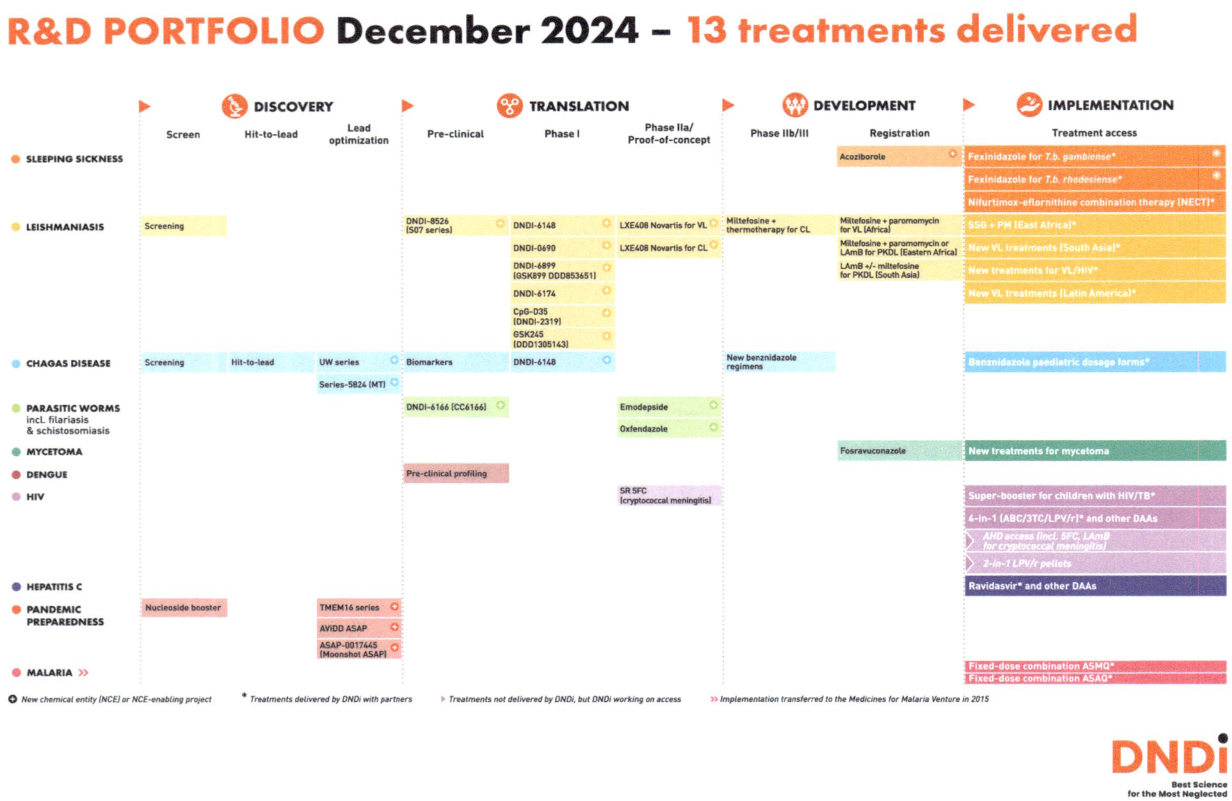

R&D Portfolio December 2024: 13 treatments delivered. (Note: This figure is reproduced from the DNDi website [16], with their authorization. These numbers are periodically updated and can be accessed at: https://dndi.org/research-development/portfolio/)

With this portfolio, a description of the scale of ongoing projects is provided as "*We are committed to developing safe, effective, and affordable treatments for people with neglected diseases and neglected patients. Together with our partners, we are working on over 40 projects, including more than 20 focused on identifying or developing new chemical entities. We are also running over 25 clinical trials*".

6 Overcoming Mental Health and Stigmatization

People living with NTDs are often affected by mental health conditions due to the suffering associated not only with physical symptoms of neglected diseases, but also with discrimination and stigmatization at the individual or community level [17]. There are some research programmes that focus on this.

One such programme is "Empowering people with cutaneous leishmaniasis: Intervention programme to improve patient journey and reduce stigma via community education" (ECLIPSE) [18] (more details of the programme can be found on their website [19] including a series of publications [20]). This interdisciplinary programme focuses on people living with cutaneous leishmaniasis (CL) and their communities in Brazil, Ethiopia, and Sri Lanka. It received £4.6 million from the United Kingdom's National Institute for Health and Care Research (NIHR) from November 2019 to October 2024 [21]. The objectives of the project presented in 2021 [22] were "*to explore and improve the psychosocial aspects associated with CL; to reduce the burden of stigma and its effects on individuals, families and communities; to capture the patient journey in its entirety to better identify gaps and draw upon areas of strength*".

According to their publication in 2022 [18], they established Community Advisory Groups (CAGs), a total of 13, with 4 in Brazil, 6 in Ethiopia, and 3 Sri Lanka. The ECLIPSE team adopted a broad community participation strategy, with a decolonial approach towards communities, differently according to the social and cultural background of each region, setting two types of groups: policy level and community level. In Brazil, it was important to consider gender, race, class, and age power relations in CAG interactions, and to minimize power dynamics while respecting diversity. The CAG membership included "*community health workers, members working in education or agriculture, people with CL, and community leaders*". In Ethiopia, a democratic approach was taken, involving meetings, discussions, nomination of people, and voting. The final CAG membership included "*community members and elders, health extension workers, people with CL and their families, local administrators, religious leaders, and traditional healers*". In Sri Lanka, through both top-down and bottom-up ways, members were formed including "*community members, religious leaders, traditional healers, people with CL and their families, teachers, local administrators, and representatives of key community groups*".

Although the COVID-19 pandemic and the armed conflict that suddenly broke out in Ethiopia hindered some of the planned activities of the project, a detailed description of their community engagement strategies and process has been described as one of their achievement [18]. This could be a model to show how the principle of the DoH could be implemented in practice. Their ethos is stated as:

no research about us without us

This phrase was proposed by the community members who guided the project and was adapted from the slogan "Nothing About us without us", as advocated during the development of the United Nations Convention on the Rights of Persons with Disabilities [23]. The latter was also led by people with disabilities.

7 Conclusion

The R&D strategies for the common goal to eliminate NTDs are aligned with the new directives of the 2024 DoH on community participation and the inclusion of vulnerable populations in research that may be of value to them, with a guarantee of post-trial access. These constructive changes in the DoH could help drive further innovation in high disease-burden areas, including NTDs, where investment in R&D has been insufficient.

The combined efforts to put into practice the principles in the DoH, as evidenced by DNDi and other private and public organizations, including pharmaceutical companies, take innovative strategies of project development partnership (PDP) ethically forward. In this way, the DoH can certainly contribute to achieving equitable and sustainable global health while in parallel safeguarding vulnerable individuals and communities from exploitation and other forms of harm, assuring equitable benefit sharing. Global collaborative efforts are expected to maximize the impact of the changes in paragraphs 6, 19, and 20 of the DoH, aiming to facilitate access to essential medicines for those most in need.

Acknowledgement We appreciate Professor Simon L. Croft, London School of Hygiene and Tropical Medicine, London, the United Kingdom, who has significantly committed to DNDi's activity, that he provide us valuable advice. We also appreciate Professor Lisa Dikomitis, University of Warwick (UK), and Professor Helen Price, Keele University (UK), co-leaders of the ECLIPSE programme, that they reviewed the part of ECLIPSE in this chapter and provided valuable advice.

CK appreciates the International Federation of Associations of Pharmaceutical Physicians and Pharmaceutical Medicine (IFAPP) for granting her travel expenditure for her in-person participation in the WMA Regional Meeting in Tel Aviv (December 2022) and the WMA for granting her travel expenditure for her in-person participation and presentation as a panellist at the WMA Regional Meeting in Washington, DC (August 2024), and for a waiver of registration fee from the WMA as a journalist to attend the General Assembly in Helsinki (October 2024). She participated in all the other WMA Regional Meetings at her own expense or online.

DG was officially engaged in discussions for the revisions of the Declaration of Helsinki from 2000 to 2008 as a member of the Brazilian Medical Association and Brazilian Medical Council teams; he also participated in the CIOMS 2016 guidelines as a member of working group; he was invited to the regional meeting in Sao Paulo by the WMA to discuss the ethics of placebo-controlled trials; and he participated, with partial funding from CIOMS, in the 2024 WMA General Assembly that adopted the 2024 version of the DoH.

Conflict of Interest AD was a Board member of the South African Medical Association, which was a member of the Workgroup of the World Medical Association for the 2024 revision of the Declaration of Helsinki. The authors declare no other competing interests relevant to the content of this chapter.

References

1. The World Medical Association. WMA declaration of Helsinki – ethical principles for medical research involving human participants. First adopted in 1964, last amended in 2024. https://www.wma.net/policies-post/wma-declaration-of-helsinki/
2. World Health Organization. Ending the neglect to attain the Sustainable Development Goals; A road map for neglected tropical diseases 2021–2030. 2020. https://www.who.int/teams/control-of-neglected-tropical-diseases/ending-ntds-together-towards-2030
3. Executive summary. Ending the neglect to attain the Sustainable Development Goals: a road map for neglected tropical diseases 2021–2030. 28 January 2021. https://www.who.int/publications/i/item/WHO-UCN-NTD-2020.01
4. Weld ED, Waitt C, Barnes K, Garcia BF. Twice neglected? Neglected diseases in neglected populations. Br J Clin Pharmacol. 2022;88(2):367–73.
5. World Health Organization. Ending the neglect to attain the sustainable development goals one health: approach for action against neglected tropical diseases 2021–2030. 28 January 2022. Available from: https://iris.who.int/bitstream/handle/10665/351193/9789240042414-eng.pdf
6. Kurihara C, Greco D, Dhai A, Matsuyama K, Baroutsou V. Vulnerability, social value and the equitable sharing of benefits from research: beyond the placebo and access debates. Front Med. 2024;11:1432267. https://doi.org/10.3389/fmed.2024.1432267.
7. Croft SL, Yamada H, Kita K, Brunger C, Hirabayashi F, Imoto D, Kurihara C, Saio , Organizer. The 2nd round table discussion on development of new medicines for neglected tropical diseases: from bench to bed, achievements and challenges*1 – partnerships, treatments and NTDs: 2005 to 2018, 13 years of change. Clin Eval. 2018;46(2): W55–73. http://cont.o.oo7.jp/46_2/w55-w73.pdf
8. Kita K, Yamada H, Hirabayashi F, Croft SL. Development of Portfolios and pipelines of drugs for the treatment, prevention and control of neglected tropical diseases. In: Kurihara C, Greco D, Dhai A, editors. Ethical innovation for global health: pandemic, democracy and ethics in research. Springer; 2023. p. 283–96.
9. World Health Organization. Neglected tropical diseases. Q&A: what are neglected tropical diseases?. 8 Jan 2025. https://www.who.int/news-room/questions-and-answers/item/neglected-tropical-diseases#:~:text=NTDs%20include%3A%20Buruli%20ulcer%3B%20Chagas,rabies%3B%20scabies%20and%20other%20ectoparasitoses%3B. Accessed 30 April 2025.
10. World Health Organization. Neglected tropical diseases. Q&A: Why are they called "neglected"?. 8 Jan 2025. https://www.who.int/news-room/questions-and-answers/item/neglected-tropical-diseases#:~:text=NTDs%20include%3A%20Buruli%20ulcer%3B%20Chagas,rabies%3B%20scabies%20and%20other%20ectoparasitoses%3B. Accessed 30 April 2025.
11. United Nations. Resolution adopted by the General Assembly on 25 September 2015 – transforming our world: the 2030 agenda for sustainable development. 21 Oct 2015. https://www.un.org/en/development/desa/population/migration/generalassembly/docs/globalcompact/A_RES_70_1_E.pdf

12. London declaration on Negelected Tropical Diseases (NTDs). 30 Jan 2012. https://www.gov.uk/government/publications/london-declaration-on-negelected-tropical-diseases-ntds. Accessed 30 April 2025.
13. World Health Organization. Neglected tropical diseases. Q&A: what progress has been achieved over the past year?. 8 Jan 2025. https://www.who.int/news-room/questions-and-answers/item/neglected-tropical-diseases#:~:text=NTDs%20include%3A%20Buruli%20ulcer%3B%20Chagas,rabies%3B%20scabies%20and%20other%20ectoparasitoses%3B. Accessed 30 April 2025.
14. World Health Organization. The global health observatory: neglected tropical diseases. https://www.who.int/data/gho/data/themes/neglected-tropical-diseases. Accessed 30 April 2025.
15. World Health Organization. Global report on neglected tropical diseases 2023. Geneva. 2023. https://www.who.int/teams/control-of-neglected-tropical-diseases/global-report-on-neglected-tropical-diseases-2023
16. DNDi. Research & development portfolio. Portfolio December 2024. https://dndi.org/research-development/portfolio/. Accessed 30 April 2025.
17. World Health Organization. Mental health of people with neglected tropical diseases – towards a person-centred approach. 6 Oct 2020. Mental health of people with neglected tropical diseases towards a person-centred approach. Accessed 30 April 2025.
18. Polidano K, Parton L, Agampodi SB, Agampodi TC, Haileselassie BH, Lalani JMG, Mota C, Price HP, Rodrigues S, Tafere GR, Trad LAB, Zerihun Z, Dikomitis L. Community engagement in Cutaneous Leishmaniasis research in Brazil, Ethiopia, and Sri Lanka: a decolonial approach for global health. Front Public Health. 2022;10:823844. https://doi.org/10.3389/fpubh.2022.823844. PMID: 35242734; PMCID: PMC8885625
19. ECLIPSE. https://www.eclipse-community.com/. Accessed 8 May 2024.
20. ECLIPSE. ECLIPSE Publications. https://www.eclipse-community.com/publications/. Accessed 8 May 2024.
21. National Institute for Health Research. Empowering people with Cutaneous Leishmaniasis: intervention programme to improve patient journey and reduce Stigma via community Education (ECLIPSE). https://www.fundingawards.nihr.ac.uk/award/NIHR200135. Accessed 30 April 2025.
22. Dikomitis L. Cutaneous Leishmaniasis and mental health: exploring psychosocial impact and CL-associated stigma (the ECLIPSE programme). Clin Eval. 2021;49(2):201–9. http://cont.o.oo7.jp/49_2/s-p208.pdf
23. United Nations. Convention on the rights of persons with disabilities. 13 Dec 2006. https://social.desa.un.org/issues/disability/crpd/convention-on-the-rights-of-persons-with-disabilities-crpd. Accessed 30 April 2025.

Chapter 16
Reformation of Global Research Ethics Principles from the Perspective of Patient Public Involvement

Kotone Matsuyama

Abstract In the context of the 2024 revision of the World Medical Association (WMA)'s Declaration of Helsinki (DoH), I have been involved in the revision process as the Chair of the Ethics Working Group of the International Federation of Associations of Pharmaceutical Physicians and Pharmaceutical Medicine (IFAPP), with its members, and finally attended the WMA General Assembly, celebrating the remarkable improvements that have been made. This revision is a historic change, particularly from the perspective of patient and public involvement.

However, some unresolved and priority issues remain. In April 2025, I was designated to be the new President-Elect of the IFAPP. As I have worked over the past 5 years together with outstanding experts and patient–public colleagues, I wish to dedicate myself to further developing collaboration and cocreation to address the remaining challenges for improvement of health and well-being in the view of a community and global scale.

Keywords Patient and public involvement · Vulnerability · Cutting-edge technology · Disaster settings · Research Ethics Committee · Informed consent

---

K. Matsuyama (✉)
Department of Health Policy Management, Nippon Medical School, Tokyo, Japan

International Federation of Associations of Pharmaceutical Physicians and Pharmaceutical Medicine (IFAPP), Woerden, The Netherlands

C. Kurihara et al. (eds.), *The 2024 Declaration of Helsinki*,
https://doi.org/10.1007/978-981-96-9294-1_16

1 Introduction

1.1 Reformation of Ethical Principles Responding to Emerging Situations

The tenth revision of the Declaration of Helsinki (DoH) [1] was adopted at the World Medical Association (WMA) General Assembly in Helsinki, Finland, in October 2024. About 60 years after the first Declaration, the revision of this commemorative Declaration took place in the same city of Helsinki, where the first version was adopted. I attended this General Assembly [2] as the Chair of the Ethics Working Group, the International Federation of Associations of Pharmaceutical Physicians and Pharmaceutical Medicine (IFAPP), which is an affiliated organization of the WMA under the Memorandum of Understanding for collaboration between the two organizations, formed in 2017.

This time, the revisions were based on the higher ethical principles, in the midst of overcoming the public health crisis of the coronavirus pandemic and welcoming the rapid changes in the medical research environment due to many technological innovations such as artificial intelligence (AI) and digital health technologies utilizing real-world data.

1.2 From "Subject" to "Participant"

First of all, I would like to point out that the most important revision is that the term "subject" has been changed to "participant" throughout the Declaration. In the preamble, paragraph 1, the DoH is defined as "*a statement of ethical principles for medical research involving* ***human participants****, including research using identifiable human material or data.*" In the following section, I will highlight some of the points of the revised DoH based on the current situation and issues related to the promotion of patient and public participation.

As of April 2025, this change has not yet been incorporated into regulations in Japan. However, in the academic communities in which I am involved, in clinical trial settings in practice, I have confirmed that some Japanese branches of global pharmaceutical companies are already translating the term "participant" into Japanese relevant term (changing from "hikenja" to "sankasha") and incorporating it into protocols and informed consent forms for the trials conducted in Japan. It would be influenced by guidance documents from the United States, Europe, and most prominently by global agreements such as the CIOMS Ethical Guidelines [3] and the ICH-GCP (R3) [4]. This means that the use of the term "participant" in official documents for clinical trials is becoming a global standard, even if it has not yet been incorporated into legally binding regulations in all countries.

The change of the term in the DoH reflects such global trends and will expand the impact more widely, both in terms of jurisdictions and research areas not limited to GCP-regulated clinical trials. In Japan, the Ministry of Health, Labour, and Welfare's decision to change this terminology is likely to occur in a short period, and it further expands the collaborative partnership with patients and research participants.

2 Meaningful Engagement Respecting Vulnerability

2.1 Engagement Is Not a Mere Formal Procedure

Paragraph 6 states: "*Meaningful engagement with potential and enrolled participants and their communities should occur before, during, and following medical research.*" The 2024 revision is groundbreaking in that it has positioned consideration of regional and cultural backgrounds as a part of research ethics, in addition to patient and public involvement. In the real world, however, there is still a lot of meaningless patient involvement. It has become a mere formality, where the collection of opinions is limited to be in a formal process. When this happens, it is difficult to reflect patient and public perspectives in the research design and decision-making process. With the new principles presented in the DoH, the social legitimacy of research could be strengthened, the quality and diversity of participation will be improved, and we can expect to see a move away from research designs that are biased toward specific groups.

This will have a significant impact on building relationships of trust with local communities, particularly in the context of international collaborative research and research in the areas of emerging infectious diseases and health crises. Moreover, as it is clarified in paragraph 8, "*While new knowledge and interventions may be urgently needed during public health emergencies, it remains essential to uphold the ethical principles in this Declaration during such emergency.*"

2.2 Community Engagement in Disaster Settings in Japan

Japan is located in an area prone to frequent earthquakes and other natural disasters. In response to the large-scale disasters that have occurred repeatedly in recent years, Japanese universities and academic societies have strengthened their support systems. In this context, the boundaries between disaster response and research have become blurred, and complaints have often been made that people receiving support while being under survey or research feel that their health and rights are not being respected. Therefore, research conducted in disaster situations requires well-equipped, careful, sensitive community engagement.

One of the commendable characteristic of the Japanese people is their patience and capacity to be gentle without causing looting or violence, which is highly evaluated internationally. However, in the recent earthquake relief efforts, I received reports from disaster relief workers affiliated with the Pharmacists Association that conflicts arose between local communities due to a lack of understanding of the historical backgrounds of these communities. In such cases, relief activities may be deemed to be careless imposition. In one case, people who were proud of their local history resisted being ordered to participate in activities to receive support together with the people in other regions, which hindered the progress of relief efforts. Under such circumstances, the members of the Pharmacists Association visited each house in the disaster area to listen to the voices of residents, which led to the finding of a shortage of medicines, which were subsequently supplied. This was a successful case of community engagement, learning from some mistakes.

The above case clearly demonstrates that, especially in crisis situations, community engagement requires medical teams to foster mutual understanding with the locally affected communities and to work together to achieve mutually desirable goals, respecting the social value and dignity of each community.

2.3 Vulnerability and Ensuring the Right and Opportunity to Participate in Research

The new statements in the DoH regarding vulnerability added to paragraphs 19 and 20 are extremely important. They are intended to prevent the inequitable distribution of benefits for certain vulnerable groups fixed to be excluded from participation in research, and blocking them from receiving the social value of research results. The harm caused by excluding them from research must be considered and weighed up against any harm caused by including them in research. It is stated that such people should receive special consideration, support, and protection in order to be included in research in a fair and responsible manner. In relation to the issue of free and informed consent, it is very important to recognize that this statement has been added in terms of ensuring that vulnerable people are not disadvantaged, that they are not harmed, and that consideration is given to enable them to share the benefits [5], as well as posttrial access of the products proven to be safe and effective.

I attended the WMA General Assembly in Helsinki held October 16—19, 2024, and the day before the new DoH was adopted, with international collaborators, organized and issued the "Helsinki Statement" [6] with endorsers from around the world. This statement celebrated the ten points of brilliant revisions of the new DoH while also expressing our continuous collaboration for remaining challenges. The position of our Statement supports the motions submitted by the Uruguayan Medical Union to the General Assembly, arguing "limiting placebo-controlled trials when there is a proven intervention" to be conducted without an increase of risk in the placebo arm and "guaranteeing posttrial access.".

Followingly, at the IFAPP Annual Meeting, International Conference on Pharmaceutical Medicine 2025 (ICPM2025), held in Amsterdam in April 2025, a researcher from Ukraine made a presentation to criticize global pharmaceutical companies for abruptly halting clinical trials without sufficient explanation in the midst of war. This is exactly the problem that IFAPP colleagues have argued for prevention [7, 8], through scientific publications and academic web meetings. In our initiatives to support Ukrainian clinical trials, in addition to providing support to those affected, the continuation of clinical trials is important to maintain the health of each trial participant as well as the life science infrastructure of Ukraine. In the next session, we organized, we reaffirmed that "posttrial access" must also be interpreted as "ensuring access to necessary care for research participants after the completion of a trial, even in the midst of a war situation.".

The title of the session we organized was "Vulnerability, Social Value, and Community Engagement: Voices from Patients and the Global South," based on an IFAPP member's paper collaborating with external experts [9]. This session featured an excellent presentation on vulnerability by Dr. Lujain Alquodmani, the Immediate Past President who led the effort for 2 years leading up to the adoption of the DoH. On the same day, Dr. Otmar Kloiber, Secretary General of the WMA, and Dr. Lembit Rago, Secretary General of the Council for International Organizations of Medical Sciences (CIOMS), also gave presentations, marking the culmination of the collaboration between the three organizations.

2.4 The Necessity of Systematic Education for Patients and the Public

Patients are always vulnerable because of the information asymmetry between them and researchers or pharmaceutical companies and because they are in a position to seek treatment [9]. To maintain an equal relationship between patients and researchers and to achieve truly meaningful engagement, a systematic educational program for patients and the public is needed.

In Europe and in the United States, there are some educational programs provided for patients and the public to support their engagement, including the famous one provided by EUPATI [10]. This is an internationally well-known patient education program developed and initiated by patient groups collaborating with experts. Many members involved in the creation of PharmaTrain, a pharmaceutical education program mainly for experts, have participated in EUPATI programs, which are mainly for patients. EUPATI's training is very comprehensive and enables patients and the public to systematically learn the knowledge they need to know about drug development.

At the same time, in order to carry out practical activities, it is desirable to develop training tailored to local conditions. An example of a localized program can be found in Japan. YORIAILab [11], a general registered association for which I am one of the executive board members, has adapted the EUPATI program to local conditions. YORIAILab is implementing a patient education program for Japan as a "clinical trial ambassador training project" [12]. In addition, a Japanese translation of the Patient Engagement Guidelines [13] developed by the Patient Focused Medicines Development (PFMD) [14], which is an initiative to facilitate patient engagement in medicines development, is prepared by the JPPaC [15]. JPPaC is an organization that facilitates patient-centred health care, and is a Japanese collaborator of the PFMD. Among other initiatives, workbooks tailored to the Japanese context are being made available to reflect regional needs. Such hands-on training, which addresses the challenges of local public health administration, will become increasingly important in all patient and citizen communities in the future.

3 Research Ethics Governance

3.1 Public and Community Engagement in Research Ethics Committee

In paragraph 23 on Research Ethics Committees (RECs), a prominent new statement was added: "*The committee must have sufficient familiarity with local circumstances and context, and include at least one member of the general public*" and "*the research protocol must be approved by research ethics committees in both the sponsoring and host countries.*" When communities participate, it leads to considerations on how to receive the cultural background of the community in question. For example, in Asian countries, there are differences in dietary habits and fasting customs sometimes due to religious backgrounds that differ from those in Europe and the United States. In addition, it is necessary to fully consider the differences in metabolism from a genetic pharmacology perspective of pharmaceuticals and the resulting safety issues that may arise.

In addition, in Japan, and presumably in non-Western regions, there is a tendency for subjective assessment data in patient-reported outcomes (PROs) to be biased toward the preferences of the treating physician. Patients may unknowingly transform and report their own desires to match those imagined by their attending physicians. Therefore, when introducing research protocols from Western regions, where respect for individual preferences is emphasized, a simple verbatim translation of the questionnaire will not generate data that accurately reflects the true subjective experiences of participants. Such issues require that RECs when evaluating protocols developed by global companies in Western countries, require modifications to reflect the reality that Japanese participants may find it difficult to express their true opinions. Implementing such governance measures is essential to ensuring the integrity of research data.

3.2 Informed Consent and Lay Communication

In paragraph 26 for the Informed Consent, I would like to stress that a new notable requirement was added: "*each potential participant must be adequately informed in plain language.*" The fundamental question of public involvement is to discover new value in pharmaceuticals and other medicinal products that patients could receive and make them available for use: posttrial access for all those in need. Nevertheless, for the creation of social value through research, the efforts are insufficient to provide easy-to-understand, plain-language communication that will help research participants and local communities understand its implications. The practice of such lay communication is very useful, but it is still in process.

It is important to note that in regions such as Asia and areas where democracy has not been fully established, women often lack the substantive right to consent to participate in research. This issue is strongly emphasized in the CIOMS Ethical Guidelines for Research Involving Humans [3]. In Japan, there are cases where elderly people who wish to participate in clinical trials are discouraged from doing so by their daughters or sons who care for them, citing concerns about their safety. The wishes of older people are often not respected, making it difficult to promote diversity in clinical trials. On the other hand, the global clinical trial's explanation sheet sometimes insensitively states the life expectancy of cancer patients.

There are also legal barriers to sharing information on research results with patients after a trial completion. In Japan, the Pharmaceuticals and Medical Devices Act prohibits the advertisement of information about unapproved drugs to medical staff and patients, although it is not prohibited to inform according to the clinical trial results disclosed in public databases. Due to the ambiguity of the boundary between lawful and inappropriate information disclosure, this situation forms a serious problem, particularly in the case of rare diseases. Despite the efforts by patient groups to increase patient participation in clinical trials, only the research participants themselves can receive the results of the trial in plain language summary form, and pharmaceutical companies cannot share the results information with patient groups. In terms of ensuring patients' right to know, the DoH's 2024 revision is insufficient. However, it should be noted that paragraph 6 of the DoH recommends the involvement of participants and their communities in "understanding and disseminating results.".

3.3 Secondary Use of the Personal Data and AI Development

In paragraph 32, the principles of data governance are described, stating that the secondary use of data and biological samples obtained through research should basically follow the WMA Declaration of Taipei on health databases and biobanks [16]. The open sourcing of data and the utilization of past data through open innovation are also closely related to the participation of patients and the public. While the utilization of research results is also important for the development of medicine, we must not neglect the fundamental principle that the unlimited use of data that were obtained in the past without explicitly informed purpose may be unethical from the perspective of protecting human dignity, even if the data are completely anonymized. In addition, we must also be prepared for the possibility of conflicts between new technologies and ethical issues, such as the utilization of real-world data and the creation of models using AI. This means that completely anonymized personal data are repeatedly used for the development of AI, beyond an individual "broad informed consent," which is described in the CIOMS guidelines [3], meaning adequately informed consent to multiple uses of personal data. The new DoH has struggled with this situation by strengthening the authority of RECs, and paragraph 32 also recommends that research using data without informed consent should be reviewed by REC.

We have to deliberate more in-depth on the possibility of AI contributing to health as well as its threat of technological singularity beyond the ability to be controlled by human beings and society. Such negative aspects of AI should be sufficiently assessed by the REC and, if approved by the REC, adequately explained to each participant in the process of "broad informed consent." Such an explanation should be provided not only to each individual participant but also to society as a whole. This is an accountability of the researcher engaged in AI development, which could lead to cocreation with the public to generate social value. In Japan, the concept of a "healthcare data biotope" has been promoted by the Ministry of Health Labour and Welfare under the concept of "the Japan Vision: Health Care 2035." [17] They explain the vision using the analogy of biotope that they will accelerate prevention and disease management through the strategic use of ICT, promoting public engagement and maintaining an ecosystem of research and development.

4 Lessons Learned from the DoH Amendment

4.1 As the Chair of the Ethics Working Group of the IFAPP

Since the time I was designated as the Chair of the IFAPP Ethics WG in 2021, I have been committed to international collaboration with various interested parties to ensure that the revision of the DoH could be in a better direction. This has been the WG's first priority until its adoption in 2024, as well as continuing efforts for implementation in practice. IFAPP has been collaborating several times with the WMA, having meetings both online and in-person to discuss various topics of research ethics. Our members published several articles in our official journal, IFAPP TODAY. During these 5 years, we also published seven peer-reviewed papers to discuss the following topics:

1. Linking the DoH to the DoT for data-driven research [18].
2. Pricing of gene therapy under cooperation between industry and society [19].
3. Theorizing benefit sharing focusing on genetics [5].
4. Ethics of research in times of war: continuation of studies [7].
5. Ethics of research in times of war: re-conceptualization of adaptive design [8].
6. General opinions for the revision of the DoH [20].
7. Reconceptualizing vulnerability and social value beyond the placebo and access debate [9].

4.2 As the President-Elect of the IFAPP

I am proud to be designated to be the President-Elect at the General Assembly of the IFAPP on April 11, 2025, during its Annual Meeting, ICPM 2025 in Amsterdam. Based on my experience as the Chair of the Ethics Working Group, I have been discussing with our colleagues the next revision of the Ethics Framework of the IFAPP [21]. IFAPP has developed our principles from the Physicians' Code of Conduct to the Ethics Framework of shared responsibility of multidisciplinary teams. Now, it is time to update our framework to more involvement of patients and the public.

There are several pressing issues in health and medicine, from the local to the global level, seeing the emergence of new technologies such as AI along with multiple utilizations of data, ensuring diversity, and emerging burdens and tasks in the super-aging society, with strengthened respect for elderly people and end-of-life care. We will conduct horizon scanning on the direction of science and technology, identify ethical issues in the use of new tools, and seek solutions in collaboration with patients, the public, and society, both respecting community values and considering the health and well-being of the global society.

5 Conclusion

In this new DoH, many valuable, significant amendments have been made that relate to patient and public involvement in medical research. Active, meaningful involvement would enable the maximized social value of the research, ensuring the fair implementation of research results and equitable benefit sharing.

Considerations for diversity, equity, and inclusiveness are essential to promote community engagement. It is important to maintain participatory practices, considering diversities in gender, religion, language, values, and cultural or social background, to enhance the benefit of research in each of them.

Working together with the international communities to achieve true patient and public participation will ultimately ensure that the benefits of research are available to patients, which will be the key to building up the research ecosystem for a better future.

Acknowledgments Some parts of this study, including attendance of the WMA General Assembly in Helsinki funded by a survey of ethical issues related to the use of advanced medical technology and bioresources according to the Japan Agency for Medical Research and Development (AMED) grant (no. 24bm1523009s0801).

Conflict of Interest The author declare no conflict of interest related to the publication of this report.

References

1. The World Medical Association. WMA Declaration of Helsinki—Ethical Principles for Medical Research Involving Human Participants. First adopted in 1964, last amended in 2024. https://www.wma.net/policies-post/wma-declaration-of-helsinki/.
2. Kurihara C, Matsuyama K, Baroutsou V. World medical association's declaration of Helsinki, 2024 revision: celebrating the 60th anniversary, at Helsinki. Clin Eval. 2025;52(3):W1–W27. http://cont.o.oo7.jp/52pop/52pop_contents_e.html.
3. Council for International Organizations of Medical Sciences. International ethical guidelines for health-related research involving humans. 2016. https://cioms.ch/publications/product/international-ethical-guidelines-for-health-related-research-involving-humans/
4. International Council for Harmonisation of Technical Requirements for Pharmaceuticals for Human Use. ICH Harmonised Guideline: Guideline for Good Clinical Practice E6(E3). Adopted January 6, 2025.
5. Matsuyama K, Kurihara C, Crawley FP, Kerpel-Fronius S. Utilization of genetic information for medicines development and equitable benefit sharing. Front Genet. 2023;14(14):1085864. https://doi.org/10.3389/fgene.2023.1085864.
6. Helsinki Statement Stakeholders. Helsinki statement. Clin Eval. 2024;52(3) http://cont.o.oo7.jp/52pop/HelsinkiStatement_18Oct24_final.pdf
7. Kerpel-Fronius S, Kurihara C, Crawley FP, Baroutsou V, Becker S, Franke-Bray B, Matsuyama K, Naseem S, Schenk J. The ethical responsibility to continue investigational treatments of research participants in situation of armed conflicts, economic sanctions or natural catastrophes. Front Med. 2022;9:950409. https://doi.org/10.3389/fmed.2022.950409.
8. Kurihara C, Crawley FP, Baroutsou V, Becker S, Franke-Bray B, Granville CA, Matsuyama K, Naseem S, Schenk J, Kerpel-Fronius S. The continuation of clinical trials in times of war: a need to develop ethics and situationally adaptive clinical research guidelines. Front Med. 2022;9:966220. https://doi.org/10.3389/fmed.2002.966220.
9. Kurihara C, Greco D, Dhai A, Matsuyama K, Baroutsou V. Vulnerability, social value and the equitable sharing of benefits from research: beyond the placebo and access debates. Front Med. 2024;11:1432267. https://doi.org/10.3389/fmed.2024.1432267.
10. EUPATI. https://eupati.eu. Assessed 28 Apr 2025.

11. YORIAILab. https://www.yoriailab.com/en. Assessed 28 Apr 2025.
12. Clinical Trial Ambassador Project. https://www.yoriailab.com/en/clinical-trial-ambassador. Assessed 28 Apr 2025.
13. Patient Engagement Management Suite. https://pemsuite.org. Assessed 28 Apr 2025.
14. PFMD. https://patientfocusedmedicine.org. Assessed 28 Apr 2025.
15. JPPaC. https://www.jppac.or.jp. Available in Japanese only. Assessed 28 Apr 2025.
16. The World Medical Association. The Declaration of Taipei on ethical considerations regarding Health Databases and Biobanks. 2002, last revised 2016. https://www.wma.net/policies-post/wma-declaration-of-taipei-on-ethical-considerations-regarding-health-databases-and-biobanks/.
17. Ministry of Health, Labour and Welfare. The Japan Vision: Health Care 2035. https://www.mhlw.go.jp/seisakunitsuite/bunya/hokabunya/shakaihoshou/hokeniryou2035/future/en/. Assessed 28 Apr 2025.
18. Kurihara C, Baroutsou V, Becker S, Brun J, Franke-Bray B, Carlesi R, Chan A, Collia LF, Kleist P, Laranjeira LF, Matsuyama K, Naseem S, Schenk J, Silva H, Kerpel-Fronius S. Linking the declarations of Helsinki and of Taipei: critical challenges of future- oriented research ethics. Front Pharmacol. 2020;11:579714. https://doi.org/10.3389/fphar.2020.579714.
19. Kerpel-Fronius S, Baroutsou V, Becker S, Carlesi R, Collia L, Franke-Bray B, Kleist P, Kurihara C, Laranjeira LF, Matsuyama K, Naseem S, Schenk J, Silva H. Development and use of gene therapy orphan drugs-ethical needs for a broader cooperation between the pharmaceutical industry and society. Front Med (Lausanne). 2020;7:608249. https://doi.org/10.3389/fmed.2020.608249.
20. Kurihara C, Kerpel-Fronius S, Becker S, Chan A, Nagaty Y, Naseem S, Schenk J, Matsuyama K, Baroutsou V. Declaration of Helsinki: ethical norm in pursuit of common global goals. Front Med (Lausanne). 2024;11:1360653. https://doi.org/10.3389/fmed.2024.1360653.
21. Kerpel-Fronius S, Becker S, Barrett J, Brun J, Carlesi R, Chan A, Collia LF, Dubois DJ, Kleist P, Koski G, Kurihara C, Laranjeira LF, Schenk J, Silva H. The shared ethical responsibility of medically and non-medicallly qualified experts in human drug development teams. Front Pharmacol. 2018;9:843. https://doi.org/10.3389/fphar.2018.00843.

Chapter 17
Global Efforts Towards the Highest Ethical Standards: Horizon for the Future

Chieko Kurihara, Dirceu Greco, and Ames Dhai

Abstract This chapter provides an overview of the book in which the 2024 Declaration of Helsinki (DoH), adopted by the World Medical Association (WMA) on 19 October 2024, is discussed. This book is planned and conducted as a voluntary initiative, independent of the WMA, with the goal of promoting the Highest Ethical Standards to protect research participants and ensure the social value of research.

It is highly commendable that the WMA leaders involved in the 2024 revision of the DoH have provided two Forewords for this book. Chapter authors participated in the review process for the 2024 DoH in a variety of ways. Editors and some authors presented alternative views that were critical of the DoH, while recognising the unique value of the DoH as a driving force of global deliberations.

This chapter also presents challenges for future revisions of the DoH as well as perspectives for a strengthened global research ethics framework established based on the Human Rights Covenant.

Keywords Research ethics · Human rights · Equity · Protecting human participants · Social value · Human Rights Covenant

---

C. Kurihara (✉)
Kanagawa Dental University, Yokosuka, Kanagawa, Japan
e-mail: chieko.kurihara@nifty.ne.jp

D. Greco
School of Medicine, Federal University of Minas Gerais, Belo Horizonte, Brazil

A. Dhai
School of Clinical Medicine, University of the Witwatersrand,
Johannesburg, Gauteng, South Africa

C. Kurihara et al. (eds.), *The 2024 Declaration of Helsinki*,
https://doi.org/10.1007/978-981-96-9294-1_17

1 An Initiative of Commitment to the Highest Ethical Standards

This book aims to promote the highest ethical standards for the protection of research participants and to ensure the social value of research through discussions in view of the 2024 revision of the Declaration of Helsinki (DoH) [1]. This is the tenth amendment since its first adoption in 1964. While this initiative is independent of the World Medical Association (WMA), which is the steward of the DoH, the authors of the chapters have been involved in its revision process in various ways, jointly or separately, by influencing, networking and collaborating with each other. They presumably share the following two objectives:

- Aspiration and commitment to "the highest globally agreed ethical standards for research involving human participants".
- Willingness to commit to maximising the protection of research participants and the social value of research based on the new principles in the 2024 DoH, together with proposed but not adopted essential principles confirmed in the revision process, to be applied in any research practice and research review, or in the policies of organisations, communities, regions or countries.

We are deeply grateful for the two Forewords contributed by four WMA authors, who played leading roles in the 2024 DoH revision process. The first Foreword is provided by Dr. Lujain Alqodmani, who served as WMA President (2023–2024), covering the most critical period of the 2 years preceding the 2024 amendment. She stated that the DoH is the "*cornerstone of our decisiveness to act together which crystallise our shared commitment and vision to uphold the highest ethical standards in medical research discussed and agreed among all the interested parties from a number of regions including conflict-affected countries through regional consultations and global dialogue*". These words on the DoH come from her continuous humanitarian dedication to global health, not limited to the questions of research ethics, but also to various agendas for human rights, equality, peace, and ecosystem sustainability.

The second Foreword is co-authored by the three WMA leaders: Dr. Ramin W. Parsa-Parsi, a member of the 2024 DoH Workgroup, who previously chaired the 2013 version of the DoH, as well as the latest revisions of the two fundamental documents defining the duties of physicians: the Declaration of Geneva [2] and the International Code of Medical Ethics [3]; Dr. Jack Resneck, Jr., who, after serving as the Workgroup Chair for the 2024 DoH, was appointed Chair of the WMA Council in April 2025; and Dr. Otmar Kloiber, who has dedicated his two decades as Secretary General of the WMA, committed to the integrity of all the WMA documents. The second Foreword introduces the topics that were the focus of discussions at each of the WMA's eight regional meetings around the world. All the regional meetings were open to the public and could be attended in person or remotely online, with recordings of some of them still available. Topics that were the focus of discussions included data-driven research; placebo use; emerging designs of clinical trials; research in resource-limited settings, including disaster settings; post-trial access; vulnerability and finally, expanding the impact of the new DoH.

2 Engagement of Chapter Authors in the Revision Process

Although the publication of this book was planned and carried out independently of the WMA, chapter authors participated, directly or indirectly, in the DoH revision process in various ways: some participated as Workgroup members representing national member associations for the 2024 or previous revisions of the DoH, or engaged in the discussions for other WMA documents; as official adviser to the WMA; members of collaborator organizations (the Council of International Organizations of Medical Sciences (CIOMS); the International Federation of Associations of Pharmaceutical Physicians and Pharmaceutical Medicine (IFAPP), under a Memorandum of Understanding; and the United Nations Educational, Scientific and Cultural Organization (UNESCO)); participants as speakers or observers at the WMA regional meetings; participants in meetings, publications or other initiatives organised by the editors of this book, independently of the WMA; and contributors of opinions to the WMA in response to the two public consultations for the 2024 DoH, organised by the WMA.

Some of the views expressed by the authors have been incorporated into the new DoH, others have not. The passionate arguments expressed by each of the authors have been directed toward promoting the achievement of the highest global ethical standards. Accordingly, this publication does not constitute an official commentary on the DoH from the perspective of the WMA. The authors' shared trajectory and common goal is to improve the protection of research participants, scientific integrity and the social value of research in the practical implementation of the 2024 DoH, and to contribute to the next revision of the DoH and other global norms of research ethics.

3 Essence of Revision and Pros and Cons

On the same date as the adoption of the revised DoH on 19 October 2024, a series of already prepared articles were published in *JAMA*, including one by the chair of the Workgroup [4–10]. Subsequently, commentaries were published in several prestigious journals [11–13]. These may provide some dominant opinions on the new DoH. In addition, some of the authors of this volume published interviews [14] with key experts who played leading roles in the 2024 DoH, which are published in *Clinical Evaluation* Vol. 52, No. 3 [15]:

http://cont.o.oo7.jp/52pop/52pop_contents_e.html

Many of the essential elements that form the foundations of this book are present in the same issue of the journal mentioned above. The last part of this issue of the journal presents an alternative world of the global research ethics community to express concerns, or add principles not adopted by the 2024 DoH. Consolidating all these perspectives, the highlights of the 2024 revision can be summarised as follows:

1. The key term has been changed from "research subject" to "research participant", promoting meaningful engagement of participants and their communities in all processes of research, with the recognition of various structural inequities in the research context.
2. Vulnerability should be reconsidered as it is not always a fixed situation, but rather as contextual and dynamic in nature, and the exclusion of vulnerable people from research may perpetuate or exacerbate their disparities, thus, inclusion comes to be promoted with strengthened safeguards and benefit sharing.
3. The WMA Declaration of Taipei [16] on health databases and biobanks is cited in the DoH; thus, secondary uses of data and materials obtained in research have to comply with the governance framework endorsed by the DoT.
4. Most of the controversial paragraphs on placebo-controlled trials and post-trial access have not undergone significant changes, as the Uruguayan Medical Union's motions [17, 18] were voted down. However, substantive discussions on these points must be seriously considered to avoid exploitation, especially of vulnerable research participants, and to reinforce their rights.

4 Helsinki Statement as an Alternative

The editors of this book were motivated to expand the impact of points 1–3 above, and as a primary agenda, point number 4, which means that the rights of the research participants to the best proven intervention, even in a control arm and after the completion of the study, must be guaranteed, in line with the fundamental principle of the DoH to prioritising the rights and interests of research participant over the purpose of the research. In the same context, the five of the authors (Kurihara, Greco, Baroutsou, Matsuyama, and Saio) planned, organised and issued on 18 October 2024, 1 day before the adoption of the 2024 DoH, the "Helsinki Statement" [19], appraising the ten items of improvement of the new DoH and expressing concerns about five remaining challenges. The Statement was issued in English, and translations in Arabic and Japanese are also available on the above website, in the same issue of the journal *Clinical Evaluation*, Vol. 52, No. 3. A Greek translation has also been prepared and will be linked on the same website. The chapters contributed by these organisers of the Helsinki Statement mention, in part, the contents of this Statement. In addition, the author of Chap. 3 is a notable endorser of the Helsinki Statement, as he represents the Egyptian Network of Research Ethics Committees and describes in this chapter the background history of the consensus-building process.

The fundamental agenda shared by the editors of this book and the Helsinki Statement could be reworded here from point 4 above:

- A fundamental principle of the DoH is that the rights and interests of research participants take precedence over the goals of research.
- Derived from this fundamental core principle, any research participant, including those in controlled trials and after the completion of research, has the right to receive the best proven intervention, if there is any. This right can be paraphrased in the language of the set of principles outlined in the 2000 version of the DoH [20], with some updates: placebo-controlled trials are not acceptable if there is a best proven intervention and post-trial access to the best proven intervention must be guaranteed to research participants who need it, as well as to the study host community and those in need worldwide.

There are certainly many difficulties in the practical implementation of these principles, but that does not mean it is impossible. The difficulties must be faced and solved by each of the parties involved. By disseminating information and taking concrete actions based on this belief, research practice could be improved both ethically and scientifically, thus generating social value.

5 Horizon for the Future: Collaboration with Patients, the Public, and Medical Students

One of the main outcomes of our voluntary initiative on the DoH revision was the consolidation and dissemination of patients' and public opinions on the DoH. The first outcome for the global audience was presented as a chapter in the editors' previous publication of 2023, entitled "*Ethical innovation for global health: pandemic, democracy and ethics in research*" [21]. The perspectives of these authors went well beyond the vision of the DoH. They presented future-looking views aimed at achieving the Sustainable Development Goals (SDGs) [22], based on the responsible engagement of patients and the public. During a webinar to disseminate the publication of this book [23], two Japanese authors, representing other authors, gave presentations in English [24]. In response to these, one of the editors, Dhai, encouraged them to develop the "Patient and Public Declaration of Research Ethics". In response, a "Patient and Public Declaration of Research Ethics" was published shortly before the adoption of the DoH 2024, and is reproduced with minor updates in this book chapter.

Additionally, some authors of this book organised webinars shortly before and after the last regional meeting of the WMA in August 2024. The proceedings are also published in the above-mentioned issue of the journal *Clinical Evaluation*. It was co-organised by the Brazilian Society of Bioethics, IFAPP and Clinical Evaluation, in collaboration with other organisations, and supported by the Japan Association for Bioethics, titled: "*The 2024 Declaration of Helsinki: Taking Forward Bioethics and Human Rights*" [15].

In this webinar, a group of medical university students called Universities Allied for Essential Medicines (UAEM), who work to ensure access to essential medicines, were invited. They have contributed one chapter to this volume. Collaborations with patient and public groups, as well as with medical university students are a precious gift that builds a bridge to the future.

Notably, in Brazil, a new patient and public initiative is being formed along similar lines to the Japanese group.

6 Highest Ethical Standards Versus Minimum Requirements

The value of the DoH lies in the fact that it is most widely known, recognised and incorporated into policies and regulations in many jurisdictions. It is a centripetal, driving force of consolidated actions seeking a global agreement to achieve the highest ethical standards for research involving human participants. This is a unique value of the DoH.

There are limitations due to its nature. The transparency of the discussion process has been improved by innovations in information technology. However, decisive agreements are made in closed meetings by the Workgroup composed exclusively of physicians and are voted on in the open General Assembly, where only physicians from associated national medical associations have the right to vote. Furthermore, the fact that DoH is embedded in various national regulations may prompt the WMA to be prudent in articulating the high ethical ideals that DoH should represent. As a result, the DoH is positioning itself to follow what has already been agreed upon by national regulations or other international organisations. This nature was called "minimal requirements" or the "lowest common denominator" in the webinar we organised [15]. Such an evaluation is a corollary of the ethical aspiration for the DoH to achieve the highest global standards.

Notably, the new principles articulated in the 2024 DoH, such as community engagement and a shift in thinking about vulnerability, could have a significant global impact, especially because they were incorporated into the DoH. We must recognise the value of the DoH and continue the deliberations aimed at improving the next amendment.

At the same time, we need to address the contradiction that a set of principles decided exclusively by physicians is binding on research communities that include non-physician members. For this reason, we must strive for an international norm that is developed by all the interested parties, with equal right to participate.

7 Research Ethics Norms as an Expansion of the Human Rights Covenant

The ideal international norm in this context would be an international convention that extends Article 7 of the United Nations (UN) International Covenant on Civil and Political Rights: "*no one shall be subjected without his free consent to medical or scientific experimentation*" [25]. Examples of such extension in national legal systems of research governance are introduced in this volume in Chaps. 2 and 3

(South Africa and Egypt). Other examples are presented to establish the framework of bioethics within the legal framework, such as in Brazil (Chap. 1) and South Korea (Chaps. 5 and 6), acquired through struggles against military regimes, as introduced in this and previous [21] books from our initiative. Similarly South Korea and, even more stringently, Taiwan have established their legal frameworks that are not limited to research but also include biobanks (Chaps. 4 and 5).

The International Covenant on Civil and Political Rights [25], together with the Covenant on Economic, Social and Cultural Rights [26], was issued based on the UN Declaration of Human Rights [27] with the commitment of the international society TO NEVER repeat the atrocities of World War II, and formed the bases of the Bill of Human Rights in the constitution of each sovereign state. The right to health is one of them, which is already included in the Constitution of the World Health Organization (WHO) [28]. The Nuremberg Code [29] provided a set of norms considered in the history of the establishment of the DoH, but it was not developed as a legal basis for the human right of research participants. Because this right conflicts with another right, the freedom of scientific research, the rights of research participants have not been well established globally. However, contemporary health research projects have expanded into enterprises with huge investment of funds and human resources, far beyond the actions of individual physicians motivated by interests in caring for individual patients.

Other examples of international agreements are the Council of Europe Convention on Human Rights and Biomedicine [30] and its Protocols. This would be a model but is limited to European jurisdictions. UNESCO's Declaration on Bioethics and Human Rights [31], unanimously approved by 195 countries in 2005, is a model, although not legally binding, and has broader content and scope, with reference to human rights and bioethical issues. It is also worth mentioning the CIOMS Ethical Guidelines for Health-related Research [32], a collaborative effort with the WHO, which have provided interpretations for the application of the DoH to regions with limited resources. The CIOMS Ethical Guidelines had already changed the term "subject" to "participant" in 2016 and anticipated the incorporation of the concept of community engagement and contemporary ideas of vulnerability. A notable point is that the term in the title has been changed from "biomedical research" to "health-related research" inferring a shift from medical studies to broad-based health studies. Therefore, the scope of the CIOMS Guidelines extends far beyond physicians.

The scope of an international norm on research ethics is a fundamental prerequisite. The central focus could be "permissible medical scientific experimentation", as suggested by the Nuremberg Code. But what about the extensional definition? Due to the difficulty of definition, many jurisdictions limit the scope of their legal systems mainly to clinical trials of drugs, or to broader views that include medical or health research. However, the disciplines of scientific research have been expanding, e.g. with the development of artificial intelligence (AI) in various fields, space development, environmental health, including the assessment of climate change and air pollution, education, sports, to name just a few. In several of these fields, research participants are exposed to risks, inequalities, and disparities in benefit sharing. This is the landscape which the DoH did not expect 60 years ago. Nevertheless, several areas of research enterprises have already begun to refer to the DoH, which has strengthened its recommendations to the external communities beyond the discretion of physicians.

This chapter concludes the discussions contained in this book, leaving further deliberations to an initiative involving the entire international community. On this final occasion, a set of principles to be considered for the next initiative is presented in three dimensions, and posted and updated on a dedicated website (Table 17.1):

http://cont.o.oo7.jp/53_2/global_research_ethics.pdf

Table 17.1 A set of principles for consideration for global Research Ethics Norm as an expansion of the Human Rights Covenant

(1) Fundamental norms referred to in the context of research ethics:

- Human beings must not be used as a mere means to an end (Kant) [33]
- No one shall participate, without free consent, in medical or scientific experimentation (International Covenant) [25]
- Health is an equal right for all (WHO) [28]
- The rights and interests of research participants prevail over the research objective of generating scientific knowledge (DoH) [1]
- Respect for dignity, autonomy, beneficence, and justice all apply in the research context (Beauchamp & Childress [34], Belmont Report [35])

(2) Fundamental principles derived from the discussions on the DoH:

- Research participants are entitled to determine their own best interests and rights, which should be shared equally between participants and those who chose not to participate in research. There must be equality of rights before inclusion and after completion of participation
- The integrity of research data must be ensured and results must be shared equitably, with transparency, and accountability. Access to research results must be guaranteed for all those in need worldwide
- Continuous capacity building and response to the emerging research landscape must be ensured by an already established research governance framework consisting of a research ethics committee, informed consent procedures; disseminating results to each participant respecting their cultural background; fair, timely disclosure of study information through public databases, adhering to international standards; and robust governance of health databases and biobanks

(3) Forward-looking perspectives presented as patient and public opinions:

- Early evaluation of the research and development plan, considering the impact on study target populations, society, global contexts and future generations:
  - Avoid discrimination or stigmatization against the target populations of the study
  - Aim for the united nation's sustainable development goals (SDGs)
  - Assess the impact on future generations, the environment, and society
- Strengthening patient/public perspectives on the ethical principles beyond the scope of autonomous norms established by and addressed to physicians:
  - Use plain language, e.g. avoiding "double-negative" phrases
  - The scope of the declaration needs to adequately address respect for embryos, foetuses, and deceased persons on which studies are performed
  - Ensure the participation of patients and the public in all processes of research
  - Multidisciplinary collaboration, including patient peer supporters
  - Shared decision-making based on informed consent
  - Diversity, fairness, and equity in the composition of the research ethics committee
  - Protect the dignity and rights of an individual who is incapable of consent and has no available proxy
  - Develop international research ethics norms that are agreed upon by all the interested parties

As of May 2025. Contents will be updated in: http://cont.o.oo7.jp/53_2/global_research_ethics.pdf

(1) **Fundamental norms referred to in the context of research ethics:** Frequently referred to in this volume and supported by well-established reasoning.
(2) **Fundamental principles derived from the discussions on the DoH**: Drawn from the arguments in this book.
(3) **Forward-looking perspectives presented as patient and public opinions**: Reproduction of the items in the table 13.1 in Chap. 13 by the patient and public group.

We hope that the challenges presented in this book will be considered and addressed by emerging initiatives in the future, and that these global efforts will culminate in the highest standards of ethical research to protect the rights of participants and to ensure that the social value of research is equitably shared among the global community.

Acknowledgement CK appreciates the International Federation of Associations of Pharmaceutical Physicians and Pharmaceutical Medicine (IFAPP) for granting her travel expenditure for her in-person participation in the WMA Regional Meeting in Tel Aviv (December 2022), and the WMA for granting her travel expenditure for her in-person participation and presentation as a panellist in WMA Regional Meeting in Washington, DC (August 2024), and for a waiver of registration fee from the WMA as a journalist to attend the General Assembly in Helsinki (October 2024). She participated in all the other WMA Regional Meetings at her own expense or online.

DG was officially engaged in discussions for the revisions of the Declaration of Helsinki from 2000 to 2008 as a member of the Brazilian Medical Association and Brazilian Medical Council teams; he also participated in the CIOMS 2016 guidelines as a member of working group; he was invited to the regional meeting in Sao Paulo by the WMA to discuss the ethics of placebo-controlled trials; and he participated, with partial funding from CIOMS, in the 2024 WMA General Assembly that adopted the 2024 version of the DoH.

Conflict of Interest AD was a Board member of the South African Medical Association, which was a member of the Workgroup of the World Medical Association for the 2024 revision of the Declaration of Helsinki. The authors declare no other competing interests relevant to the content of this chapter.

References

1. The World Medical Association. WMA Declaration of Helsinki—ethical principles for medical research involving human participants. First adopted in 1964, last amended in 2024. https://www.wma.net/policies-post/wma-declaration-of-helsinki/
2. The World Medical Association. The Declaration of Geneva. 1948. Last amended 2017. https://www.wma.net/policies-post/wma-declaration-of-geneva/
3. The World Medical Association. The International Code of Medical Ethics. 1949. Last revised in 2022. https://www.wma.net/policies-post/wma-international-code-of-medical-ethics/
4. [Editor's Note] Bibbins-Domingo K, Brubaker L, Curfman G. The 2024 revision to the Declaration of Helsinki: modern ethics for medical research. JAMA. 2024. https://doi.org/10.1001/jama.2024.22530. Epub ahead of print.
5. Resneck JS Jr. Revisions to the Declaration of Helsinki on its 60th anniversary: a modernized set of ethical principles to promote and ensure respect for participants in a rapidly innovating medical research ecosystem. JAMA. 2024. https://doi.org/10.1001/jama.2024.21902. Epub ahead of print.

6. Shaw JA. The revised Declaration of Helsinki-considerations for the future of artificial intelligence in health and medical research. JAMA. 2024. https://doi.org/10.1001/jama.2024.22074. Epub ahead of print.
7. Bloom L. Revisiting the Declaration of Helsinki – a patient-Centered perspective. JAMA. 2024. https://doi.org/10.1001/jama.2024.22077. Epub ahead of print.
8. Reis AA, Upshur R, Moodley K. Future-proofing research ethics-key revisions of the Declaration of Helsinki 2024. JAMA. 2024. https://doi.org/10.1001/jama.2024.22254. Epub ahead of print..
9. Saenz C, Carracedo S. The revision of the Declaration of Helsinki viewed from the Americas-paving the way to better research. JAMA. 2024. https://doi.org/10.1001/jama.2024.22270. Epub ahead of print..
10. Bierer BE. Declaration of Helsinki – revisions for the 21st century. JAMA. 2024. https://doi.org/10.1001/jama.2024.22281. Epub ahead of print..
11. [News] O'Grady C. Cornerstone medical ethics guidelines get a major update. Science. 2024;386(6721):473–4. https://doi.org/10.1126/science.adu2384. Epub 2024 Oct 31.
12. [Editorial] Declaration of Helsinki embraces health equity. Nat Med. 2024. https://doi.org/10.1038/s41591-024-03433-5. Epub ahead of print.
13. Sheather J. Declaration of Helsinki puts global justice up front. BMJ. 2024;(387):q2405. https://doi.org/10.1136/bmj.q2405.
14. Kurihara C, Matsuyama K, Baroutsou V. World Medical Association's Declaration of Helsinki, 2024 revision: celebrating the 60th anniversary, at Helsinki. Clin Eval. 2025;52(3):W1–27. http://cont.o.oo7.jp/52pop/52pop_contents_e.html
15. Clin Eval. 2025; 52(3). http://cont.o.oo7.jp/52pop/52pop_contents_e.html
16. The World Medical Association. The Declaration of Taipei on ethical considerations regarding Health Databases and Biobanks. 2002. Last revised 2016. https://www.wma.net/policies-post/wma-declaration-of-taipei-on-ethical-considerations-regarding-health-databases-and-biobanks/
17. Sindicato Médico del Uruguay. World Medical Association Submission of a Motion to Amend (on Paragraph 33, submitted to) 227th Council Session and General Assembly, Helsinki 2024, 10 Oct 2024.
18. Sindicato Médico del Uruguay. World Medical Association Submission of a Motion to Amend (on Paragraph 34, submitted to) 227th Council Session and General Assembly, Helsinki 2024, 10 Oct 2024.
19. Helsinki Statement Stakeholders. Helsinki statement 2024. Clin Eval. 2025;52(3) http://cont.o.oo7.jp/52pop/52pop_contents_e.html
20. World Medical Association. Declaration of Helsinki: ethical principles for medical research involving human subjects. 2000. Rescinded. https://www.wma.net/wp-content/uploads/2018/07/DoH-Oct2000.pdf
21. Kurihara C, Greco D, Dhai A, editors. Ethical innovation for global health: pandemic, democracy and ethics in research. Singapore: Springer; 2023.
22. United Nations. Resolution adopted by the General Assembly on 25 September 2015 – Transforming our world: the 2030 Agenda for Sustainable Development. 21 Oct 2015. https://www.un.org/en/development/desa/population/migration/generalassembly/docs/globalcompact/A_RES_70_1_E.pdf
23. Clinical Evaluation. Ethical innovation for global health. http://cont.o.oo7.jp/sympo/eigh.html. Accessed 15 May 2025.
24. Clinical Evaluation. Patient and public perspectives on the WMA Declaration of Helsinki. http://cont.o.oo7.jp/grareco.html. Accessed 15 May 2025.
25. United Nations. International Covenant on Civil and Political Rights. Adopted by General Assembly resolution 2200A (XXI) of 16 December 1966, entry into force 23 March 1976. https://www.ohchr.org/en/instruments-mechanisms/instruments/international-covenant-civil-and-political-rights

26. United Nations. International Covenant on Economic, Social and Cultural Rights. Adopted by General Assembly resolution 2200A (XXI) of 16 December 1966, entry into force 3 January 1976. https://www.ohchr.org/en/instruments-mechanisms/instruments/international-covenant-economic-social-and-cultural-rights
27. United Nations. The Universal Declaration of Human Rights. Adopted December 10, 1948. https://www.un.org/en/about-us/universal-declaration-of-human-rights
28. World Health Organization. Constitution of the World Health Organization. 1946 July 22. https://www.who.int/about/governance/constitution.
29. International Military Tribunal. Trials of war criminals before the Nuremberg Military Tribunals under Control Council law no. 10 Nuernberg, October 1946–April 1949. Vol. 2, p. 181–2. (The Nuremberg Code.)
30. Council of Europe. Convention for the protection of human rights and dignity of the human being with regard to the application of biology and medicine: Convention on human rights and biomedicine (European Treaty Series No. 164) 1997. https://www.coe.int/en/web/conventions/full-list?module=treaty-detail&treatynum=164
31. United Nations Educational, Scientific and Cultural Organization. Universal Declaration on Bioethics and Human Rights 19 October 2005. https://www.unesco.org/en/legal-affairs/universal-declaration-bioethics-and-human-rights?hub=66535
32. Council for International Organizations of Medical Sciences. International ethical guidelines for health-related research involving humans. 2016. https://cioms.ch/publications/product/international-ethical-guidelines-for-health-related-research-involving-humans/
33. Kant I. Groundwork of the metaphysics of morals. 1785.
34. Beauchamp TL, Childress JF. Principles of biomedical ethics. Oxford: Oxford University Press; 1979.
35. The National Commission for the Protection of Human Subjects of Biomedical and Behavioral Research. The Belmont Report: ethical principles and guidelines for the protection of human subjects of research. 1979. https://www.hhs.gov/ohrp/regulations-and-policy/belmont-report/read-the-belmont-report/index.html